T0296357

A COMPANION TO ANIMAL PHYSIOLOGY

*Papers from the Fifth International Conference on Comparative
Physiology held at Sandbjerg, Denmark, July 22–26, 1980*

The conference was sponsored by the Interunion Commission
on Comparative Physiology representing the International
Unions of Biological Sciences, Physiological Sciences, and
Pure and Applied Biophysics. The conference was made
possible through the generous support of the Cocos
Foundation, Indianapolis, Indiana; Danish Research Council
for Natural Sciences; Novo Foundation, Copenhagen,
Denmark; and University of Aarhus, Denmark.

Conference logo design by Margaret L. Estey.

A companion to
ANIMAL
PHYSIOLOGY

Editors

C. RICHARD TAYLOR
Harvard University

KJELL JOHANSEN
University of Aarhus

LIANA BOLIS
University of Messina

CAMBRIDGE
UNIVERSITY PRESS

CAMBRIDGE UNIVERSITY PRESS
Cambridge, New York, Melbourne, Madrid, Cape Town,
Singapore, São Paulo, Delhi, Tokyo, Mexico City

Cambridge University Press
The Edinburgh Building, Cambridge CB2 8RU, UK

Published in the United States of America by
Cambridge University Press, New York

www.cambridge.org
Information on this title: www.cambridge.org/9780521286855

First published 1982
Reprinted 1984, 1993,1994
Re-issued 2013

A catalogue record for this publication is available from the British Library

ISBN 978-0-521-24437-4 Hardback
ISBN 978-0-521-28685-5 Paperback

Contents

v

Participants

R. McN. AND ANN ALEXANDER, Department of Pure and Applied Zoology, The University of Leeds, Leeds LS2 9JT, England

JÜRGEN AND HILDE ASCHOFF, Max-Planck-Institut für Verhaltensphysiologie, 8131 Erling-Andechs, West Germany

RUSSELL AND KAYE BAUDINETTE, School of Biological Sciences, Flinders University, Bedford Park, South Australia 5042, Australia

CLAUS BECH, Department of Zoology, University of Aarhus, DK-8000 Aarhus C, Denmark

REINIER AND NANCY BEEUWKES, Department of Physiology, Harvard Medical School, 25 Shattuck Street, Boston, Massachusetts 02115, U.S.A.

DENIS BELLAMY, University College, P.O. Box 78, Cardiff CF1 1XL, Wales

P. J. BENTLEY, Department of Pharmacology, Mt. Sinai School of Medicine, New York, New York 10029, U.S.A.

MARVIN H. BERNSTEIN, Department of Biology, New Mexico State University, Las Cruces, New Mexico 88003, U.S.A.

JACOB J. BLUM, Duke University, Department of Physiology, Durham, North Carolina 27710, U.S.A.

LIANA BOLIS, Istituto di Fisiologia Generale, Università di Messina, Messina, Italy

ARIEH BORUT AND DUBA YAAKOBY, Department of Zoology, The Hebrew University, Jerusalem, and School of Education, Tel Aviv University, Tel Aviv, Israel

PIERRE BOUVEROT, Laboratoire de Physiologie Respiratoire, Centre National de la Recherche Scientifique, 23 Rue Becquerel, 67087 Strasbourg, France

OLE BRIX, Department of Zoology, University of Aarhus, DK-8000 Aarhus C, Denmark

FRANCIS G. CAREY, Woods Hole Oceanographic Institution, Woods Hole, Massachusetts 02543, U.S.A.

SHOU-TEH CHIANG, Department of Physiology, National Yang-Ming Medical College, Taipei, Taiwan, Republic of China

EUGENE C. CRAWFORD, JR., University of Kentucky, School of Biological Sciences, Lexington, Kentucky 40506, U.S.A.

TERENCE DAWSON, School of Zoology, The University of New South Wales, P.O. Box 1, Kensington, New South Wales 2033, Australia

PIERRE DEJOURS, Laboratoire de Physiologie Respiratoire, Centre National de la Recherche Scientifique, 23 Rue Becquerel, 67087 Strasbourg, France

RAZI AND OFRA DMI'EL, Department of Zoology, Tel Aviv University, Tel Aviv, Israel

RAGNAR FÄNGE, Institute of Zoophysiology, University of Gothenburg, Gack, S-400 33 Goteborg, Sweden

MICHAEL A. FEDAK, Sea Mammal Research Unit, British Antarctic Survey, Madingley Road, Cambridge CB3 0ET, England

MALCOLM S. GORDON, Department of Biology, University of California, Los Angeles, California 90024, U.S.A.

CLYDE AND JANET HERREID, Department of Biology, State University of New York, Buffalo, New York 14214, U.S.A.

P. W. AND BRENDA HOCHACHKA, Department of Zoology, University of British Columbia, Vancouver, British Columbia V6T 1W5, Canada

MARYANNE R. AND GILBERT HUGHES, Department of Zoology, University of British Columbia, Vancouver, British Columbia V6T 1W5, Canada

DONALD C. AND DIANA JACKSON, Division of Biology and Medicine, Brown University, Providence, Rhode Island 02912, U.S.A.

KJELL JOHANSEN, Department of Zoophysiology, University of Aarhus, DK-8000 Aarhus C, Denmark

C. BARKER JØRGENSEN, Zoolysiologisk Laboratorium, Kobenhavns Universitet, August Krogh Institut, Universitetsparken 13, DK-2100 Kobenhavn, Denmark

DARWIN JORGENSEN, Department of Zoology, Duke University, Durham, North Carolina 27706, U.S.A.

RICHARD KEYNES, Department of Physiology, Cambridge University, Cambridge CB2 3EG, England

G. L. AND MELBA KOOYMAN, Scripps Institution of Oceanography, University of California, San Diego, LaJolla, California 92093, U.S.A.

JOHN KROG, Institute of Zoophysiology, University of Oslo, Postbaks 1051, Blindern, Oslo 3, Norway

GEORGE LAPENNAS, Duke University Marine Laboratory, Beaufort, North Carolina 28516, U.S.A.

JACQUES LAROCHELLE, Department de Biologie, Université Laval, Quebec G1K 7P4, Canada

CLAUDE LENFANT, Division of Lung Diseases, NHLBI, National Institutes of Health, Westwood Building, Bethesda, Maryland 20205, U.S.A.

PETER LUTZ, Division of Biology and Living Resources, University of Miami, School of Marine and Atmospheric Science, Miami, Florida 33149, U.S.A.

GUNNAR LYKKEBOE, Department of Zoophysiology, University of Aarhus, DK-8000 Aarhus C, Denmark

SIMON MADDRELL, Department of Zoology, Cambridge University, Cambridge CB2 3EJ, England

GEOFFREY M. O. MALOIY, University of Nairobi, Faculty of Veterinary Medicine, P.O. Box 29053, Kabete, Kenya

ALIE MINK, Department of Zoophysiology, University of Aarhus, DK-8000 Aarhus C, Denmark

M. PEAKER, Physiology Department, The Hannah Research Institute, Ayr KA6 5HL, Scotland

JOHANNES AND ILSE PIIPER, Max-Planck-Institut für experimentelle Medizin, Hermann-Rein-Strasse 3, D-3400 Göttingen, West Germany

BEROLD PINSHOW, Ben-Gurion University of the Negev, Sde Boker Campus, P.O. Box 653, Beersheva 84210, Israel

C. LADD AND HAZEL PROSSER, Department of Physiology and Biophysics, University of Illinois at Urbana-Champaign, Urbana, Illinois 61801, U.S.A.

HERMANN RAHN, Department of Physiology, Schools of Medicine and Dentistry, State University of New York at Buffalo, Buffalo, New York 14214, U.S.A.

BLAKE REEVES, Department of Physiology, State University of New York at Buffalo, Buffalo, New York 14214, U.S.A.

PETER SCHEID, Max-Planck-Institut für experimentelle Medizin, Abteilung Physiologie, Hermann-Rein-Strasse 3, D-3400 Göttingen, West Germany

BENT SCHMIDT-NIELSEN, 21 Beechglen Road, Roxbury, Massachusetts 02119, U.S.A.

KNUT AND MARGARETA SCHMIDT-NIELSEN, Duke University, Department of Zoology, Durham, North Carolina 27706, U.S.A.

ROBERT SCHROTER, Physiological Flow Studies Unit, Imperial College, London, SW7 2AZ, England

AMIRAM SHKOLNIK, Department of Zoology, Tel Aviv University, Tel Aviv, Israel

JOHN FLENG STEFFENSEN, Department of Zoophysiology, University of Aarhus, DK-8000 Aarhus C, Denmark

C. RICHARD TAYLOR, CFS-Museum of Comparative Zoology, Harvard University, Old Causeway Road, Bedford, Massachusetts 01730, U.S.A.

VILHELNS TEXENS, Department of Zoophysiology, University of Aarhus, DK-8000 Aarhus C, Denmark

STEPHEN THESLEFF, Farmakologiska Institutionen, Lunds Universitet, Lund, Sweden

JOSÉ TORRE-BUENO, Department of Zoology, Duke University, Durham, North Carolina 27706, U.S.A.

HANS AND ANNEMARIE USSING, University of Copenhagen, Institute of Biological Chemistry A, 13 Universitetsparken, DK-2100 Copenhagen Ø, Denmark

STEPHEN A. AND RUTH WAINWRIGHT, Department of Zoology, Duke University, Durham, North Carolina 27706, U.S.A.

EWALD AND VRENY WEIBEL, Anatomisches Institut der Universität Bern, Bern 9, Postfach 139, Bühlstrasse 26, Bern, Switzerland

STEPHEN C. WOOD, Department of Physiology, University of New Mexico, School of Medicine, Albuquerque, New Mexico 87131, U.S.A.

About this book

A *companion to animal physiology* is designed as a supplement to Knut Schmidt-Nielsen's *Animal physiology* text. It attempts to provide students and professional biologists with the opportunity to pursue some of the topics introduced in *Animal physiology*. It brings together a series of papers prepared to honor Knut Schmidt-Nielsen on the occasion of his sixty-fifth birthday. The papers were presented at the Fifth International Conference on Comparative Physiology, Comparative Physiology: Perspectives, a Satellite Symposium of the XXVIII International Physiological Congress. The conference was held in Sandbjerg, Denmark, in July 1980 and was sponsored by the Interunion Commission on Comparative Physiology, representing the International Unions of Biological Sciences, Physiological Sciences, and Pure and Applied Biophysics. It brought together a group of scientists who have made important contributions to different areas in comparative physiology, and asked them to provide a perspective on what they thought was interesting and important in their particular areas.

Vikings have always sensed discovery and discovered with sensitivity. Knut Schmidt-Nielsen, a modern-day Viking of science, and his wife, Margareta, are flanked by two members of the Danish Philharmonic at the symposium banquet celebrating his birthday. The ancient Viking horns, lurs, provided the music for the banquet.

About Knut Schmidt-Nielsen

Knut Schmidt-Nielsen was born in Trondheim, Norway, in 1915. His early schooling took place in Norway. He studied under August Krogh and obtained his Doctor of Philosophy degree under Kaj Linderstrøm-Lang at the Carlsberg Laboratories in Denmark.

Knut Schmidt-Nielsen's books and his research reveal his curiosity, enthusiasm, and originality as a scientist. He enjoys physiology and is able to convey this enjoyment in his books and papers. He and his collaborators have helped us understand how animals survive in deserts; discovered salt glands in birds and reptiles; unraveled the mysteries of the bird lung; and shown that countercurrent exchanges can be organized temporally as well as spatially. He has written not only about what we know, but about what we do not as yet understand. His books and his research have stimulated students and professional biologists to try to find the answers to how animals work, and perhaps this is why Knut Schmidt-Nielsen has had such an enormous impact on comparative physiology.

PART ONE

Oxygen

Overview

Part One considers the exchange of the respiratory gases.

In Chapter 1 Peter Scheid develops a simple model of respiratory exchange. The model is used to define and evaluate the gas exchange at the respiratory surface. It compares the efficiency of three mechanisms that vertebrates use for bringing oxygen in contact with the respiratory surfaces: the countercurrent mechanism of fish gills, the crosscurrent mechanism of bird lungs, and the pool mechanism of the mammalian lung.

In Chapter 2 the pressure gradients and conductances for oxygen and carbon dioxide (defined in Scheid's model) are analyzed by Pierre Dejours to examine whether humans can climb Mount Everest without supplemental oxygen. He extends the analysis to birds, and concludes that it is the greater gas-exchange efficiency of their lungs that enables them to fly over Mount Everest.

In Chapter 3 Ewald Weibel extends the model for oxygen transport to include the pathway from the lung to the mitochondria of muscles. He divides the pathway into a number of steps and proposes that the structural design of each step is quantitatively matched. Allometry and morphometric analysis of structure are used as tools to test this hypothesis.

In Chapter 4 Johannes Piiper develops a theoretical model to evaluate the extent to which diffusion limits the exchange of O_2 and CO_2 in vertebrate respiratory organs. The model predicts a strong diffusion limitation for O_2 across amphibian skin, an intermediate limitation across fish gills, and a weak limitation across the human lung.

In Chapter 5 Peter Lutz discusses some unusual ways in which vertebrates increase the capacity of blood for carrying oxygen during exercise. He also discusses vertebrate mechanisms for the effective use of oxygen stores in the lungs during breath holds and/or dives.

In Chapter 6 Donald Jackson discusses the problems for CO_2 elimination and acid–base regulation posed by the variable temperature of ectothermic vertebrates. He describes the physiological mechanisms used by a number of ectotherms for solving these problems.

In Chapter 7 Kjell Johansen asks whether the incomplete separation between pulmonary and systemic circulation found in air-breathing fish, in amphibians, and in reptiles is advantageous. Available data reveal clear advantages in retaining an incomplete separation between pulmonary and systemic circuits in these animals.

1 A model for comparing gas-exchange systems in vertebrates

PETER SCHEID

Like all animals with aerobic metabolism, vertebrates obtain the necessary O_2 from their environment, which can be water or air. External gas exchange is aided by specialized organs – gills for water breathing in fish and some amphibians and lungs for air breathing in reptiles, birds, and mammals. Skin may aid gills and lungs as a gas-exchange organ, and in some amphibian species it represents the only route for gas exchange (see Chapter 4). In this chapter a model is developed that defines the parameters involved in gas exchange. The model is then used to compare three of the most highly structured vertebrate gas-exchange organs: gills in fish, parabronchial lungs in birds, and alveolar lungs in mammals. Most of the other gas-exchange systems can be viewed as various steps in the transition between these types and are thus more complex mixtures of the basic types.

General model for gas-exchange organs

Schema

A general schema for a gas-exchange model is depicted in Figure 1.1 (Piiper and Scheid, 1975, 1977). Medium (air or water) and blood enter the system, in which they reach an intimate contact for gases to be exchanged, separated by a tissue barrier. Whereas both medium and blood act as vehicles to convectively transport respiratory gases to the site of gas exchange, diffusion is the sole mechanism by which these gases are transferred from medium to blood or from blood to medium.

Basic parameters

For a quantitative treatment of gas exchange, a number of basic parameters must be known (see Figure 1.1). These are listed below, with typical units given in parentheses:

1. Gas-exchange rate, $\dot{M}$ (mmol $\cdot$ min^{-1})
2. Flow rates of medium and blood, $\dot{V}_m$, $\dot{V}_b$ (liter $\cdot$ min^{-1})
3. Capacitance coefficients in medium and blood, β_m, β_b (mmol $\cdot$ liter^{-1} $\cdot$ Torr^{-1}) (see below)

3

4. Partial pressures of the gas under consideration, in the inflowing medium (P_i) and blood (P_v) and in the outflowing medium (P_e) and blood (P_a) (Torr)
5. Diffusive conductance of the blood–medium barrier, G_{diff} (mmol · min^{-1} · Torr^{-1}). G_{diff} is often referred to as the diffusing capacity of the barrier, D (see below)

Basic overall equations

In order to arrive at simple equations for gas exchange, the following assumptions will be made; their validity will be discussed later:

1. The gas-exchange system is in steady state; that is, all variables are constant in time.
2. The β_m and β_b are constant over the partial pressure range considered.
3. No concentration gradients exist inside the gas-exchange system in the direction perpendicular to the medium–blood barrier.

The mass balance provides some simple links between the basic parameters:

$$\dot{M} = \dot{V}_m \cdot \beta_m \cdot (P_i - P_e) = \dot{V}_b \cdot \beta_b \cdot (P_a - P_v) \tag{1}$$

FIGURE 1.1 General schema for gas-exchange systems in vertebrates. The figure shows the various basic parameters (defined in the text) that determine gas-exchange function. No commitment is made, however, as to the specific arrangement of medium flow to blood flow at the site of gas exchange. This specific arrangement is the basis for differences among the various types of gas-exchange organ in vertebrates and is determined entirely by the structure of the system.

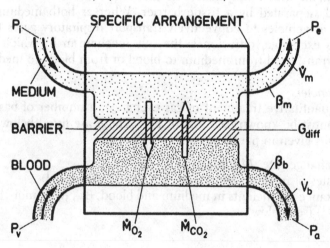

In analogy with electric circuits, the ratio of $\dot{M}/\Delta P$ may be defined as a conductance. Equation 1 thus allows definition of a convective conductance by ventilation:

$$G_{vent} = \dot{V}_m \cdot \beta_m \tag{2}$$

and of a convective conductance by perfusion:

$$G_{perf} = \dot{V}_b \cdot \beta_b \tag{3}$$

Similarly, for diffusive transport across the medium–blood barrier

$$\dot{M} = G_{diff} \cdot (\overline{P_m} - \overline{P_b}) \tag{4}$$

where $\overline{P_m}$ and $\overline{P_b}$ are spatial mean values of partial pressures in medium and blood.

The diffusive conductance, G_{diff}, is equal to the diffusing capacity (D), provided the diffusive resistance is concentrated into the medium–blood barrier. In this case, G_{diff} is related to area (A) and thickness (L) of the barrier and to the solubility (α) and diffusivity (d) of the gas in this membrane:

$$G_{diff} = \frac{A}{L} \cdot d \cdot \alpha \tag{5}$$

Because the parameters on the right-hand side of equation 5 are not easily determined for biological systems, G_{diff} is usually obtained from equation 4, when $\overline{P_m}$ and $\overline{P_b}$ are known for the gas-exchange organ in question. Knowledge of these values entails knowledge of the specific arrangement of the gas-exchange system.

The total gas-exchange conductance, G_{tot}, may be defined from the total partial pressure difference from inspired medium to inflowing blood:

$$G_{tot} = \dot{M}/(P_i - P_v) \tag{6}$$

Capacitance coefficient and differences between air and water breathing

The capacitance coefficient, β, is defined as the slope in the relationship between gas content (C) and partial pressure (P) in a given liquid or gas mixture (Piiper et al., 1971):

$$\beta = dC/dP \tag{7}$$

External medium: water or air

The principal difference between water and air as breathing media concerns the capacitance coefficient (Dejours, 1975; Piiper and Scheid, 1977; Schmidt-Nielsen, 1979; see Figure 1.2). For air, the relationship of

6   PETER SCHEID

equation 7 is linear and at any given temperature identical for all ideal gases (Piiper et al., 1971):

$$\beta_g = 1/RT \tag{8}$$

For (fresh) water, the relation between C and P is also linear, and β is usually termed the physical solubility coefficient, but β differs among gases and among solvents. Thus β_{CO_2} is about equal for fresh water and for air, but β_{O_2} is about 30 times smaller in water than in air (Figure 1.2). For sea water, the C/P relationship is alinear, and β is thus a function of P. This is due to HCO_3^- formation in the presence of bicarbonate/carbonate and boric acid/borate (see Piiper and Scheid, 1977). However, the value of about 30 for the β_{CO_2}/β_{O_2} ratio applies to sea water as well.

FIGURE 1.2 Differences in capacitance coefficients, β, of O_2 and CO_2 in water and air. The upper diagram shows β as the slope in the diagram of content (C) against partial pressure (P). In the lower, P_{O_2}–P_{CO_2} diagram, the lines connect corresponding P_{O_2} and P_{CO_2} values in the medium with those encountered in ambient air or water (e.g., P_{iO_2} = 147 Torr, P_{iCO_2} = 0) at RQ = 1.0. Arterial values (a) are close to expired values (e). Skin breathers show similar values to those of gill breathers because gas transport is limited by diffusion through the liquid medium (provided by the skin tissue). Arterial values for animals with exchange through both lung and gill/skin lie in an intermediate range.

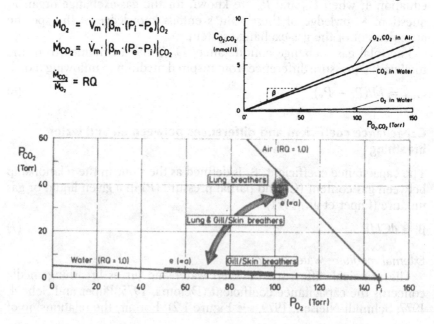

The result of these differences in β among the respiratory media, air, and water become apparent in the CO_2/O_2 diagram, which constitutes a plot of corresponding P_{CO_2} and P_{O_2} values anywhere in the organism. Equation 1 predicts:

$$\frac{(P_e - P_i)_{CO_2}}{(P_i - P_e)_{O_2}} = \frac{(\beta_m)_{O_2}}{(\beta_m)_{CO_2}} \cdot RQ \tag{9}$$

where the respiratory quotient, $RQ = \dot{M}_{CO_2}/\dot{M}_{O_2}$, lies between 0.65 and 1.0 for most steady-state conditions. Equation 9 represents the slope of the line connecting corresponding P values in the outflowing medium, P_e, to those in the inflowing medium (= inspiratory), P_i. For water as the breathing medium, the slope is about 30 times smaller than for air (Figure 1.2), a direct consequence of the differences in medium solubilities, β_m. Because arterial P_{O_2} and P_{CO_2} values are similar to expired values in all animals, equation 9 and Figure 1.2 predict much lower arterial P_{CO_2} values in water breathers than in air breathers, and this is in fact observed (see Piiper and Scheid, 1977). It could be said that the water breather must provide a very high ventilatory flow rate to extract sufficient O_2 from water in view of the low water solubility for O_2; this high flow means hyperventilation for CO_2 with ensuing low body P_{CO_2} values (see Dejours, 1975).

Internal medium: blood

For inert gases, blood solubility is generally assumed to be constant, independent of P, the absolute value being similar to that for water. Owing to the reversible binding of O_2 to hemoglobin (Hb), the C/P relation (= dissociation curve) is not linear for O_2, and β_{bO_2} thus is not constant. Likewise for CO_2, where the chemical reaction to form HCO_3^- accounts for a nonlinear dissociation curve and hence for the dependence of β_{bCO_2} on P_{CO_2} (see Piiper and Scheid, 1977).

Specific arrangement: anatomy and models

The basic system of Figure 1.1 fits all gas-exchange organs of vertebrates. However, a quantitative analysis requires information about the specific arrangement of medium flow to blood flow at the site of gas exchange. This information must be provided by the anatomical structure of the systems.

Fish gills: countercurrent arrangement

The fish gill apparatus (Hughes and Shelton, 1962; Randall, 1970; Johansen, 1971; Hughes and Morgan, 1973) comprises gill arches with gill filaments and secondary lamellae (Figure 1.3). These secondary lamellae form a sieve through which water flows in a direction opposite to that of blood flow in the secondary lamellae. A secondary lamella with water flowing past it may be regarded as the gas-exchange unit. The counter-current model for gas exchange in this unit takes account of (1) the blood–water contact all along the separating tissue barrier, and (2) the opposite directions of water and blood flow.

Although the schema of Figure 1.3 applies to teleost fish, there is now evidence that the countercurrent model is appropriate in elasmobranch fish as well (see Piiper and Scheid, 1977).

Avian parabronchial lungs: crosscurrent arrangement

Air sacs and parabronchi constitute the basic structures of the respiratory system in birds. Whereas the air sacs, through the action of respiratory muscles, provide respiratory gas flow in the breathing cycle, gas exchange takes place virtually exclusively in the parabronchial lung (see Scheid, 1979). The parabronchi constitute tubes, of several centimeters in length, that are open at both ends (Figure 1.4). Air capillaries depart from the parabronchial lumen, or from atria and infundibula (Duncker, 1971), in radial directions into the periparabronchial tissue. Blood capillaries, on the other hand, which originate from arterioles in the periparabronchial tissue, run toward the parabronchial lumen to drain into venules there. Thus air capillaries and blood capillaries intertwine intimately to exchange gases across a very thin blood–gas separating membrane.

FIGURE 1.3 Schema of respiratory apparatus in teleost fish.

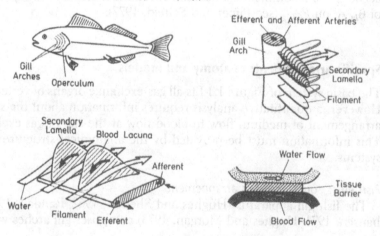

To understand the functional model underlying parabronchial gas exchange, it is important to note that any given blood capillary contacts air capillaries originating from only a short fraction of the total parabronchial length. The parabronchus thus represents a serial arrangement (with respect to parabronchial air flow) of a multitude of blood capillaries, and blood flow (at the site of gas exchange) and parabronchial gas flow form a crosscurrent arrangement (Scheid and Piiper, 1970; see Scheid, 1979).

When air traverses the parabronchial tube, it is continually depleted of O_2 and enriched in CO_2 as it contacts the blood capillaries in serial order. The blood in the blood capillaries, in turn, will be arterialized to varying degrees, depending on their location along the parabronchial tube; the arterialization will be best at the gas inflow end of the parabronchus, and worst on its outflow end. Systemic arterial blood thus constitutes a mixture of blood from all capillaries, and it is easy to conceive that the P_{O_2} in this arterial blood can exceed the P_{O_2} in end-par-

FIGURE 1.4 Schema of respiratory apparatus in birds. The figure shows the various air sacs, secondary bronchi, and tertiary bronchi (= parabronchi). In the periparabronchial tissue, air capillaries, radiating from the parabronchial lumen, meet blood capillaries. These blood capillaries originate from arterioles, in the periphery of the periparabronchial tissue, and drain into venules, located close to the lumen.

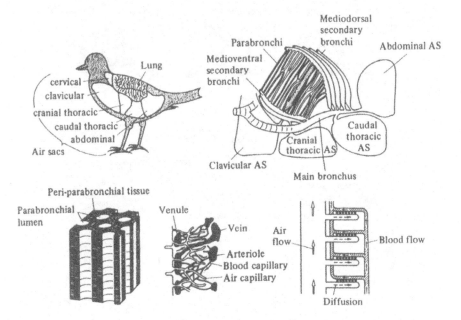

abronchial gas, which represents the lowest P_{O_2} in the gas phase (see Figure 1.5).

This overlap of partial pressures is similarly observed in the counter-current model (Figure 1.3). However, unlike the situation in the countercurrent system, the gas profiles in the crosscurrent system do not depend on the direction of parabronchial gas flow. The experimentally observed phenomenon that respiratory air flows through the parabronchus in the same direction during both inspiration and expiration thus can not be easily explained as a result of gas-exchange needs (Scheid and Piiper, 1972; Scheid, 1979).

Mammalian lungs: pool arrangement

The bronchial tree of the mammalian lung displays a pattern of dichotomous branching. The first approximately 20 generations of bronchi serve as conducting airways, whereas alveoli arise only from the last generations. In an arrangement unlike the avian respiratory system, alveoli are the distensible structures that accommodate tidal fluctuations of lung volume and also constitute the site of gas exchange.

FIGURE 1.5 Schema of respiratory apparatus of fish gills (left), avian lungs (middle), and mammalian lungs (right). (See text for discussion.)

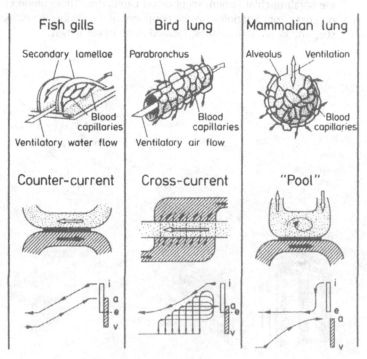

The functional unit of the alveolar lung is the alveolus, surrounded by blood capillaries. In the pool model (Figure 1.5, right), which is generally assumed to be the appropriate gas-exchange model, transfer of gas between the well-mixed alveolar compartment and capillary blood occurs across a blood–gas separating tissue membrane.

Quantitative relationships

The functional models defined in the preceding section are highly simplified approximations of anatomical reality. The key advantage of this simplification is a quantitative analysis of gas-exchange function, which forms the basis for a comparison between the systems with regard to their gas-transfer efficiency.

Gas-exchange conductances and relative partial pressure differences

Piiper and Scheid (1975) have analyzed the three models by defining relative partial pressure differences for ventilation (Δp_{vent}), perfusion (Δp_{perf}), and transfer (Δp_{tr}):

$$\Delta p_{vent} = \frac{P_i - P_e}{P_i - P_v} \tag{10}$$

$$\Delta p_{perf} = \frac{P_a - P_v}{P_i - P_v} \tag{11}$$

$$\Delta p_{tr} = \frac{P_e - P_a}{P_i - P_v} \tag{12}$$

Equations 1 to 3 and 9 show that

$$\Delta p_{vent} = G_{tot}/G_{vent} \tag{13}$$

$$\Delta p_{perf} = G_{tot}/G_{perf} \tag{14}$$

denote the fractional resistance (inverse of conductance), relative to the total gas-transfer resistance, provided by the component processes ventilation and perfusion.

The term Δp_{tr} is more complex as it depends not only on the diffusive resistance provided by the blood–medium separating membrane, but also on the specific arrangement of the system.

The Δp values for each system can be expressed as functions of two ratios of the three component conductances defined above; for example, $X = G_{vent}/G_{perf}$ and $Y = G_{diff}/G_{perf}$ (see Table 1.1).

Real system and idealized model

The equations of Table 1.1, which quantitatively describe gas exchange in the three basic models, may be applied to experimental data, particularly to assess G_{diff}, which is difficult to determine otherwise (see Piiper and Scheid, 1975, 1977; Chapter 4). However, a number of deviations in gas-exchange performance of the real organ from the simplified model must be appreciated (see assumptions listed above), some of which are listed below (for details, see Piiper and Scheid, 1977).

Anatomical arrangement. Medium dead space is particularly present in tidally ventilated lungs, but not in the unidirectionally irrigated fish gills. Arterialized blood leaving the gas-exchange organ may be substantially modified by venous admixture before entering the systemic arterial system.

Regional inhomogeneities. Regional variation in the component conductances have particularly been shown in mammalian lungs but appear to be present in bird lungs and in fish gills as well (see Scheid and Piiper, 1981). Their effect is to reduce the gas-exchange efficiency.

Temporal variations. Temporal variations in blood flow exist in all vertebrates as a result of the rhythmic action of the heart. In mammals and birds the tidal ventilation results in temporal variations in P_{O_2} and P_{CO_2} in medium and blood within the gas-exchange organ. These temporal variations reduce the gas-exchange efficiency in all models, but theoretical and experimental evidence suggests that this reduction is quantitatively not very pronounced (Scheid, 1979; Piiper and Scheid, 1980).

Diffusion in medium. Imperfect mixing of gas in the alveolar region results in concentration gradients in alveolar gas. This stratified inhomogeneity would constitute an additional gas-transfer resistance and would thus diminish the gas-exchange efficacy below that of the idealized system. However, in alveolar lungs stratification appears to be of minor

TABLE 1.1 Relative partial pressure differences in the three systems: $X = G_{vent}/G_{perf}$; $Y = G_{diff}/G_{perf}$; $Z = Y(1 - 1/X)$; $Z' = [1 - \exp(-Y)]/X$

	Countercurrent	Crosscurrent	Pool
Δp_{vent}	$\dfrac{1 - \exp(-Z)}{X - \exp(-Z)}$	$1 - \exp(-Z')$	$\dfrac{1 - \exp(-Y)}{X + 1 - \exp(-Y)}$
Δp_{perf}	—[a]	$X \cdot \Delta p_{vent}$	—
Δp_{tr}	—	$1 - \Delta p_{vent} - \Delta p_{perf}$	—

[a] No data.

importance (Scheid and Piiper, 1980). Gas transport inside the air capillaries of avian lungs takes place by diffusion alone. Thus stratification in these structures may be expected to represent a significant resistance to parabronchial gas exchange. Whereas this may be so under flight conditions, both theoretical and experimental evidence shows that the gas phase of air capillaries imposes no major obstacle to gas exchange (see Scheid, 1979). When water flows past the secondary lamellae in fish gills, O_2 has to diffuse from central regions of this flow toward the secondary lamellar surface to be transferred into blood. As the diffusivity of O_2, and that of other gases, is about 100 000 times smaller in water than in air, a significant diffusion resistance is expected to exist in the water irrigating the secondary lamellae. Calculations suggest in fact that about one-half of the total O_2 uptake resistance in the gills of the elasmobranch fish *Scyliorhinus stellaris* resides inside the secondary lamellar water, the remaining resistance being mainly in the water–blood separating tissue barrier (Scheid and Piiper, 1976; Chapter 4).

Kinetics of chemical reactions. In particular the time needed for equilibration of CO_2, HCO_3^-, and H^+ in plasma and red cells, and for the associated exchange processes across the red cell membrane, is comparable with the time the red cell spends in the gas-exchange region. Hence, part of the limitation of CO_2 exchange may derive from finite equilibration kinetics (Piiper and Scheid, 1980). For ectothermic animals, this factor may be larger than for mammals or birds because the reactions involved become slower with lowered temperature. Thus, in fish even O_2 uptake may be limited by the chemical reaction with hemoglobin.

Linearity of blood dissociation curves. In the analysis, β_b has been assumed to be constant. For CO_2 and for O_2 in hypoxia this assumption constitutes a valid approximation; however, for O_2 in normoxia special methods must be employed to allow for the alinearity of the blood O_2 dissociation curve (Piiper and Scheid, 1977).

Bohr and Haldane effects. These effects describe the interactions among O_2 and H^+ (and CO_2) in binding to hemoglobin. These effects may be described as a dependence of β_{bO_2} not only on P_{O_2} but also on P_{CO_2}; and vice versa for β_{bCO_2}. Both effects have been shown to be favorable for gas exchange. A peculiar enhancement of CO_2 exchange by the Haldane effect in the crosscurrent system has recently been shown experimentally and theoretically (Meyer et al., 1976).

Comparison of the gas-exchange organs: gas-exchange efficiency

Figure 1.5 summarizes the structure of the three basic gas-exchange organs and the models used for their functional analysis. In this section we shall attempt to identify the differences in gas-exchange efficiency among the three types.

Gas-exchange efficiency

Various factors may be relevant for the efficiency of a gas-exchange organ. In particular, mechanical properties of the breathing apparatus in producing ventilatory flow and flow resistances at the medium and blood sites may be as important as the physical properties of the respired medium and of the blood. Gas-exchange efficiency in this discussion shall be used to identify differences in the gas-exchange performance among the functional models shown in Figure 1.5. We shall thus consider these systems under identical conditions, whereby the only difference concerns their structural arrangement.

The conductances defined in the model offer the possibility of quantitatively defining gas-exchange efficiency as the total conductance, G_{tot}, of a given model for given values of G_{vent}, G_{perf}, and G_{diff}.

Figure 1.6 shows G_{tot} against G_{diff} for $G_{vent} = G_{perf} = 1$ (left), and G_{tot} against G_{vent} for $G_{diff} \rightarrow \infty$ and $G_{perf} = 1$ (right). Both diagrams apply to the more general case in which G_{tot}/G_{perf} is plotted against G_{perf}/G_{diff} (for $G_{vent}/G_{perf} = 1$; left) or against G_{vent}/G_{perf} (for $G_{diff}/G_{perf} \rightarrow \infty$; right).

It can be seen from Figure 1.6 that there is a sequence in efficiency between the systems:

countercurrent $\geq$ crosscurrent $\geq$ pool

FIGURE 1.6 Comparison of gas-exchange efficiency in the three basic types of gas-exchange systems. (For details, see text.)

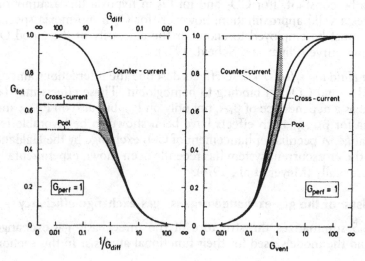

It should be noted that the extraction coefficient, defined as $(P_i - P_e)/P_i$, relates to the ratio of $\dot{M}/G_{vent}$ only (equations 1 and 2) and does not allow any conclusion as to the gas-exchange efficiency of the underlying gas-exchange system.

Gas-exchange efficiency as survival factor

It is tempting to assume that the gas-exchange efficiency constituted a significant parameter in the evolutionary development of the gas-exchange organs. It has in fact been argued that the higher tolerance of hypoxia of birds compared with mammals was due, in part or mainly, to the higher gas-exchange efficiency of the crosscurrent compared with the pool system.

Figure 1.6 shows that this view is at least doubtful. The plot shows that the large difference in gas-exchange efficiency among the systems is achieved only when G_{diff} is large. With decreasing G_{diff}, the differences disappear. It can be shown that a number of the deviations of real organs from the idealized model are equivalent to a reduction in G_{diff} and act thus in attenuating the differences in gas-exchange efficiency. Hence, in real life these differences are expected to remain significantly below their upper, theoretical limits.

Moreover, under conditions of high gas transfer (i.e., during exercise) all vertebrates appear to increase their ventilation almost proportionally with O_2 uptake, whereas the concomitant increases in cardiac output are usually limited. Thereby, G_{vent}/G_{perf} generally increases, from around unity during rest (shaded area in Figure 1.6, right) to significantly larger values, where differences in gas-exchange efficacy among the systems become increasingly attenuated.

We conclude that the large differences that theoretically exist in gas-exchange performance among the basic types of gas-exchange organs are probably much less pronounced in real organs and under physiological conditions. It seems thus not very likely that these differences can be regarded as significant determinants in the evolutionary development of the particular gas-exchange organs.

References

Dejours, P. (1975) *Principles of comparative respiratory physiology*. Amsterdam: North-Holland.

Duncker, H.-R. (1971) The lung air sac system of birds. *Ergebn. Anat. Entwickl.-Gesch.* 45, pt. 6.

Hughes, G. M., and Morgan, M. (1973) The structure of fish gills in relation to their respiratory function. *Biol. Rev.* 48:419–475.

Hughes, G. M., and Shelton, G. (1962) Respiratory mechanisms and their nervous control in fish. *Adv. Comp. Physiol. Biochem.* 1:275–364.

Johansen, K. (1971) Comparative physiology: Gas exchange and circulation in fishes. *Annu. Rev. Physiol.* 33:569–612.

Meyer, M., Worth, H., and Scheid, P. (1976) Gas–blood CO_2 equilibration in parabronchial lungs of birds. *J. Appl. Physiol.* 41:302–309.

Piiper, J., Dejours, P., Haab, P., and Rahn, H. (1971) Concepts and basic quantities in gas exchange physiology. *Respir. Physiol.* 13:292–304.

Piiper, J., and Scheid, P. (1975) Gas transport efficacy of gills, lungs, and skin: Theory and experimental data. *Respir. Physiol.* 23:209–221.

Piiper, J., and Scheid, P. (1977) Comparative physiology of respiration: Functional analysis of gas exchange organs in vertebrates. In *International review of physiology, respiration physiology II*, vol. 14 (J. G. Widdicombe, ed.), pp. 219–253. Baltimore: University Park Press.

Piiper, J., and Scheid, P. (1980) Blood–gas equilibration in lungs. In *Pulmonary gas exchange*, vol. I (J. B. West, ed.), pp. 131–171. New York: Academic Press.

Randall, D. J. (1970) Gas exchange in fish. In *Fish physiology*, vol. IV (W. S. Hoar and D. J. Randall, eds.), pp. 253–292. New York: Academic Press.

Scheid, P. (1979) Mechanisms of gas exchange in bird lungs. *Rev. Physiol. Biochem. Pharmacol.* 86:137–186.

Scheid, P., and Piiper, J. (1970) Analysis of gas exchange in the avian lung: Theory and experiments in the domestic fowl. *Respir. Physiol.* 9:246–262.

Scheid, P., and Piiper, J. (1972) Cross-current gas exchange in avian lungs: Effects of reversed parabronchial air flow in ducks. *Respir. Physiol.* 16:304–312.

Scheid, P., and Piiper, J. (1976) Quantitative functional analysis of branchial gas transfer: Theory and application to *Scyliorhinus stellaris* (Elasmobranchii). In *Respiration of amphibious vertebrates* (G. M. Hughes, ed.), pp. 17–38. London: Academic Press.

Scheid, P., and Piiper, J. (1980) Intrapulmonary gas mixing and stratification. In *Pulmonary gas exchange*, vol. I (J. B. West, ed.), pp. 87–130. New York: Academic Press.

Scheid, P., and Piiper, J. (1981) Functional inhomogeneities in gas exchange organs of non-mammalian vertebrates. In *Advances in Physiological Sciences*, vol. 10, *Respiration* (I. Hutás and L. A. Debreczeni, eds.). Budapest: Akadémiai Kiadó and Pergamon Press.

Schmidt-Nielsen, K. (1979) *Animal physiology: Adaptation and environment*, 2nd ed. Cambridge University Press.

2 Mount Everest and beyond: breathing air

PIERRE DEJOURS

Exercise at very high altitude, above 7000 m, is the greatest respiratory challenge that animals can encounter. Indeed, in this situation, the body, viewed as a system for the transfer of O_2 from the ambient air to the cells, must transport O_2 molecules at a rather high rate to cover the energy requirement of exercise in the face of an unfavorable ambient condition, namely the low O_2 partial pressure.

It has long been known that some birds fly or soar near 10 000 m. The highest recorded altitude for an airborne bird is that of a large vulture, Ruppel's griffon, which collided with an airplane at 11 280 m (Laybourne, 1974). The bar-headed goose can fly from near sea level to elevations as high as 9200 m within a few hours (Black and Tenney, 1980).

Birds can fly very high, but not for very long; they fly over a mountain range, then down for rest or search for food, and their extreme altitude excursion, at least in active flight, is brief. But a high-altitude excursion is another story for mammals. Because of their slow progression, mammals must carry a store of food sufficient to ensure their survival for days or weeks. Presumably only humans undertake such journeys, and to the load of food must be added all the equipment necessary for combatting extreme cold, blizzards, eventually very intense radiation, and so on. Yet man, thanks to his technical advancement, has been able to climb the highest summit in the world, Mount Everest, 8848 m (Figure 2.1). To date, five men – two in May and three in October 1978 – have climbed Mount Everest, by "fair means" (Messner, 1980, p. 174), that is, breathing air, without supplemental oxygen.

I well remember that 20 years ago, some physiologists, including myself, greatly doubted that humans could ever get to the top of Mount Everest without some oxygen breathing, either continuous or intermittent, the latter technique, the so-called oxygen shower, being that used by the Chinese expedition in 1975. Indeed about 80 humans have climbed Mount Everest breathing some oxygen. But how may we account for the achievement without oxygen breathing (Dejours, 1979, 1981; Ölz, 1979; West and Wagner, 1980)? In fact, we may only speculate, because there are no sure data concerning the barometric pressure, the work rate, the O_2 consumption, and the respiratory and circulatory variables near the

17

top. It is clear that what has to be explained is how one can *climb* the last leg of Mount Everest without breathing O_2. Once on the summit, many Mount Everest O_2-breathing climbers have removed their oxygen masks, but then their O_2 consumption was again relatively low.

Let us return to the two key factors: (1) low barometric pressure and (2) relatively high O_2 consumption required for the exercise of climbing.

FIGURE 2.1 Relationship between altitude and barometric pressure. On the abscissa, the partial pressure of inspired oxygen, P_{IO_2}, saturated with water vapor at 37 °C, is given for the various barometric pressures, P_B. At an ambient pressure of about 47 mm Hg (19 100 m), water boils. Arrow 1: altitude and barometric pressure at which most acutely exposed subjects lose consciousness or are on the verge of doing so; this altitude corresponds to inhalation of 10% oxygen at sea level. Arrow 2: the highest recorded permanent human habitat. Arrow 3: the maximal altitude (Mount Everest) that air-breathing subjects previously acclimatized to a lower altitude can tolerate for a few hours. Arrow 4: the maximal altitude that can be reached, without immediate death hazard, by a pilot breathing pure oxygen in a nonpressurized plane. Arrow 5: the maximal altitude for a pilot breathing oxygen at positive pressure (+ 20 mm Hg). Higher altitudes can be attained only in pressurized cabins or pressure suits.

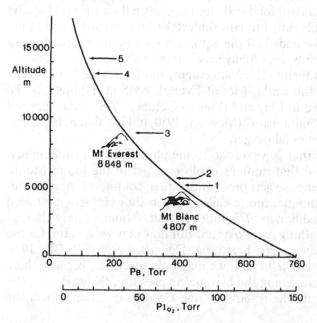

Barometric pressure

The table in the *Handbook of respiratory data on aviation* (1944) gives the barometric pressure, P_B, at 8848 m as 236 Torr. However, there is great doubt that the pressure is this low. According to Pugh (1957) and to West and Wagner (1980) it is higher. In 1979, I published some calculations concerning respiratory variables near the Mount Everest summit, in which I made two assumptions: (1) I used a value for atmospheric pressure of 247 Torr, which can be calculated from Figure 2 of Keys's paper (1936), and (2) I took as the pulmonary temperature a value of 37°C, and as the water vapor tension, 47 Torr. The wet air entering the lung would then have a partial pressure of O_2 of 42 Torr [(247 − 47) × 0.2095]. In the calculation of West and Wagner (1980) the ambient pressure is taken to be 250 Torr, and S. Lahiri (pers. commun.) thinks that the lung temperature may be lower than 37°C, and that, consequently, the water vapor tension at saturation would also be lower; this would allow a higher tracheal O_2 pressure. In fact, the hypothesis of S. Lahiri is important; because the percentage of water vapor in the alveolar gas at high altitude is large (19% at 37°C), a lower lung temperature would markedly influence the tracheal oxygen pressure and concentration (Figure 2.2). One must also consider that barometric pressure fluctuates, as it does at sea level; and the inspired O_2 tension being a key factor, these P_B variations may not be negligible (West and Wagner, 1980). Of course, on the final slope of Mount Everest, the atmospheric pressure is slightly higher than at the summit, but the pressure difference between a point 10 m below the summit and the summit itself is small because at this altitude the "barometric slope" is 29 m · Torr^{-1} instead of the 11 m · Torr^{-1} at sea level. In this text, except at one place, we shall assume that the tracheal inspired P_{O_2} is 42 Torr.

Oxygen consumption

One may attempt to estimate the O_2 consumption necessary for climbing the upper slopes of Mount Everest in two ways: (1) from an extrapolation of the oxygen consumption in prolonged exercise at lower altitudes, and (2) from the work rate on the slope of Mount Everest (Dejours, 1981).

Figure 2.3 shows the maximal O_2 consumption and the O_2 consumption during normal climbing (60% of maximal $\dot{M}_{O_2}$) at various altitudes from sea level to 7440 m (from Pugh et al., 1964). At 7440 m, the maximal oxygen consumption for short-term exercise was about 62 mmol · min^{-1}. Considering that the O_2 consumption in sustained exercise is 60% of that in maximal exercise (see discussion of this assumption in Åstrand and

Rodahl, 1977, p. 302), O_2 consumption during climbing at 7440 m, barometric pressure 300 Torr, would be about 37 mmol $\cdot$ min^{-1}. If an extrapolation down to 242 Torr is made (dotted line on Figure 2.1), the O_2 consumption for sustained exercise near the top of Mount Everest would be 21 mmol $\cdot$ min^{-1} (470 ml STPD $\cdot$ min^{-1}).

On Mount Everest, it takes over 8 h to climb from an altitude of about 8000 m to the top at 8848 m, and about 1 h for the last 60 m, that is, 1 m $\cdot$ min^{-1} (see Messner, 1980, pp. 137–144). If the subject with his load weighs 80 kg, his mechanical work rate is 80 kg $\cdot$ m $\cdot$ min^{-1}, that is, 785 J $\cdot$ min^{-1} or 13 W. To perform this mechanical work at an assumed efficiency of 20%, the rate of total energy expenditure would be 3925

FIGURE 2.2 Fractions (upper curves) and concentrations (lower curves) of nitrogen, oxygen, and water in a volume of air BIPS at 37 °C ($P_{H_2O} =$ 47 Torr) as functions of the change of barometric pressure with altitude. The bottom line gives the O_2 concentration in inspired tracheal air at 37 °C at sea level and at the top of Mount Everest.

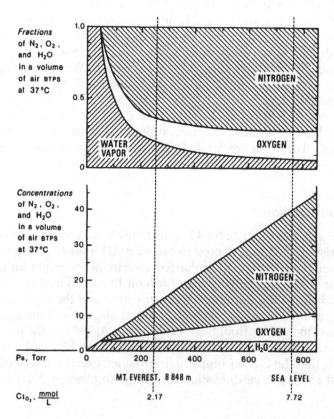

$J \cdot \min^{-1}$, which requires an O_2 consumption of about 8.7 mmol $\cdot$ min^{-1}. Adding this O_2 consumption to the O_2 consumption of the subject at rest ($\approx$ 12 mmol $\cdot$ min^{-1}), one obtains a total O_2 consumption for the last 60 m on Mount Everest of about 21 mmol $\cdot$ min^{-1}. But the efficiency of the mechanical work may be lower than 20% because of the technical difficulties and eventually the load for thermal regulation; 25 mmol $\cdot$ min^{-1} (560 ml STPD $\cdot$ min^{-1}) may be a more acceptable value of $\dot{M}_{O_2}$ near the summit. This value is more than the extrapolated value of Figure 2.1, but those very fit mountaineers who climbed Mount Everest without oxygen breathing may well have had a higher maximal O_2 consumption than the subjects of the experiment of Figure 2.1.

Reactions of the respiratory system

How is it possible to transfer 25 mmol $\cdot$ min^{-1} of oxygen from the air to the mitochondria when the pressure head of O_2 in the ambient air is only 42 Torr? Obviously, the conductances at the various stages of O_2 transfer

FIGURE 2.3 Top curve is maximum oxygen consumption at the end of a few minutes' exercise, in which fit young adults worked on a bicycle ergometer, as a function of the barometric pressure from sea level (760 Torr) to 7440 m (300 Torr). Bottom curve is derived and represents 60% of the maximal $\dot{M}_{O_2}$, a rate of oxygen consumption that may be sustained for several hours. Dotted lines are these curves extrapolated to lower barometric pressures in order to estimate $\dot{M}_{O_2}$ of maximal and sustained exercises near the top of Mount Everest. [Drawn from data given in Pugh et al., 1964]

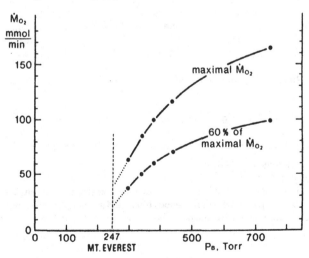

must increase greatly. At sea level, in the example of Table 2.1 (columns 1 and 2), the overall I, $\bar{v}$ difference for O_2 is 108 Torr at rest and 126 Torr during exercise. If we assume that $P\bar{v}_{O_2}$ at very high altitude cannot be lower than 17 Torr ($S\bar{v}_{O_2} = 36\%$), then the P_{O_2} difference between inspired gas and mixed venous blood is reduced to 25 Torr (42 − 17). How is this difference distributed at the intermediate transfer steps: (1) inspired air to alveolar gas, (2) alveolar gas to arterial blood, and (3) arterial blood to mixed venous blood?

Alveolar gas composition

The P_{O_2} difference between inspired and alveolar gas is.51 Torr at sea level, and 37 Torr at 4510 m (Table 2.1, columns 3 and 4). The highest

TABLE 2.1 Respiratory data in inspired air, in alveolar gas, and in arterial and mixed venous blood, for subjects at rest or exercising at sea level (columns 1 and 2); at rest at sea level and at 4510 m (columns 3 and 4); and during climbing near the summit of Mount Everest (column 5).

	1[a]	2[a]	3[b]	4[b]	5	6[c]
Altitude (m)	Sea level	Sea level	Sea level	4510	8848	≈9500
P_B (Torr)	760	760	760	446	247	225
$\dot{M}_{O_2}$ (mmol · min^{-1})	12.5	108	12.6	9.7	25	12.5
Condition	Rest	Exercise	Rest	Rest	Exercise	Rest
$P_{I_{O_2}}$ (Torr)	150	149	150	83	42	37
$P_{A_{O_2}}$ (Torr)	97	108	99	46	30	24
Pa_{O_2} (Torr)	91	92	94	44	22	19
$P\bar{v}_{O_2}$ (Torr)	42	23	39	33	17	17.3
$C_{HbO_2}{}^{max}$ (mmol · L^{-1})	9.1	—	9.4	12.6	13.5	13.5
Sa_{O_2} (%)	96	96	97	73	50	62
$S\bar{v}_{O_2}$ (%)	75	24	74	55	36	36
Pa_{CO_2} (Torr)	40	36	40	32	11	11.5
$P\bar{v}_{CO_2}$ (Torr)	—[d]	—	47	36	13.7	12.9
pH_a	7.42	7.36	7.40	7.40	7.58	7.58
pH_v	—	—	—	—	7.56	7.56

[a] Data from Ekelund and Holmgren (1964).
[b] Data from Torrance et al. (1970/71).
[c] Like column 5, column 6 is hypothetical; it indicates the lowest barometric pressure attainable by the subject of column 5 having about the same venous blood and a resting O_2 consumption of 12.5 mmol · min^{-1} (see text). Values of partial pressures in columns 3, 4, and 5 are illustrated in Figures 2.5 and 2.6.
[d] No data.

possible alveolar ventilation, $\dot{V}_A$, would seem to be the most advantageous because the ($P_{I_{O_2}} - P_{A_{O_2}}$) difference would be less, and there would be a higher pressure head to work with beyond the alveolar gas. However, such an increase of alveolar ventilation entails a fall of alveolar P_{CO_2}, which, if it is not accompanied by a proportionate fall of [HCO_3^-], leads to a hypocapnic alkalosis. For this reason, it seems improbable that alveolar P_{CO_2} would be lower than 10 Torr, that is, about one-fourth its sea-level value. Because, presumably, [HCO_3^-] does not fall proportionately to the fall of $P_{A_{CO_2}}$, then a state of hypocapnic alkalosis should exist in air breathers on the upper slopes of Mount Everest. Even at a lower altitude (5800 m), an alkalosis has been reported in two out of three exercising subjects (Pugh et al., 1964). Furthermore, Messner (1980, p. 141) reported that his teammate, P. Habeler, who also climbed Mount Everest without breathing oxygen, had two attacks of involuntary flexions of the fingers. Another Mount Everest climber, breathing air on the summit, had facial muscle cramps and a lip tremor, and could not close his mouth (Messner, 1980, p. 116). These symptoms presumably were tetanic spasms, a common symptom of marked alkalosis.

If the alveolar ventilation were 40 liters $\cdot$ min^{-1}, the O_2 consumption 25 mmol $\cdot$ min^{-1}, and the respiratory quotient, R, 0.9, the alveolar P_{CO_2} on Mount Everest would be 11 Torr, and the alveolar P_{O_2} 30 Torr.

Why assume a respiratory quotient of 0.9? It is an important assumption, for if $P_{A_{CO_2}}$ is unchanged at 11 Torr, $P_{A_{O_2}}$ would be 32 Torr for $R = 1$ (oxidation of carbohydrates) and 27 Torr for $R = 0.7$ (oxidation of lipids). But both these R values are quite improbable (Åstrand and Rodahl, 1977, p. 487 et seq.); R never reaches 1 in exercise even if the subject has ingested a large quantity of carbohydrates in the hours preceding the exercise; and an R value of 0.7, typical of lipid catabolism, is observed only after several hours of exercise without carbohydrate ingestion when the fuel for energy metabolism is provided by the fatty acids. There is no detailed description of the nature of the food intake during the 8-h climbing of the last slopes of Mount Everest, but certainly the climbers drank something sweet, like sugared tea. In the absence of more information, I feel that an R value of about 0.9 is the most reasonable assumption.

Alveolar–arterial P_{O_2} difference: arterial blood

The evaluation of the ($A - a)P_{O_2}$ difference is crucial. West et al. (1962) reported alveolar–arterial P_{O_2} differences of 8–20 Torr during mild exercise at 5800 m (380 Torr). Calculations show that a difference of much more than 8 Torr is very improbable, if not impossible. Assuming an ($A - a)P_{O_2}$ difference of 8 Torr on the upper slopes of Mount Everest, the arterial O_2 pressure would be 22 Torr. Then the O_2 and CO_2 concentrations and the acid–base balance in the arterial blood would depend on the physicochemical properties of the blood, for which there are apparently no more recent complete data than those of Dill et al. (1937b), which I use here. These authors studied the blood of 10 male high-altitude natives living at 5.34 km and climbing every day to 5.80 km to work in the mines (Keys, 1936). One of the main characteristics of their blood was the high [Hb], 13.5 mmol · liter^{-1}, whereas the blood [Hb] of sea-level natives examined at sea level was 9.0 mmol · liter^{-1} (Dill et al., 1937a) (Figure 2.4). Higher [Hb] seems to have no adaptive value; instead it is a characteristic of illness, chronic mountain polycythemia. For the blood of high-altitude and sea-level natives, Dill et al. (1937a) measured an oxygen affinity at pH 7.4 of about 25–26 Torr. Since 1937, many authors have measured the blood O_2 affinity, stating precisely the conditions of measurement, yet the existence of change or its direction in high-altitude natives is not clear; any change there is does not seem to be very important in humans (see Torrance et al., 1970/71) and may be negligible (Samaja et al., 1979). Anyway it is not obvious what the physiological consequence of a shift of the O_2 dissociation curve would be: Whereas a decrease of O_2 affinity favors O_2 unloading in the systemic capillaries, at very high altitude it impairs O_2 loading by blood in the pulmonary capillaries (see Eaton et al., 1974; Petschow et al., 1977). As

FIGURE 2.4 Oxygen concentration of blood vs. P_{O_2} in sea-level subject at rest, and in a subject climbing the last meters to the top of Mount Everest (see text). Plot shows the arterial and mixed venous blood points in both subjects and the higher maximal concentration of HbO_2, $C_{HbO_2}^{max}$ (see Table 2.1).

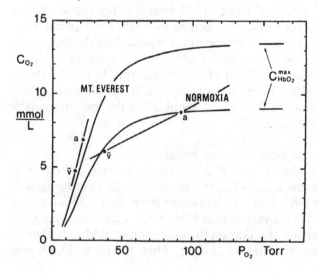

we shall see later, the main characteristic of the excellent high-altitude flyer, the bar-headed goose, is its very high blood O_2 affinity, and calculations by West and Wagner (1980) show also that a leftward shift of the human O_2 dissociation curve favors O_2 uptake.

Obviously the difference between the alveolar and arterial P_{O_2} values depends essentially on the values given to the alveolar air–blood O_2 transfer factor, or pulmonary diffusing capacity for oxygen, $D_{L_{O_2}}$. For our calculation we have taken a relatively high $D_{L_{O_2}}$ value of 2.4 mmol $\cdot$ min^{-1} $\cdot$ Torr^{-1}, which is nonetheless lower than the value of 2.8 assumed by Piiper (1980). For either value, arterial blood P_{O_2}'s are in fact similar. However, with the $D_{L_{O_2}}$ value calculated on morphological data by Weibel's group (Gehr et al., 1978), which is about three times higher than the value we used, the alveolar–arterial P_{O_2} diffusion is reduced to a few Torr. In all these calculations we assume that (1) the blood oxygen dissociation curve is linear over the range that is concerned, as is approximately the case in fact, and (2) the unevenness of the $\dot{V}_A/\dot{Q}$ distribution, and the venous admixture, have no important effect on the (A − a)P_{O_2} difference at high altitude, as shown by Rahn and Farhi (1964). To summarize, it is clear that an irritating unknown remains concerning the alveolar–capillary O_2 transfer factor, and that, as pointed out also by Piiper (1980) and by West and Wagner (1980), this factor is crucial.

Finally, taking values of arterial blood P_{O_2} and P_{CO_2} of 22 Torr and 11 Torr respectively, the use of the Dill et al. (1937b) nomogram shows that arterial C_{O_2} is 6.7 mmol $\cdot$ liter^{-1}, S_{O_2} is about 50%, and pH is 7.58. One will note (Figure 2.4) that the O_2 concentration in arterial blood on Mount Everest is not much lower than at sea level, as a result of the high [Hb] and the increase of affinity resulting from the arterial blood alkalosis (Bohr effect).

Arterial–mixed venous blood differences

The O_2 and CO_2 status in the mixed venous blood depends on the composition of the arterial blood, on the physicochemical properties of the blood, and on the cardiac output. It seems obvious that the higher the cardiac output, the lower would be the arteriovenous differences and the higher the oxygenation of the venous blood and the mean systemic capillary P_{O_2}. However, the cardiac output at high altitude cannot be very high (Cerretelli, 1976; Åstrand and Rodahl, 1977, p. 631). The value of 12 liters $\cdot$ min^{-1} is the most probable for a sustained exercise near the top of Mount Everest. With this figure we obtain the following values: $P\bar{v}_{O_2} = 17$ Torr; $C\bar{v}_{O_2} = 4.8$ mmol $\cdot$ liter^{-1}; $S\bar{v}_{O_2} = 36\%$; $P\bar{v}_{CO_2} = 13.7$ Torr; pH$\bar{v} = 7.56$.

For an oxygen consumption of 25 mmol $\cdot$ min^{-1}, calculations with a cardiac output higher than 12 liters $\cdot$ min^{-1} do not lead to a much less hypoxic mixed venous blood. Also we should consider that a higher cardiac output or a higher ventilation would mean an increase of the air and blood convection rates. As Otis pointed out in 1954, either rate increase would cost energy, and the seeming advantage – an increase of the O_2 convective transport – might very well be offset, or more than offset, by the concomitant increase of the energy requirement and the O_2 consumption.

Table 2.1 summarizes certain respiratory variables at various altitudes, at rest and in exercise; column 5 gives the hypothetical figures for the air-breathing Mount Everest mountaineer. Figure 2.5 is drawn from columns 3 and 4 (Torrance et al., 1970/71) and column 5. The data of Figure 2.5 are replotted on Figure 2.6, a P_{CO_2} vs. P_{O_2} diagram of Fenn et al. (1946).

As for the next step of the O_2 transfer, namely from capillary blood to mitochondria, no pertinent information exists. No doubt, the fit subjects in question have an excellent skeletal muscle capillary supply, and their enzymatic machinery is at its optimum development. But presumably the limits to exercise are not at the tissue level, but in the concentration and pressure of O_2 in the systemic capillaries.

In my opinion, the two main limiting factors to be considered are: the danger of pathological hypocapnic alkalosis due to a relative hyperven-

FIGURE 2.5 Partial pressures of O_2 and CO_2, P_{O_2} and P_{CO_2}, in ambient air and in the various compartments of the gas-exchange system of a man at rest, at sea level; at rest, at 4510 m; in climbing, near the top of Mount Everest. The partial pressures of O_2 and CO_2 in the interstitial fluid and inside the cells are not well known.

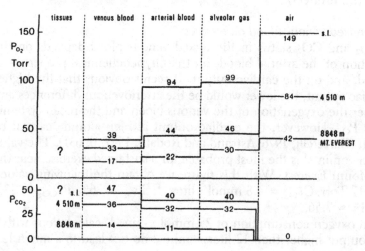

tilation in spite of some metabolic compensation of the alkalosis, and the O_2 transfer factor from alveolar gas to pulmonary capillary blood ($D_{L_{O_2}}$).

These speculations, based on what data we have, show how the problem of O_2 and CO_2 transfer and acid–base balance in exercising humans at very high altitude can be approached. We now look forward to the pertinent measurements to be made during the forthcoming American medical research expedition to Everest (fall 1981) headed by John West.

Beyond Mount Everest

Man

Could man climb higher than Mount Everest? Maybe a little higher, if the slope were moderate and the mountaineering conditions excellent, but certainly not much higher, in view of the fact that it takes more than 10 min to climb the last 10 m on Mount Everest. However, it is known that man adapted to a lower altitude, at rest, breathing air, can be brought by plane or in a pressure chamber (Velasquez, 1959) to a higher altitude and lower pressure than Mount Everest's. Actually one can calculate what barometric pressure the subject of Table 2.1, column 5, could bear having about the same mixed venous blood, but a resting O_2 consumption

FIGURE 2.6 Partial pressures of O_2 and CO_2 in inspired (*I*) and alveolar (*A*) gas and in arterial (*a*) and mixed venous blood (*v̄*) in resting sea-level natives at sea level, in resting high-altitude natives (4510 m) at high altitude, and in air-breathing climbers near the top of Mount Everest. Values from columns 3, 4, and 5 of Table 2.1. The Mount Everest figures are speculative (see text).

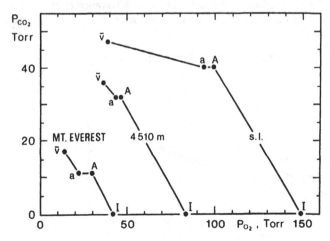

of 12.5 mmol · min^{-1}, an R value of 0.85, and assumed values for other respiratory variables as in Table 2.1, column 6. The calculation yields (column 6) a P_{IO_2} value of 37.3 Torr, one that, according to Keys (1936), is available in inspired BTPS (37°C) air at an altitude of about 9500 m ($P_B \simeq 225$ Torr).

Birds

Although man is most probably the only mammal to have walked to the summit of Mount Everest, many species of birds have been seen flying at this altitude and above, as mentioned at the beginning of this chapter. Black and Tenney (1980) have made a systematic comparative study of bar-headed geese, which have been observed crossing directly over the summit of Mount Everest, and Pekin ducks. In particular, after preadapting both species at 5640 m for 4 weeks, they compared their respiratory reactions when exposed to marked hypoxia, down to an oxygen pressure of 23 Torr, a pressure prevailing at 11 520 m. The most remarkable characteristic of the bar-headed goose was its relatively high blood O_2 affinity (P_{50} is 14 Torr lower in the bar-headed goose than in the duck). However, even Pekin ducks tolerate very severe hypoxia, although less well. Presumably another factor, related to the characteristics of the air-to-blood O_2 transfer in birds, must be taken into consideration to account for their high tolerance to marked hypoxia compared to mammals: In all observations at very high altitude reported by Black and Tenney, the difference between inspired and arterial P_{O_2} was only a few Torr (Figure 2.7), whereas it is 20 Torr in our example of man on Mount

FIGURE 2.7 Partial pressures of oxygen in inspired air and in arterial blood, P_{IO_2} and Pa_{O_2}, in the bar-headed goose and in man, as functions of the barometric pressure, P_B. [Data for goose, from Black and Tenney, 1980; data for man, from Velasquez, 1959, and Torrance et al., 1970/71; for the data from Velasquez, an alveolar–arterial P_{O_2} difference of 6 Torr assumed]

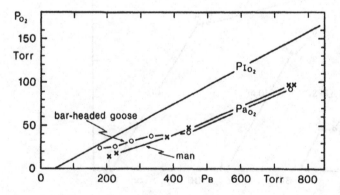

Everest, a figure that is possibly too small. This peculiarity of the inspired-air-to-arterial-blood O_2 transfer in birds may be related to the efficiency of the crosscurrent disposition in the parabronchial lung of these animals.

References

Åstrand, P.-O., and Rodahl, K. (1977) *Textbook of work physiology.* New York: McGraw-Hill. 681 pp.

Black, C. P., and Tenney, S. M. (1980) Oxygen transport during progressive hypoxia in high-altitude and sea-level waterfowl. *Respir. Physiol.* 39:217–239.

Cerretelli, P. (1976) Limiting factors to oxygen transport on Mount Everest. *J. Appl. Physiol.* 40:658–667.

Dejours, P. (1979) L'Everest sans oxygène: le problème respiratoire (abstract). *J. Physiol. (Paris)* 75:43A.

Dejours, P. (1981) *Principles of comparative physiology of respiration,* 2nd ed. Amsterdam and New York: North-Holland/American Elsevier.

Dill, D. B., Edwards, H. T., and Consolazio, W. V. (1937a) Blood as a physicochemical system. II. Man at rest. *J. Biol. Chem.* 118:635–648.

Dill, D. B., Talbott, J. H., and Consolazio, W. V. (1937b) Blood as a physicochemical system. XII. Man at high altitude. *J. Biol. Chem.* 118:649–666.

Eaton, J. W., Skelton, T. D., and Berger, E. (1974) Survival at extreme altitude: positive effect of increased hemoglobin oxygen affinity. *Science* 183:743–744.

Ekelund, L. G., and Holmgren, A. (1964) Circulatory and respiratory adaptation during long-term, non-steady state exercise, in the sitting position. *Acta Physiol. Scand.* 62:240–255.

Fenn, W. O., Rahn, H., and Otis, A. B. (1946) A theoretical study of the composition of the alveolar air at altitude. *Am. J. Physiol.* 146:637–653.

Gehr, P., Bachofen, M., and Weibel, E. R. (1978) The normal human lung: Ultrastructure and morphological estimation of diffusion capacity. *Respir. Physiol.* 32:121–140.

Handbook of respiratory data on aviation. (1944) Washington, D.C.: Committee on Medical Research.

Keys, A. (1936) The physiology of life at high altitudes. The international high altitude expedition to Chile, 1935. *Sci. Monthly* 43:289–312.

Laybourne, R. C. (1974) Collision between a vulture and an aircraft at an altitude of 37,000 feet. *Wilson Bull.* 86:461–462.

Messner, R. (1980) *Everest. Expedition zum Endpunkt.* München: Dromer Knaur. 200 pp.

Ölz, O. (1979) Everest ohne Sauerstoff: die medizinischen Grundlagen. In *Der einsame Sieg* (P. Habeler, ed.), pp. 215–223. München: Wilhem Goldmann. 223 pp.

Otis, A. B. (1954) The work of breathing. *Physiol. Rev.* 34:449–458.

Petschow, D., Würdinger, I., Baumann, R., Duhm, J., Braunitzer, G., and Bauer, C. (1977) Causes of high blood O_2 affinity of animals living at high altitude. *J. Appl. Physiol.* 42:139–143.

30 PIERRE DEJOURS

Piiper, J. (1980) Oxygen uptake at high altitude: Limiting role of diffusion in lungs. In *High altitude physiology and medicine*, vol. I, *Physiology of adaptation* (W. Brendel and R. A. Zink, eds.), New York: Springer.

Pugh, L. G. C. E. (1957) Resting ventilation and alveolar air on Mount Everest: with remarks on the relation of barometric pressure to altitude in mountains. *J. Physiol. (London)* 135:590–610.

Pugh, L. G. C. E., Gill, M. B., Lahiri, S., Milledge, J. S., Ward, M. P., and West, J. B. (1964) Muscular exercise at great altitudes. *J. Appl. Physiol.* 19:431–440.

Rahn, H., and Farhi, L. E. (1964) Ventilation, perfusion, and gas exchange – the $\dot{V}A/\dot{Q}$ concept. In *Handbook of physiology*, section 3, Respiration, vol. 1 (W. O. Fenn and H. Rahn, eds.), pp. 735–766. Washington, D.C.: American Physiological Society. 926 pp.

Samaja, M., Veicsteinas, A., and Cerretelli, P. (1979) Oxygen affinity of blood in altitude Sherpas. *J. Appl. Physiol.* 47:337–341.

Torrance, J. D., Lenfant, C., Cruz, J., and Marticorena, E. (1970/71) Oxygen transport mechanisms in residents at high altitude. *Respir. Physiol.* 11:1–15.

Velasquez, T. (1959) Tolerance to acute anoxia in high altitude natives. *J. Appl. Physiol.* 14:357–362.

West, J. B., Lahiri, S., Gill, M. B., Milledge, J. S., Pugh, L. G. C. E., and Ward, M. P. (1962) Arterial oxygen saturation during exercise at high altitude. *J. Appl. Physiol.* 17:617–621.

West, J. B., and Wagner, P. D. (1980) Predicted gas exchange on the summit of Mount Everest. *Respir. Physiol.* 42:1–16.

3

The pathway for oxygen: lung to mitochondria

EWALD R. WEIBEL

In most terrestrial mammals aerobic metabolism plays a dominant role in generating the energy (in the form of adenine triphosphate, ATP) used by the muscles during prolonged running. The rate of O_2 consumption is therefore a good measure of an animal's energy metabolism (Chapter 12).

The oxidative generation of ATP is performed in the mitochondria, specifically on the respiratory chain units packed into their inner mitochondrial membranes. Oxygen is delivered to the cells from the capillary blood, and the blood is continuously replenished in the lung with O_2 from the store in environmental air. The pathway for O_2 can therefore be broken up into a number of steps, and one can define the flow rate at each step as the product of a prevailing driving force and an associated conductance for O_2 (Figure 3.1). In the model shown in Figure 3.1 (Weibel and Taylor, 1981) the driving forces are O_2 partial pressure differences. The conductances are of three kinds: In air and in the blood they depend on the mass flow rate of the carrier, multiplied by its O_2 capacitance (the quantity of O_2 that can be loaded onto the carrier for a unit increase in partial pressure); the transfer conductances from air to blood or from blood to cells are diffusing capacities; whereas the final conductance in the mitochondria is an as yet ill-defined "molecular" conductance, related to the enzyme kinetics of the respiratory chain units.

Some of these conductances are determined by structural design properties of the system. Branched airway tubes guide inspired air into the lung, delivering O_2 to the alveoli. A close contact between air in alveoli and blood in capillaries is established across a very thin but extensive tissue barrier through which O_2 is transferred from air to blood. The blood is distributed to the various parts of the body by means of a network of blood vessels to the expansive capillary network that pervades all tissues. The transfer of O_2 from the blood to the cells depends on the distance from the capillaries to the mitochondria, where O_2 is consumed, and this is related to the density of capillaries in the tissue. Finally, the O_2 consumption in the cells must be related in some manner to the number of mitochondria in the cells.

31

The flow of O_2 across this respiratory system is essentially determined by the O_2 required for energy conversion in the respiratory chains in the mitochondria. In order to maintain steady-state conditions the flow of O_2 at each level of the respiratory system must equal the O_2 flow into the mitochondrial sink.

Structural design as limiting factor of O_2 flow rate

The basis of the present considerations is the hypothesis that the structural design of the pathway for O_2 is quantitatively matched to the O_2 flow rates. If this hypothesis is valid, one would expect that the structural properties defining the various conductances can have a limiting effect on the O_2 flow rate at each step of the pathway. In other words, we would expect the lung to maintain just enough alveolar surface, the tissues just enough capillaries, and the muscle cells just enough mitochondria to satisfy the O_2 flow requirements under conditions of maximal aerobic metabolism.

How can we test this hypothesis? The general approach is to compare animals that differ sufficiently in terms of their maximal rate of O_2 consumption, $\dot{V}_{O_2max}$, measured, for example, while they run on a treadmill (Seeherman et al., 1981), and to estimate by morphometry the magnitude

FIGURE 3.1 Model of respiratory system indicating O_2 flow rates, driving forces, and conductances. At levels A (air) and B (blood), the conductances are related to mass flow rates ($\dot{V}$) multiplied by the capacitance (β) of the medium for O_2. At the transfer levels A-B and B-C, the conductances are "diffusing capacities" determined in part by morphometric properties. In the mitochondria (C), the conductance is related to the molecular makeup (respiratory chains) of the mitochondrial membranes. [From Weibel and Taylor, 1981]

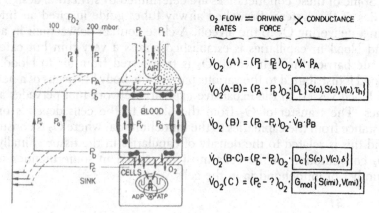

of the structural parameters that determine the conductances at the various levels of the pathway for O_2. The hypothesis is valid if the conductances are proportional to the maximal O_2 flow rates measured.

A particularly promising approach is to look at animals of largely varying body mass, M_b, and to exploit the fact that maximal $\dot{V}_{O_2}$ varies nonlinearly with M_b, namely about with $M_b^{0.8}$ (Chapter 12). The mass-specific $\dot{V}_{O_2max}$ of a 20-g mouse is about 12 times as high as that of a 500-kg cow. If the structural parameters determining the conductances are matched to the O_2 flow requirements, then we would expect them to scale also to $M_b^{0.8}$. This experiment of nature is particularly useful if we are interested in looking at the entire pathway from the lung to the mitochondria.

The following account is based primarily on a study of a group of African mammals on which measurements of $\dot{V}_{O_2max}$ were made (Chapter 12), followed by a morphometric evaluation of the lungs and various muscles (Weibel and Taylor, 1981).

The muscles' O_2-consuming machinery: mitochondria

The respiratory chain units, where O_2 is consumed, are densely packed into the inner mitochondrial membranes which form the characteristic cristae (Figure 3.2). The structural parameter that determines the number of such units present is the surface area of the inner mitochondrial membranes. It appears that mitochondria of skeletal muscles have a characteristic density in the inner membranes (Hoppeler et al., 1981a), and that, as a first approximation, the volume of mitochondria can serve as a morphometric correlate to the muscle cell's potential to generate ATP aerobically by consuming O_2. The validity of this concept is seen by comparing different types of muscle fibers (Figure 3.3). Some muscle fibers generate their ATP mostly anaerobically, that is, directly from glycolysis ("white fibers"), whereas others ("red fibers") depend on aerobic metabolism. It is seen from Figure 3.3 that the aerobic fibers contain a large mitochondrial complement, whereas the "white fibers" have only a few very small mitochondria (Hoppeler et al., 1981a).

In order to quantitatively relate the mitochondrial content of muscle cells to the level of oxidative metabolism we can compare the total mitochondrial volume density, $V_V(mt)$ (the volume of mitochondria contained in 1 cm^3, or g, of muscle tissue), to the maximal O_2 consumption per gram body weight, $\dot{V}_{O_2max}/M_b$. It is known from studies on man that in well-trained athletes $V_V(mt)$ is increased in proportion to the increase in $\dot{V}_{O_2max}/M_b$ (Hoppeler et al., 1973). How does $V_V(mt)$ vary with animal size? Does it scale similarly to $\dot{V}_{O_2max}/M_b$, that is, about to $M_b^{-0.2}$?

In our African study we have examined four functionally different

FIGURE 3.2 Mitochondrion of skeletal muscle fiber showing inner mito-chondrial membranes forming cristae.

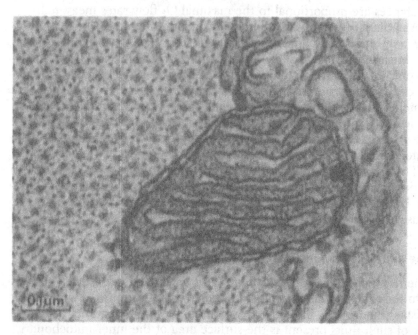

muscles that are highly active when animals run at top speed (Mathieu et al., 1981): the diaphragm as the main inspiratory muscle, a flexor and an extensor of the hind leg, the semitendinosus and vastus medialis muscles, and an extensor of the spine, the longissimus dorsi. Figure 3.4 shows that the mitochondrial densities scale very differently for the four muscles, and only that of the semitendinosus appears to be proportional to the mass-specific $\dot{V}_{O_2max}$. Does that mean that the mitochondria of muscles are not related to the muscles' capacity for aerobic metabolism? The results from the study of athletes quoted above do not justify such a conclusion.

It may be that the volume *density* of mitochondria is not a sufficient parameter to fully describe the relationship between $\dot{V}_{O_2max}$ and mito-chondrial O_2 consumption. It would be sufficient if the total volume of the muscles considered increased in simple proportion to body mass, but

FIGURE 3.3 Comparison of cross sections of two portions of adjacent muscle fibers showing considerable differences in their content in sub-sarcolemmal (ms) and in interfibrillar or central (mc) mitochondria, in diaphragm (a) and in semitendinosus muscle (b) of dwarf mongoose. [From Mathieu et al., 1981]

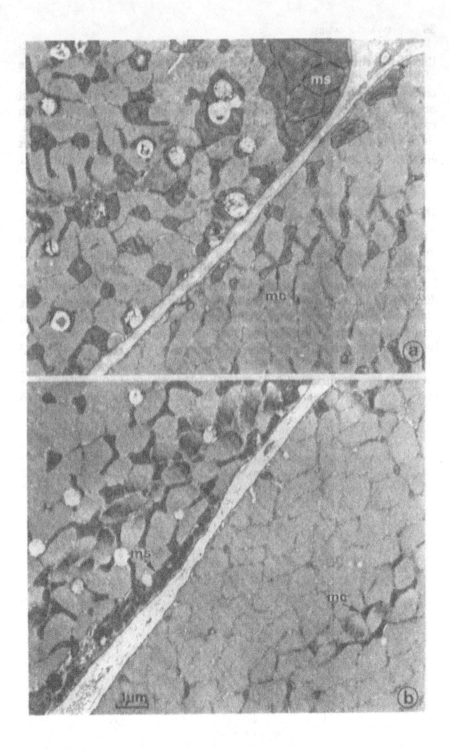

that is not the case. Alexander (Chapter 21) has measured the volume of some of the muscles on a large size range of African mammals. If we calculate the *total* mitochondrial volume of each muscle (the product of muscle volume times mitochondrial volume density), we find that in each case it scales in proportion to $\dot{V}_{O_2max}$, as shown in Figure 3.5 (Mathieu et al., 1981). Thus, the amount of mitochondria in these muscles appears closely matched to the O_2 consumed during heavy muscular work.

The O_2 supply to cells: capillaries

The general notion that the capillary density limits the O_2 supply to muscle cells goes back to the pioneering studies of August Krogh (1922). Because we have shown mitochondrial volume is related to the rate of O_2 consumption, we can now ask whether the mitochondrial volume density is matched by a proportional density of blood capillaries. The comparison of cross sections of muscle bundles from diaphragm and semitendinosus muscle of a small African gazelle in Figure 3.6 allows two observations: (1) The diaphragm has a much higher mitochondrial volume density than the semitendinosus muscle (see also Figure 3.4), and this appears matched by a larger relative number of capillary cross sections in the diaphragm; (2) the mitochondria are not evenly distributed

FIGURE 3.4 Regression lines of log $V_V(mt, f)$ vs. log M_b in diaphragm (D), vastus medialis muscle (VM), longissimus dorsi muscle (LD), and semitendinosus muscle (ST). The regression line of log $(\dot{V}_{O_2max}/M_b)$ vs. log M_b is added (dashed line) for comparison. [From Mathieu et al., 1981]

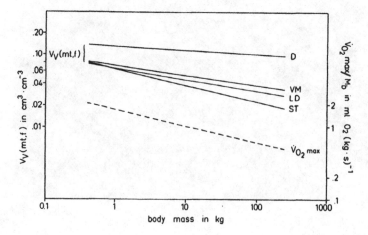

throughout the muscle fibers but rather appear to concentrate toward the surface, clustering in part close to the capillaries. This is quite suggestive of a close relation between the capillaries and the mitochondria they supply.

To obtain relevant morphometric measurements describing the capillary network is not a simple matter. For one thing, it is highly but not totally aligned with the longitudinal course of muscle fibers, and the stereological methods are not well worked out to account for that. A reasonable approach is to estimate the number of capillary cross sections per unit area of muscle or muscle fiber, N_A, which is an approximate estimate of the capillary length per unit volume of muscle, $J_V(c)$ (Hoppeler et al., 1981b). This can then be directly related to the volume density of mitochondria because the relative area of a fiber cross section covered by mitochondrial profiles is directly proportional to $V_V(mt)$ (see Weibel, 1979b).

If this is done for a large number of different muscles taken from the series of African mammals (Hoppeler et al., 1981b), one finds that, on the average, a higher mitochondrial density is associated with a higher capillary density, but that the data show a remarkable scatter (Figure 3.7). The relationship improves somewhat if one includes heart muscle and much smaller species, down to the smallest mammals (the Etruscan

FIGURE 3.5 Regression lines of log V_{mt} vs. log M_b in diaphragm (D), semitendinosus muscle (ST), and vastus medialis muscle (VM). The regression line of log ($\dot{V}_{O2max}$) vs. log M_b is added (dashed line) for comparison. [From Mathieu et al., 1981]

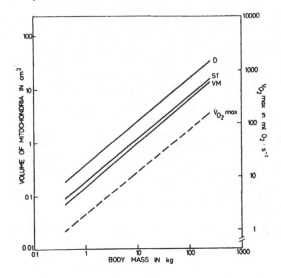

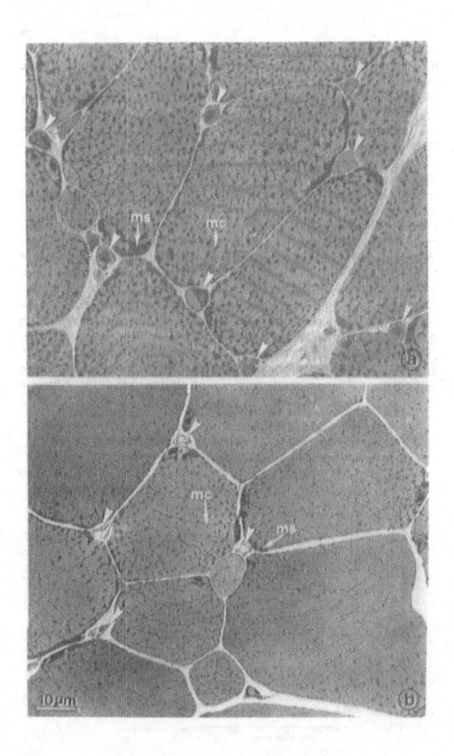

shrew with 45% mitochondria in the myocardial cells), but the scatter is still appreciable.

Thus it appears that there is a consistent trend for the capillary supply of muscle fibers to be quantitatively related to their mitochondrial content, but apparently not in a simple fashion. Two reasons can be evoked to help explain the great variability. First, it seems that the capillary density is not constant along the entire length of the capillary path. It appears to increase toward the venular end of the capillary network, possibly to compensate for the fall in capillary O_2 partial pressure. Second, and perhaps of more importance, capillaries supply not only O_2 to the cells but also nutrients. The need for substrates for glycolysis may help explain why the capillary supply of white or glycolytic fibers is much more abundant than would be predicted from their very low mitochondrial content. We must refine our model to consider the entire "energy flow" from the capillary blood into the cells.

O_2 supply to the blood: the lung's diffusing capacity

In the lung the situation is simpler. The lung's dominant function is to transfer O_2 from the environmental air to the blood's erythrocytes, and to remove CO_2. Under conditions of sustained heavy work, CO_2 production and O_2 consumption are about equal.

The first step in taking O_2 from the store in environmental air is to inspire a certain quantity of air deep into the lung, close to the alveolar surface where gas exchange between air and blood takes place (Figure 3.8). The main characteristics of this pulmonary gas-exchange region are that it establishes a very large surface of contact between air and blood – about the size of a tennis court in man – and that the tissue barrier separating air and blood is extremely thin (Figure 3.9). The diffusion conductance across this barrier, called the pulmonary diffusing capacity for oxygen, D_{LO_2}, is determined by a number of morphometric properties such as the total alveolar and capillary surface areas, the capillary blood volume, and the harmonic mean barrier thickness, together with some material constants such as permeability coefficients and the rate of O_2 binding of red blood cells. The model by which these morphometric

FIGURE 3.6 Comparison of cross sections of portions of muscle bundles showing considerably larger amounts of both capillary (arrow heads) and mitochondria profiles in diaphragm (a) than in semitendinosus muscle (b) from suni. Note that mitochondria (subsarcolemmal, ms, and interfibrillar or central, mc) are peculiarly abundant in portions of the muscle fibers located near the capillaries, in both muscles. [From Hoppeler et al., 1981b]

parameters can be combined to give an estimate of D_{LO_2} is not to be discussed here (Weibel, 1970/71, 1973). The question we want to address is whether these morphometric parameters and particularly D_{LO_2} are matched to the O_2 flow rates under conditions of maximal aerobic metabolism.

There is evidence that shows that D_{LO_2} is proportional to O_2 consumption. If one compares pairs of animals of similar body size but different O_2 needs, one finds that the animal with a higher level of O_2 consumption has a proportionately larger D_{LO_2}. This is the case if we compare horse with cow, dog with man, or Japanese waltzing mice with normal laboratory mice (Figure 3.10). It can also be shown that the lung can adapt to higher O_2 needs by increasing the area of gas-exchanging surfaces by differential adaptive growth of the lung. Thus it appears that the size of the pulmonary gas-exchange apparatus is malleable, that it can be made commensurate to the body's O_2 needs, at least if they are imposed during the active growth phase.

Because the lung appears adaptable to O_2 needs, we would expect the main morphometric determinants of diffusing capacity to scale proportionately to O_2 consumption. Tenney and Remmers (1963) have reported

FIGURE 3.7 Linear relationship between N_A(c, f) and V_V(mt, f) in diaphragm (D) and semitendinosus (ST), longissimus dorsi (LD), and vastus medialis (VM) muscles from a number of African mammals. Data from all muscles are taken together. The 95% confidence band for the regression line is limited by dashed lines ($r = 0.728$). [From Hoppeler et al., 1981b]

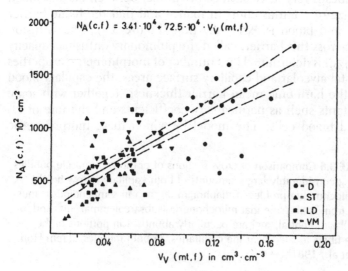

that alveolar surface area is linearly proportional to (resting) O_2 consumption in a wide range of mammalian species, from the bat (10 g) to the whale (200 kg). This indicates that the alveolar surface area scales with $M_b^{0.75}$, the scaling of standard resting metabolic rate (Kleiber, 1961). But when we started to estimate D_{LO_2} for different mammalian species,

FIGURE 3.8 Scanning electron micrograph of lung of small African gazelle (suni, *Nesotragus moschatus*) showing foamlike structure of lung parenchyma with terminal bronchiole (TB) branching into respiratory bronchioles (RB) and alveolar ducts (AD).

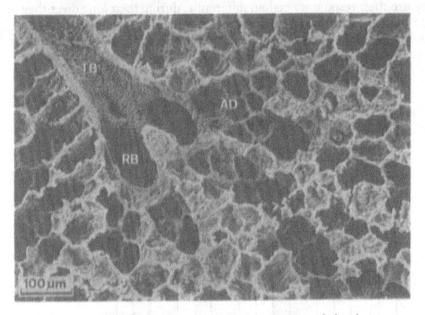

FIGURE 3.9 Electron micrograph of thin section of alveolar septum showing capillaries with erythrocytes (EC) and thin air–blood barrier (T).

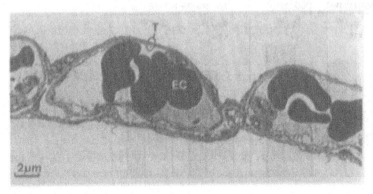

we obtained a different result (Weibel, 1972, 1973, 1979a). We found that both alveolar surface area and D_{LO_2} scale almost linearly with body mass. Thus we have on the one hand an inconsistent finding when comparing our results to those of Tenney and Remmers (1963), and on the other hand a paradox in our own results.

The inconsistent finding is easy to explain. Our own studies were restricted to terrestrial animals, whereas Tenney and Remmers (1963) considered terrestrial and marine mammals together, and all their large species were marine mammals. Marine mammals are very peculiar in various respects. First, in terms of physiology they are peculiar because they use their respiratory system differently; during their long dives they hold their breath for very long periods, reduce their circulation to the muscles, and incur important O_2 debts as they rely on anaerobic metabolism (Schmidt-Nielsen, 1979, ch. 6). Second, in terms of lung structure

FIGURE 3.10 Proportionality of pulmonary diffusing capacity and O_2 consumption in four pairs of mammals of similar size but different O_2 needs. C = control laboratory mice.

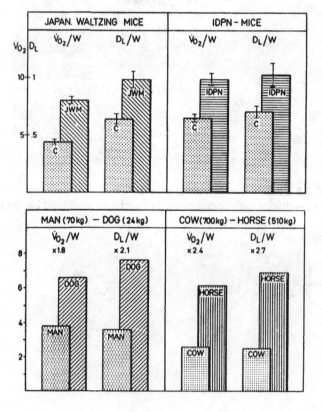

they are peculiar in that they have uncommonly large alveoli and a double capillary network in their alveolar walls in contrast to the simple network in terrestrial mammals. Therefore, these two classes of mammals use their lungs quite differently, and it is questionable whether one may indiscriminately compare their alveolar surface areas to O_2 consumption in an allometric study. Indeed, if one considers only the terrestrial mammals of the series of Tenney and Remmers (1963), one finds that their alveolar surface area scales identically to that in our series. So there is really no inconsistency and no contradiction, except with respect to the question of whether marine mammals are sufficiently comparable to terrestrial mammals to consider them jointly in an allometric study of this kind.

The paradox is more intriguing: Comparing mammals of similar body mass, D_{LO_2} is proportional to O_2 consumption, whereas comparing mammals of a large range of body mass (from the Etruscan shrew of 2 g to the cow and horse of 700 kg), we find that D_{LO_2} is not proportional to O_2 consumption. A cow therefore has a conductance for O_2 transfer from air to blood that is about 20 times that of the shrew when related to the unit volume of O_2 that has to be transferred from air to blood per minute. How can this be explained?

One possibility was that the allometric comparison was based on wrong assumptions; the scaling factor of 0.75 related to resting $\dot{V}_{O_2}$, and we would hardly believe the lung to have a limiting effect on the modest O_2 flow rate that has to be maintained to satisfy the O_2 needs of a resting animal. Clearly, we would expect D_{LO_2} to be of such magnitude as to satisfy – and perhaps limit – the *maximal* O_2 flow rate as it is required during heavy physical work such as running. Could maximal O_2 consumption, $\dot{V}_{O_2max}$, be linearly proportional to body mass in mammals of varying size? In 1976 this appeared possible, and there were some preliminary physiological data that even suggested it. This provided the motive for C. R. Taylor and me to plan an expedition to Africa, joining forces with Geoffrey M. O. Maloiy in Nairobi, in order to investigate the relationship between maximal O_2 needs of wild running animals and the quantitative design properties of the lung and other parts of the respiratory system.

The results did not come out as expected; we found that $\dot{V}_{O_2max}$ scaled to $M_b^{0.8}$ and thus was simply about 10 times resting O_2 consumption for all species (Taylor et al., 1981, and Chapter 12). But when we estimated the morphometric lung parameters, we found D_{LO_2} scaled linearly to body mass (Gehr et al., 1981). The slope of 0.95 shown in Figure 3.11 is not significantly different from 1.0, but it differs from the slope of 0.77 for $\dot{V}_{O_2max}$ obtained for the same animals, and shown as a broken line. Thus we have confirmed on this limited population of African mammals an allometric relationship that appears valid for a much broader population of terrestrial mammals (Figure 3.12), namely, that the pulmonary diffusing capacity is scaled about linearly to body mass and is hence not proportional to maximal O_2 consumption. Instead of solving the paradox, the African study confirmed it.

Does this finding mean that the lung's diffusion conductance is not matched to functional needs? Such a conclusion is not justified for the very reason that D_{LO_2} is closely matched to $\dot{V}_{O_2}$ when animals of similar size but differing O_2 needs are compared. The allometric study seems to indicate that our model for structure–function correlation is inaccurate or incomplete, specifically that we are missing additional size-dependent factors.

One possibility is that one or the other of the physical coefficients used for calculating D_{LO_2}, such as the O_2 binding rate of blood, is size-dependent, faster in small than in large animals; the data are insufficient

FIGURE 3.11 Allometric plot of pulmonary diffusing capacity. African mammals. Broken line represents slope of $\dot{V}_{O_2max}$ in the same animals. [From Gehr et al., 1981]

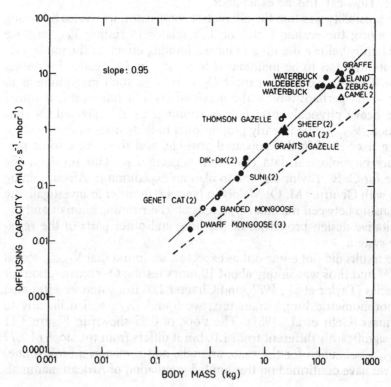

to judge the extent of this possible effect. Another possibly erroneous assumption lies in the unqualified assertion that D_{LO_2} should be directly proportional to $\dot{V}_{O_2}$, a claim that derives from Bohr's (1909) definition of D_{LO_2} by

$$\dot{V}_{O_2} = D_{LO_2} \cdot (P_A - P_{\bar{c}})_{O_2}$$

whereby one assumes the driving force, the O_2 partial pressure difference, to be constant (i.e., size-independent). But there is little evidence to support this assumption. Indeed, if we trust our estimate of D_{LO_2}, we could attempt to estimate the partial pressure gradient by dividing $\dot{V}_{O_2max}$ by D_{LO_2} for those animals where we have both values estimated (Taylor et al., 1981; Gehr et al., 1981); it turns out that the gradient, ΔP_{O_2}, is scaled to $M_b^{-0.2}$. For a large animal, similar in size to adult man, we would estimate ΔP_{O_2} to be about 10 mbar ($\sim$ 7 mm Hg), which is what one estimates by physiological methods, whereas for an animal of 500 g, it is about 40 mbar ($\sim$ 30 mm Hg).

FIGURE 3.12 Allometric plot of pulmonary diffusing capacity for all species. [From Gehr et al., 1981]

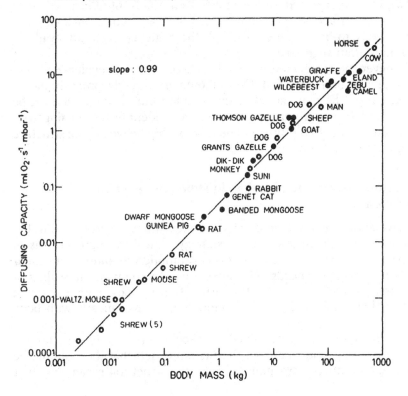

A lower alveolar P_{O_2} in large than in small mammals could result in a larger D_{LO_2} as large animals adapted to a reduced pressure head. There is morphometric evidence that lung structure can respond to alterations in alveolar P_{O_2} from studies on adaptive growth at high altitude (Bartlett and Remmers, 1971; Burri and Weibel, 1971); D_{LO_2} becomes enlarged at high altitude, (i.e., under lower inspired P_{O_2}) but reduced when an animal is given 40% O_2 to breathe. Is there any evidence that suggests a lower alveolar P_{O_2} in large as compared to small animals? Referring to Figure 3.1, we see that "alveolar P_{O_2}" is a rather ill-defined term; it may range anywhere from inspired P_{O_2} (200 mbar) to P_{O_2} in venous blood (50 mbar). But in estimating the transfer of O_2 across the air–blood barrier, P_{AO_2} has to mean the P_{O_2} prevailing at the alveolar surface of the barrier. What is its correct value or range?

We cannot answer this question because the first step of our model is incomplete; the flow of O_2 in the air phase, from environmental air to the alveolar surface, is affected not only by mass flow of air but also, and importantly, by diffusion through the air phase. Indeed, the alveolar surface is separated from inspired air by a large volume of residual air through which O_2 molecules have to diffuse. A size-dependent difference in P_{AO_2} is to be expected if the distance from the front of inspired air to the alveolar surface is considerably larger in large than in small animals. This is likely to be the case, although the evidence currently available is no more than suggestive.

Whatever the correct answer will be, there is enough evidence to suggest that the postulate that D_{LO_2} should be directly proportional to $\dot{V}_{O_2max}$ cannot be maintained in this simple form. The model must be refined, but before we can do this, we must identify the missing size-dependent factors that influence the relationship between lung structure and gas-exchange function.

Conclusions: the value of scaling in structure–function correlation

We have approached the study of structure–function correlation in the respiratory system using a double strategy. The first approach was to compare structural parameters to O_2 consumption in animals of similar size but differing O_2 needs. The second was to compare the scaling of structural parameters and of O_2 consumption to body mass, exploiting the fact that maximal O_2 consumption changes nonlinearly with body size.

In each of the levels examined – mitochondria, capillaries, lung – there was strong evidence from the first approach that there should be a close quantitative correlation between the structural parameters that determine the various conductances and the flow rate of O_2. But in none of these levels did we find, at first, an agreement between the scaling factors for maximal O_2 consumption and for the relevant structural parameters. Thus we ended up with a number of paradoxes that needed explaining. The clue was to identify additional size-dependent factors that influence the relationship between structure and function and that may well have remained undetected if we had simply compared animals differing in terms of O_2 need but matched in size. Thus one of the important insights from this kind of study is that the double strategy of varying functional loads on the system independent of body size and as a function of body size is a useful way of checking out the conception of models on which we attempt to base the structure–function correlation.

With respect to mitochondria, we found that it was the total mitochondrial volume of a muscle that had to be compared with O_2 consumption, rather than mitochondrial volume density. For capillaries we need to look at the total energy supply – O_2 and substrates – rather than only at O_2 flow. And in the lung, a system that looked simple and well worked out at first, we need to consider a number of boundary conditions that affect gas exchange between air and blood as a function of body mass. There is still hope that the structural parameters influencing the various conductances of the respiratory system are well matched to the O_2 flow requirements even in animals of different body size. But the situation is less simple than one would think at first. A thorough analysis of scaling relations, when viewed critically, is a very useful tool for improving our understanding of how structural design may affect functional performance.

References

Bartlett, D., Jr., and Remmers, J. E. (1971) Effects of high altitude exposure on the lungs of young rats. *Respir. Physiol.* 13:116–125.

Bohr, C. (1909) Ueber die spezifische Tätigkeit der Lungen bei der respiratorischen Gasaufnahme. *Scand. Arch. Physiol.* 22:221–280.

Burri, P. H., and Weibel, E. R. (1971) Morphometric estimation of pulmonary diffusion capacity. II. Effect of P_{O_2} on the growing rat lung to hypoxia and hyperoxia. *Respir. Physiol.* 11:247–264.

Gehr, P., Mwangi, D. K., Ammann, A., Maloiy, G. M. O., Taylor, C. R., and Weibel, E. R. (1981) Design of the mammalian respiratory system. V. Scaling morphometric pulmonary diffusing capacity to body mass: Wild and domestic mammals. *Respir. Physiol.* 44:61–86.

Hoppeler, H., Lüthi, P., Claassen, H., Weibel, E. R., and Howald, H. (1973) The ultrastructure of the normal human skeletal muscle: a morphometric analysis on untrained men, women and well trained orienteers. *Pflügers Arch.* 344:217–232.

Hoppeler, H., Mathieu, O., Krauer, R., Claassen, H., Armstrong, R. B., and
  Weibel, E. R. (1981a) Design of the mammalian respiratory system. VI.
  Distribution of mitochondria and capillaries in various muscles. *Respir.
  Physiol.* 44:87–111.
Hoppeler, H., Mathieu, O., Weibel, E. R., Krauer, R., Lindstedt, S. L., and
  Taylor, C. R. (1981b) Design of the mammalian respiratory system. VIII.
  Capillaries in skeletal muscles. *Respir. Physiol.* 44:129–150.
Kleiber, M. (1961) *The fire of life: an introduction to animal energetics.* New
  York: Wiley. 454 pp.
Krogh, A. (1922) *Anatomy and physiology of capillaries.* New Haven: Yale
  University Press. (2nd ed. 1929)
Mathieu, O., Krauer, R., Hoppeler, H., Gehr, P., Lindstedt, S. L., Alex-
  ander, R. McN., Taylor, C. R., and Weibel, E. R. (1981) Design of the
  mammalian respiratory system. VII. Scaling mitochondrial volume in skel-
  etal muscle to body mass. *Respir. Physiol.* 44:113–128.
Schmidt-Nielsen, K. (1979) *Animal physiology,* 2nd ed. Cambridge University
  Press.
Seeherman, H. J., Taylor, C. R., Maloiy, G. M. O., and Armstrong, R. B.
  (1981) Design of the mammalian respiratory system. II. Measuring maximum aerobic capacity. *Respir. Physiol.* 44:11–23.
Taylor, C. R., Maloiy, G. M. O., Weibel, E. R., Langman, V. A., Kamau, J.
  M. Z., Seeherman, H. J., and Heglund, N. C. (1981) Design of the mammalian respiratory system. III. Scaling maximum aerobic capacity to body
  mass: Wild and domestic mammals. *Respir. Physiol.* 44:25–37.
Tenney, S. M., and Remmers, J. E. (1963) Comparative quantitative morphology of mammalian lungs: diffusing areas. *Nature* 197:54–56.
Weibel, E. R. (1970/71) Morphometric estimation of pulmonary diffusion capacity. I. Model and method. *Respir. Physiol.* 11:54–75.
Weibel, E. R. (1972) Morphometric estimation of pulmonary diffusion capacity. V. Comparative morphometry of alveolar lungs. *Respir. Physiol.*
  14:26–43.
Weibel, E. R. (1973) Morphological basis of alveolar–capillary gas exchange.
  Physiol. Rev. 53:419–495.
Weibel, E. R. (1979a) Oxygen demand and the size of respiratory structures in
  mammals. In *Evolution of respiratory processes. A comparative approach,*
  vol. 13 of *Lung biology in health and disease* (C. Lenfant ed.), pp.
  289–346. New York: Marcel Dekker.
Weibel, E. R. (1979b) *Stereological methods,* vol. 1, *Practical methods for biological morphometry.* New York: Academic Press.
Weibel, E. R., and Taylor, C. R. (eds.). (1981) Design of the mammalian respiratory system. *Respir. Physiol.* 44:1–164 (No. 1).

4

A model for evaluating diffusion limitation in gas-exchange organs of vertebrates

JOHANNES PIIPER

This chapter develops a quantitative model for evaluating the extent to which diffusion limits the exchange of O_2 and CO_2 across vertebrate respiratory organs. The model is then applied to three types of gas-exchange organs: amphibian skin, fish gills, and human lungs. The model predicts a strong diffusion limitation for O_2 across amphibian skin, an intermediate limitation across fish gills, and a very weak limitation across the human lung. Possibilities for refining the model are then discussed.

Model

Diffusion

Diffusion designates transport of matter due to the statistical thermal motion of molecules. A net diffusive transport occurs only in the presence of a concentration (or partial pressure) gradient. Quantitative treatment of diffusion in physical and chemical engineering, and particularly in biology, is difficult owing to complexities in geometry, presence of multiple materials with differing diffusive properties, combination with convective processes, formation and consumption of substances involved, variation of all variables with time, and so on.

In many cases, however, a reasonable approximation is achieved by use of the simplest model of steady-state diffusion through a flat sheet of homogeneous material, representing the barrier to diffusion. In this case, Fick's first law of diffusion can be applied:

$$\dot{M} = d \cdot \frac{A}{l} \cdot (C_1 - C_2) \tag{1}$$

where $\dot{M}$ is the transfer rate; d, diffusion coefficient; A, surface area; l, thickness of barrier; and C_1 and C_2 are concentrations on each side of the barrier. For diffusion of gases, it is practical and customary to consider diffusion as related to partial pressure (P) differences rather than to concentration differences. After Henry's law, C is proportional to P:

$$C = \alpha \cdot P \tag{2}$$

where α is the solubility coefficient of the diffusing gas in the diffusion barrier. Thus the first Fick's law of diffusion assumes the form:

$$\dot{M} = d \cdot \alpha \cdot \frac{A}{l} \cdot (P_1 - P_2) \tag{3}$$

With the introduction of Krogh's diffusion constant, K,

$$K = d \cdot \alpha \tag{4}$$

Fick's first law of diffusion becomes:

$$\dot{M} = K \cdot \frac{A}{l} \cdot (P_1 - P_2) \tag{5}$$

Krogh's diffusion constant may be defined as diffusive flux density $(\dot{M}/A)$ per partial pressure gradient $((P_1 - P_2)/l)$.

For physiology of gas exchange it is important to view the K values comparatively for the various media (air, water, tissue) and for various gases, particularly for the two respiratory gases, O_2 and CO_2. Relative values of K, standardized to O_2 in air, are compiled in Table 4.1. The following features are of particular significance.

1. In air, the diffusivity is by an order of 10^4 to 10^5 higher than in water or tissue.
2. The K values for CO_2 in air are somewhat smaller than those for O_2 (owing to differences in size of the molecules), whereas in water and tissue K for CO_2 is about 20–30 times higher (owing to the much higher solubility coefficient of CO_2 in water).

Diffusive conductance, G, defined as diffusive transfer rate $(\dot{M})$ per effective partial pressure difference $(P_1 - P_2)$ and its reciprocal, the diffusive resistance $R \, (= 1/G)$, are particularly useful quantities for analysis of biological diffusion processes:

$$G = \dot{M}/(P_1 - P_2) \tag{6}$$

$$R = (P_1 - P_2)/\dot{M} \tag{7}$$

TABLE 4.1 Diffusion of CO_2 and O_2 in gas, water, and tissue. Approximate relative values of Krogh's diffusion coefficient $K \, (= d \cdot \alpha)$.

	Gas	Water	Tissue	Gas/water	Gas/tissue
CO_2	0.72	12×10^{-5}	6×10^{-5}	6 000	12 000
O_2	1.00	0.5×10^{-5}	0.25×10^{-5}	200 000	400 000
CO_2/O_2	0.72	24	24		

For conductance derived from physiological measurements the term "diffusing capacity" (or "transfer factor"), D, is generally used (see below). Combining equations 6 and 3 yields the following relationship:

$$D = d \cdot \alpha \cdot \frac{A}{l} \tag{8}$$

The problems involved with D and equation 8 will be discussed in the following section.

Combination of diffusion with convection (blood flow)

In many cases of vertebrate respiratory organs, blood perfusing capillary vessels has gas exchange with a more or less homogeneous and constant external medium (e.g., alveolar gas in mammalian lungs, environmental air in skin breathing of amphibians). In such cases a very simple model combining diffusion with convection is applicable (Figure 4.1).

A flat sheet of blood, flowing at a constant rate $\dot{Q}$, has gas-exchange contact with the medium phase through a (flat and uniform) barrier to diffusion. The partial pressures (of O_2, CO_2, or any other gas) in blood, starting from the mixed venous value, $P_{\bar{v}}$, approach the partial pressure

FIGURE 4.1 Diffusion–perfusion limitation in gas transfer. Left above: model and partial pressure profile in blood. Left below: blood dissociation curve (plot of concentration, C, against partial pressure, P) to visualize capacitance coefficient, β. Right: partial pressure profiles for different values of $D/(\beta\dot{Q})$.

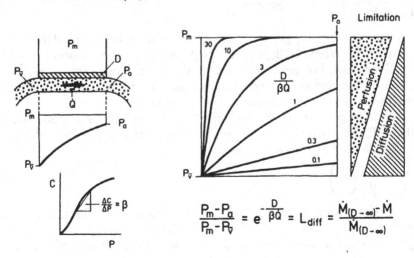

$$\frac{P_m - P_a}{P_m - P_{\bar{v}}} = e^{-\frac{D}{\beta\dot{Q}}} = L_{\text{diff}} = \frac{\dot{M}_{(D \to \infty)} - \dot{M}}{\dot{M}_{(D \to \infty)}}$$

in the medium, P_m, reaching the arterialized value, P_a, at the end of blood–medium contact. The following relationship follows from the first Fick diffusion equation for any length element of the model, dx:

$$d\dot{M}/dx = (P_A - P_c) \cdot dD/dx \tag{9}$$

($d\dot{M}$, gas transfer rate in the element; P_c, partial pressure in the capillary; dD, corresponding element of diffusing capacity). The Fick principle applied to the model element dx yields:

$$d\dot{M}/dx = \beta \cdot \dot{Q} \cdot dP_c/dx \tag{10}$$

(dP_c/dx, partial pressure gradient in blood). The quantity β is the capacitance coefficient (Piiper et al., 1971). For inert gases, β is equal to the solubility coefficient (α); for gases chemically bound in blood (O_2, CO_2, CO), β is the (mean) slope of the dissociation curve.

Combination of the differential equations 9 and 10 and subsequent integration with respect to P_c (from $P_{\bar{v}}$ to P_a), assuming β = const. (see below), yields:

$$\frac{P_m - P_a}{P_m - P_{\bar{v}}} = e^{-D/\beta\dot{Q}} \tag{11}$$

This fundamental relationship states that the diffusive equilibration deficit (left side) is a function, not of D alone, but of the $D/(\beta\dot{Q})$ ratio.

The equilibration deficit is equal to the diffusion limitation index, L_{diff}, defined as the fractional difference between the transfer rate achieved without diffusion limitation ($D = \infty$) and the actual transfer rate:

$$\frac{P_m - P_a}{P_m - P_{\bar{v}}} = L_{diff} = \frac{\dot{M}(D = \infty) - \dot{M}}{\dot{M}(D = \infty)} \tag{12}$$

An increase of $D/(\beta\dot{Q})$ means a decrease in diffusion limitation (see Figure 4.1). When $D/(\beta\dot{Q})$ is very high (above 3), the transfer is limited by blood flow, but practically not by diffusion. Conversely, a very low $D/(\beta\dot{Q})$ ratio, below 0.1, signifies that the transfer process is practically exclusively limited by diffusion, not by blood flow.

For inert gases it may be assumed that the solubilities in blood and in the diffusion barrier are equal (or that their ratio is about equal for all inert gases). Equation 8 shows that

$$\frac{D}{\beta\dot{Q}} = \frac{\alpha}{\beta} \cdot \frac{d \cdot A}{\dot{Q} \cdot l} \tag{13}$$

This relationship indicates that, because α/β is nearly identical for all inert gases, the degree of diffusion limitation should be the same for these gases. On the other hand, because for O_2, CO_2, and CO, β is markedly higher than α, the transfer of these gases must always be more markedly diffusion-limited than that of inert gases.

The more important problems and difficulties in application of equation 11 to analysis of diffusion limitation in gas-exchange organs are the following:

1. The analysis assumes steady state. However, P_m is variable (i.e., alveolar gas within the respiratory cycle), $\dot{Q}$ is generally not completely steady but pulsatile, and so on. Often it is difficult to estimate the significance of such variations.
2. The capacitance coefficients β (slope of blood dissociation curves) are not constant (i.e., independent of partial pressure, P) for the chemically bound gases O_2 and CO_2. A meaningful averaging is often problematic.
3. The barrier to diffusion is geometrically complex and composed of heterogeneous materials and media (Figure 4.2). In many cases only a fraction of the total resistance to gas transfer is located in an anatomical tissue barrier between medium and blood. A considerable fraction may be within the medium, particularly when it is water. But diffusion inside blood and red blood cells must also be considered. Moreover, the chemical reactions of O_2 binding and CO_2 release are not instantaneous and may limit the overall equilibration rate. Thus the meaning of the term diffusing capacity is complex, anatomically ill-defined, and even logically incorrect. A better term might be equilibration capacity.
4. Every respiratory organ is composed of a great number of parallel elements which may vary considerably with respect to several parameters (in terms of the model of Figure 4.1, P_m, D, and $\dot{Q}$ may vary).

FIGURE 4.2 Serial component processes in medium-to-blood transfer of O_2. Right: various terms in use and their meaning.

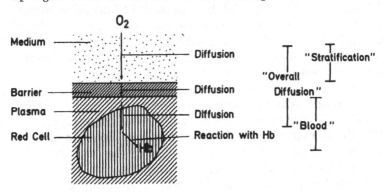

The resulting "inhomogeneity" ("nonuniformity" or "unequal distribution") effects have been studied in detail for mammalian lungs (e.g.,
West, 1977).

Diffusion limitation in various gas-exchange organs

To illustrate the application of the $D/(\beta\dot{Q})$ model concept to gas-exchange
organs, three selected examples representing widely differing systems will
be discussed in this section. The main criteria for the selection were the
relative morphological (and probably functional) uniformity of the gas-
exchange organs and, of course, availability of reliable experimental data.

Amphibian skin

In a number of studies the quantitative roles of skin, lungs, and gills
in CO_2 elimination and O_2 uptake have been analyzed in various amphibians (see Guimond and Hutchison, 1976). The general finding is that
cutaneous gas exchange is important in all extant amphibians.

Typically, the aquatic amphibian larva breathes by gills and skin, and
the (terrestrial or aquatic) adult by lungs and skin. In individual families
and species of urodelan amphibians, however, a large diversity is encountered. In many aquatic forms gills are retained in the adult, and in
the majority of such cases lungs are also present, so that three respiratory
organs function simultaneously. Of particular interest is the tendency,
obviously occurring independently in many urodelan families, of a reduction or a total suppression of lungs. In the family Plethodontidae
lungs are absent in all species, the total gas exchange being cutaneous.
Interestingly, the plethodontids are the predominant group of urodeles
in North America and the only urodeles extending their distribution into
tropical regions (South America).

In the plethodontid common dusky salamander (*Desmognathus fuscus*) the anatomy of the skin vascularization has been studied in some
detail by Czopek (1961). All the skin (and the buccopharyngeal mucosa)
contains a rather uniform dense capillary plexus, located directly under
the epithelium of 26 μm mean thickness. In the same species the gas
exchange and transport function has been studied in recent years by
Gatz and co-workers.

In a particular experimental series (Gatz et al., 1975) the salamanders
were exposed to an atmosphere containing 20% Freon 22 (chlorodifluo-
romethane) until an equilibrium had been reached (Figure 4.3). There-
after the animals were placed in a closed vessel containing room air that
was expelled every 10 min, and the test gas concentration therein was
determined by gas chromatography. A plot of the amount of Freon 22
eliminated against time usually could be analyzed as two exponential

components. The same procedure was repeated after the animal had been killed. The slower wash-out kinetics in the dead animal was attributed to the absence of convective transport by blood circulation.

For further analysis (Gatz et al., 1976), a general model for the blood circulation has to be assumed. The following model was considered in agreement with the anatomical arrangement. The cardiac output of the univentricular heart is functionally divided into two fractions, one perfusing the skin (where it is partially cleared of the inert gas), the remaining fraction being distributed to all other tissues of the body (where inert gas is taken up by the blood). In the venous system the blood returning from the skin and from internal organs is mixed, and the mixing is complete by the time the blood is ejected from the ventricle into the arterial system.

In order to estimate the cardiac output, several alternative models (with parallel or serial arrangement of the two functional compartments identified by inert gas elimination) were used. Because there was no reasonable approach allowing estimation of the two functional fractions of the cardiac output, it was assumed that 20 to 50% of the cardiac output was skin blood flow. The estimated range for values of cardiac output, cu-

FIGURE 4.3 Estimation of blood flow in the lungless salamander (*Desmognathus fuscus*). Upper part: methodology. Below left: elimination kinetics. Below right: schema of circulation and estimated range of blood flow; $\dot{Q}_{tot}$, cardiac output; $\dot{Q}_{cut}$, cutaneous blood flow; $\dot{Q}_{int}$, internal organ blood flow.

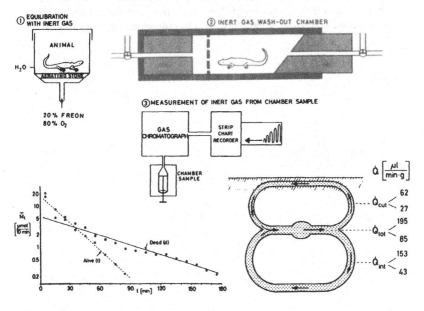

taneous flow, and internal organ flow is shown in Figure 4.3. A direct estimate of the cardiac output, obtained by inserting the cut end of the conus arteriosus into a graded capillary tube, yielded 171 $\mu l \cdot min^{-1} \cdot g^{-1}$, in good agreement with the values calculated from the inert gas elimination rate.

In further calculations the Fick principle was applied, using the O_2 uptake ($\dot{M}_{O_2}$) values measured in another series of experiments in the same salamander species (Gatz et al., 1974b):

$$\dot{M}_{O_2} = \dot{Q}_{cut} \cdot (C_a - C_m)_{O_2} \tag{14}$$

$$\dot{M}_{O_2} = \dot{Q}_{int} \cdot (C_m - C_v)_{O_2} \tag{15}$$

(a, arterialized blood; v, averaged venous blood; m, mixed cardiac blood; $\dot{Q}_{cut}$, skin blood flow; $\dot{Q}_{int}$, internal organ blood flow).

The value of $C_{m_{O_2}}$ could be obtained from P_{O_2} determined in blood samples obtained by cardiac puncture, using the blood O_2 dissociation curve of this species (Gatz et al., 1974a); $C_{a_{O_2}}$ and $C_{v_{O_2}}$ could be calculated using equations 14 and 15; and the corresponding P_{O_2} values were estimated from the O_2 dissociation curve (Figure 4.4).

FIGURE 4.4 Respiratory gas transport in the lungless salamander *Desmognathus fuscus*. Left: schema of circulation (cf. Figure 4.3). Mean experimental values for mixed (cardiac) blood (*m*) and estimates for arterialized (a) and averaged venous blood (*v*) are indicated in the table. Right: P_{O_2} and P_{CO_2} profiles in skin capillaries. $\overline{\Delta P}$, mean partial pressure difference between ambient air and cutaneous capillary blood.

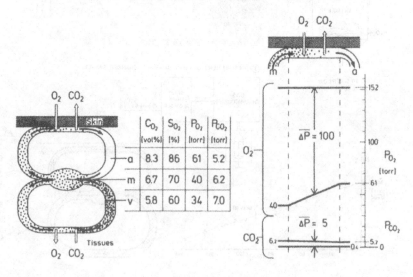

	C_{O_2} (vol%)	S_{O_2} (%)	P_{O_2} (torr)	P_{CO_2} (torr)
a	8.3	86	61	5.2
m	6.7	70	40	6.2
v	5.8	60	34	7.0

On the basis of P_{O_2} in inflowing and outflowing cutaneous blood (P_{mO_2} and P_{aO_2}), the P_{O_2} profile in cutaneous blood, shown in Figure 4.4, was estimated. The large mean P_{O_2} difference for diffusion, $\overline{\Delta P}_{O_2}$ = 100 Torr, compared to the small $(P_a - P_m)_{O_2}$ = 21 Torr, clearly demonstrated the preponderant role of diffusion limitation. The $D/(\beta \dot{Q})$ value for cutaneous O_2 transfer, which can simply be estimated as the ratio of these values, is 0.2, and the diffusion limitation index, L_{diff}, is close to 0.8. With such a high degree of diffusion limitation, the cutaneous O_2 uptake is predicted to be severely reduced in environmental hypoxia. Indeed, the O_2 uptake of D. *fuscus* has been found to be much reduced in hypoxia (Gatz et al., 1974b), the O_2 uptake deficit being in part energetically compensated by anaerobic glycolysis and splitting of high-energy phosphates (Gatz and Piiper, 1979).

The corresponding CO_2 values deserve particular attention. First, all blood P_{CO_2} values are very low for an "air-breathing" animal, about 5–7 Torr. This is due to the fact that both CO_2 and O_2 transport is mainly limited by diffusion through the skin epidermis. Because K_{CO_2} in skin is expected to be about 20 times higher than K_{O_2} (see Table 4.1), the mean air–cutaneous blood P_{CO_2} difference should be about 100/20 = 5 Torr, and this is the case. The $D/(\beta \dot{Q})$ ratio, however, comes out for CO_2 around 0.3, thus similar to that for O_2, although K_{CO_2}, and therefore D_{CO_2} should be 20 times higher than the values for O_2. This is due to the fact that β_{CO_2} is much higher than β_{O_2} (i.e., the blood CO_2 dissociation curve is much steeper than the O_2 dissociation curve).

Fish gills

It is generally accepted that the countercurrent model is the appropriate basic model for analysis of gas exchange in both teleostean and elasmobranch gills because respiratory water in the interlamellar spaces (between the secondary lamellae) flows in a direction opposite to that of blood flow in the secondary lamellae (see Chapter 1).

Estimations of the gill diffusing capacity or transfer factor, D, have been performed in the rainbow trout (Randall et al., 1967) and in the dogfish *Scyliorhinus stellaris* (Piiper et al., 1968, 1977).

When the dissociation curve can be considered straight, a simple formula involving P_{O_2} in inspired and expired water, and in arterial and mixed venous blood, may be used (Figure 4.5). Usually however, this condition cannot be met, and a particular variant of the Bohr integration technique should be employed (Figure 4.5). With this evaluation method, mean D_{O_2} values in experiments on unanesthetized S. *stellaris* of 0.34 to 0.64 μmol $\cdot$ min^{-1} $\cdot$ Torr^{-1} $\cdot$ (kg body mass)$^{-1}$ were obtained.

Because the diffusivity of O_2 in water is not much higher than in tissue, a considerable part of the resistance to diffusion may be suspected to reside in the interlamellar water. To estimate the role of diffusion limitation in interlamellar water, the diffusion limitation in a comparable system, in which the whole resistance to O_2 uptake was in water, was investigated (Scheid and Piiper, 1971, 1976).

The model used is depicted in Figure 4.6. The secondary lamellae are simulated by a series of plates. The O_2 pressure on the surface of the plates, P_o, is considered to be constant (see below).

The degree of O_2 equilibration of water is conveniently expressed as the equilibration deficit, $\epsilon = (P_{\bar{E}} - P_o)/(P_I - P_o)$, where P_I is O_2 pressure in inflowing water and $P_{\bar{E}}$, in mixed outflowing water. It has been shown that ϵ is a function of the parameter $d \cdot l/(2b \cdot \bar{u})$, where l is the length of the plate; $2b$, the interlamellar distance; d, the diffusion coefficient of O_2 in water; $\bar{u}$, the mean water velocity. Further modifying variables are the shape of the plates (pores) and the water velocity profile in the interlamellar space.

The equilibration deficit, ϵ, may be viewed as an equivalent water shunt in a model in which part of the water is assumed to equilibrate completely to P_o, the remaining part (shunt) retaining P_I (Figure 4.7). For comparison with experimental diffusing capacity measurements,

FIGURE 4.5 Determination of diffusing capacity D for fish gills on the basis of the countercurrent model. Left: for linear dissociation curves. Right: a particular Bohr integration technique for nonlinear dissociation curves.

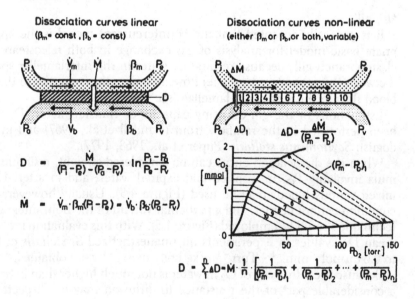

however, it is useful to compute the conductance (diffusing capacity) of a barrier on the surface of the plates that would yield the same degree of equilibration, $D_{eq} = -\dot{V} \cdot \alpha \cdot \ln \epsilon$ ($\dot{V}$, water flow; α, solubility coefficient of O_2 in water).

The morphometrical data obtained by Perry and Hughes (cf. Scheid and Piiper, 1976) were used and are in part shown in Figure 4.8, where the assumed velocity distribution is also illustrated (parabolic transversal flow profile combined with a hyperbolic vertical flow gradient produced by tapering of the plates toward the free edge). For the standard (resting) conditions the equivalent water shunt is 5%, and the equivalent diffusing capacity D_{eq} is 2.0 μmol $\cdot$ min^{-1} $\cdot$ Torr^{-1} or 0.9 μmol $\cdot$ min^{-1} $\cdot$ Torr^{-1} $\cdot$ (kg body mass)$^{-1}$. The value is 1.4 to 2.6 times the total D_{O_2} value, indicating that 70 to 40% of the resistance to O_2 uptake may be sought in interlamellar water.

These figures should be considered as estimates of the order of magnitude only, because of many problems and deficiencies regarding both the physiological and morphometrical measurements. Furthermore, the assumption of constant P_o in the model calculations certainly is unrealistic. More precise data and more adequate modeling are required for a more accurate estimation of the quantitative role of diffusion limitation in interlamellar water of fish gills.

FIGURE 4.6 Model for simulation of O_2 equilibration in gill interlamellar space. (For explanations, see text.)

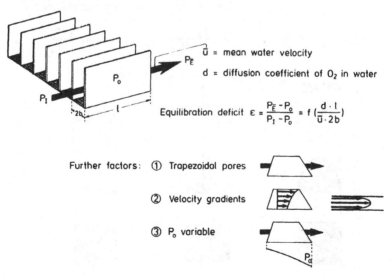

Human lungs

The concept and measurements of the diffusing capacity of human lungs were initiated by Marie Krogh and Christian Bohr more than 70 years ago, but the normal values and the nature of the limiting process in alveolar–capillary equilibration of O_2 and CO_2 are still controversial. This is so mainly because the degree of diffusion limitation in normal human lungs (and probably other mammalian lungs) is very small and therefore difficult to quantify (reviewed by Piiper and Scheid, 1980).

In our laboratory in Göttingen rebreathing techniques for determination of D_{O_2} (D_{CO} and D_{CO_2}) have been elaborated (Adaro et al., 1973). Besides their being noninvasive (requiring no blood sampling), their important advantage is that by rebreathing, lung gas is homogenized, and thus all disturbing "inhomogeneity" effects (see above, under "Combination of diffusion with convection") are reduced. In principle, for determination of D_{O_2} the time course of the exponential approach of al-

FIGURE 4.7 Equilibration of O_2 in gill interlamellar space. Left: change of P_{O_2} profile in interlamellar water (from bottom) – at entrance, inside, at exit of interlamellar space, mixed after exiting. Middle: water shunt analog. Right: equilibration conductance analog; at the wall a barrier, of conductance D_{eq}, assumed, whereas there is no transversal gradient in interlamellar water.

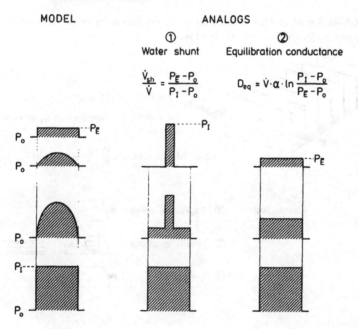

MODEL   ANALOGS

① Water shunt

$$\frac{\dot{V}_{sh}}{\dot{V}} = \frac{P_E - P_o}{P_I - P_o}$$

② Equilibration conductance

$$D_{eq} = \dot{V} \cdot \alpha \cdot \ln \frac{P_I - P_o}{P_E - P_o}$$

veolar P_{O_2} to mixed venous P_{O_2} is continuously recorded during a rebreathing maneuver. In order to obtain D, a number of other variables must be determined: lung volume, rebreathing bag volume, effective rebreathing ventilation, capillary blood flow and β, and the slope of the blood O_2 dissociation curve. The accuracy of the methods is much increased when stable isotopes of O_2 (CO, CO_2) are used (Meyer et al., 1981).

In normal healthy males D_{O_2} was found to average 54 ml $\cdot$ min^{-1} $\cdot$ Torr^{-1} (Meyer et al., 1981), whereas standard (textbook) values are about half as much. The $D/(\beta\dot{Q})$ value for hypoxia, in which condition the measurements are performed for methodological reasons, is about 2 (when using standard D_{O_2} values: about 1). Calculation of $D/(\beta\dot{Q})$ for normoxia is more problematic. Formally a value of 8 (or 4) is obtained, apparently showing that there is no diffusion limitation for O_2 exchange.

Measurements of D_{CO_2} are even more difficult than those of D_{O_2}. Piiper et al. (1980) found by particular rebreathing techniques a D_{CO_2}/D_{O_2} ratio of about 5:7. The reasons for deviation of the ratio from the Krogh diffusion constant ratio for tissue, about 20, are (1) possible diffusion limitation in alveolar gas and (2) relatively slow equilibration kinetics of the $H^+/HCO_3^-/CO_2$ system in blood. The $D/(\beta\dot{Q})$ ratio for resting man is estimated at 4, but during heavy exercise it may drop to 1.7, thus indicating diffusion limitation (better: equilibration limitation; see section on "Combination of diffusion with convection").

FIGURE 4.8 Equilibration of O_2 in interlamellar space. Values according to Scheid and Piiper (1976). Left: water velocity distribution. Right: anatomical dimensions.

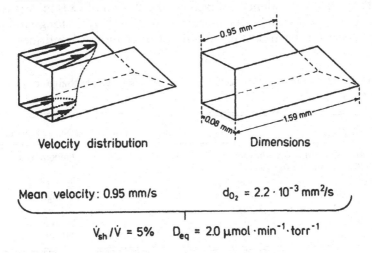

Velocity distribution   Dimensions

Mean velocity: 0.95 mm/s $d_{O_2} = 2.2 \cdot 10^{-3}$ mm^2/s

$\dot{V}_{sh}/\dot{V} = 5\%$ $D_{eq} = 2.0$ μmol $\cdot$ min$^{-1} \cdot$ torr^{-1}

Conclusions and perspectives

The results of the analysis of the role of diffusion limitation for O_2 in the three selected examples of vertebrate gas-exchange organs are summarized in Table 4.2. The $D/(\beta\dot{Q})$ ratio for O_2 and the diffusion limitation index, L_{diff}, derived therefrom, vary in a large range, from 0.2 (in amphibian skin) to 8 (in human lung), with L_{diff} decreasing from 80% (prevalent diffusion limitation) to less than 1% (practically no diffusion limitation). Most of the values are in a range, $0.1 < D/(\beta\dot{Q}) < 3$, where both diffusion and perfusion exert a limiting role.

The results should be considered preliminary, and much work on models and on experimental animals remains to be performed, using improved methods, to arrive at more accurate estimates of the role of diffusion in gas-exchange organs.

For future research efforts in this field the following areas and aspects deserve particular attention.

1. The role of diffusion limitation in the respiratory medium, not only in water breathing (fish gills) but also in mammalian lungs (here called "stratification"), requires further modeling and experimentation.
2. Investigation into structure–function relationships on the basis of quantitative morphological and physiological data is expected to yield new insights.
3. In spite of considerable effort our knowledge of reaction limitation in O_2 (and CO) uptake by blood cells is insufficient.
4. The role of the slow equilibration of the $H^+/HCO_3^-/CO_2$ system in blood (due to absence of carbonic anhydrase in blood plasma) should be studied in submammalian vertebrates, particularly water breathers.
5. There is good evidence for facilitated diffusion of CO_2 in vitro and in vivo, whereas facilitated transport of CO (and O_2) in lungs and in the placenta has been postulated but not unanimously confirmed.

TABLE 4.2 Diffusion limitation of O_2 uptake in various gas exchange organs: L_{diff} after equation 12.

Respiratory organ	$D/\beta\dot{Q}$	L_{diff}	Diffusion limitation
Amphibian skin (*Desmognathus fuscus*)	0.2	0.8	Predominant
Fish gills (*Scyliorhinus stellaris*)			
Overall	0.5	0.6	Strong
Water–gas barrier	1	0.4	Strong
Mammalian lungs (*Homo sapiens*)			
Hypoxia	2	0.1	Weak
Normoxia	8	0.0003	Absent

6. Extension of such analysis to invertebrates is desirable. A particularly promising area is that of gas-exchange mechanisms in insects.
7. The respiratory apparatus certainly is more dimensioned for gas exchange in activity than at rest. In particular the role of diffusion limitation at the upper limit of O_2 uptake is of interest.
8. The role of diffusion was considered in this paper in connection with blood flow. However, in most respiratory gas-exchange organs there is convective transport by ventilation which must be included in an integrated view of the performance of a gas-exchange organ. Moreover, anatomical/functional nonuniformity (which was considered a disturbing factor in this analysis) must be incorporated in a more comprehensive model for gas exchange.

References

Adaro, F., Scheid, P., Teichmann, J., and Piiper, J. (1973) A rebreathing method for estimating pulmonary D_{O_2}: Theory and measurements in dog lungs. *Respir. Physiol.* 18:43–63.
Czopek, J. (1961) Vascularization of the respiratory surfaces of some Plethodontidae. *Zool. Pol.* 11:131–148.
Gatz, R. N., Crawford, E. C., Jr., and Piiper, J. (1974a) Respiratory properties of the blood of a lungless and gill-less salamander, *Desmognathus fuscus*. *Respir. Physiol.* 20:33–41.
Gatz, R. N., Crawford, E. C., Jr., and Piiper, J. (1974b) Metabolic and heart rate response of the plethodontid salamander *Desmognathus fuscus* to hypoxia. *Respir. Physiol.* 20:43–49.
Gatz, R. N., Crawford, E. C., Jr., and Piiper, J. (1975) Kinetics of inert gas equilibration in an exclusively skin-breathing salamander, *Desmognathus fuscus*. *Respir. Physiol.* 24:15–29.
Gatz, R. N., Crawford, E. C., Jr., and Piiper, J. (1976) Gas transport characteristics in an exclusively skin-breathing salamander, *Desmognathus fuscus* (*Plethodontidae*) In *Respiration of amphibious vertebrates* (G. M. Hughes, ed.), pp. 339–356. New York: Academic Press.
Gatz, R. N., and Piiper, J. (1979) Anaerobic metabolism during severe hypoxia in the lungless salamander *Desmognathus fuscus* (*Plethodontidae*). *Respir. Physiol.* 38:377–384.
Guimond, R. W., and Hutchison, V. H. (1976) Gas exchange of the giant salamanders of North America. In *Respiration of amphibious vertebrates* (G. M. Hughes, ed.), pp. 313–338. New York: Academic Press.
Meyer, M., Scheid, P., Riepl, G., Wagner, H.-J., and Piiper, J. (1981) Relationship between pulmonary diffusing capacities for O_2 and CO measured by a rebreathing technique. *J. Appl. Physiol.* (In press.)
Piiper, J., and Baumgarten-Schumann, D. (1968) Effectiveness of O_2 and CO_2 exchange in the gills of the dogfish (*Scyliorhinus stellaris*). *Respir. Physiol.* 5:338–349.

Piiper, J., Dejours, P., Haab, P., and Rahn, H. (1971) Concepts and basic quantities in gas exchange physiology. *Respir. Physiol.* 13:292–304.

Piiper, J., Meyer, M., Marconi, C., and Scheid, P. (1980) Alveolar–capillary equilibration kinetics of $^{13}CO_2$ in human lungs studied by rebreathing. *Respir. Physiol.* 42:29–41.

Piiper, J., Meyer, M., Worth, H., and Willmer, H. (1977) Respiration and circulation during swimming activity in the dogfish *Scyliorhinus stellaris.* *Respir. Physiol.* 30:221–239.

Piiper, J., and Scheid, P. (1980) Blood–gas equilibration in lungs. In *Pulmonary gas exchange*, vol. I (J. B. West, ed.), pp. 131–171. New York: Academic Press.

Randall, D. J., Holeton, G. F., and Stevens, E. D. (1967) The exchange of oxygen and carbon dioxide across the gills of rainbow trout. *J. Exp. Biol.* 46:339–348.

Scheid, P., and Piiper, J. (1971) Theoretical analysis of respiratory gas equilibration in water passing through fish gills. *Respir. Physiol.* 13:305–318.

Scheid, P., and Piiper, J. (1976) Quantitative functional analysis of bronchial gas transfer: Theory and application to *Scyliorhinus stellaris* (*Elasmobranchii*). In *Respiration of amphibious vertebrates* (G. M. Hughes, ed.), pp. 17–38. New York: Academic Press.

West, J. B. (1977) *Regional differences in the lungs.* New York: Academic Press.

5

Oxygen transport in vertebrate blood: challenges

PETER LUTZ

The most important challenge for a blood gas system is its ability to cope with the demand of flight or fight. Oxygen supply can be represented by the Fick equation:

$$\dot{V}_{O_2} = \dot{Q}_h(Ca_{O_2} - Cv_{O_2})$$

and increased oxygen consumption ($\dot{V}_{O_2}$) can be accomplished by a rise in cardiac output ($\dot{Q}_h$) and/or by taking more oxygen from the blood (i.e., by increasing the difference between the oxygen contents of arterial and venous blood, $Ca_{O_2} - Cv_{O_2}$). Characteristically, increases in oxygen extraction from the blood are due to a fall in Cv_{O_2} rather than an increase in Ca_{O_2}, as discussed in Knut Schmidt-Nielsen's (1979) text, *Animal physiology*. This chapter considers a few "uncharacteristic" or unusual ways vertebrate blood has met the increased demands for O_2.

Oxygen content of arterial blood during exercise

Optimal numbers of red blood cells

Although the oxygen-carrying capacity of blood could be increased by raising the proportion of red blood cells (the hematocrit), there is a trade-off, as the consequent increase in viscosity results in a severe increase in resistance to flow. There is therefore, under normal circumstances, an optimal hematocrit where oxygen delivery ($Ca_{O_2} \times \dot{Q}_h$) is maximal, the actual value of which presumably depends on the hemodynamic characteristics of the circulatory system. For man this has been calculated to be about 40% (Crowell and Smith, 1967), close to that actually found, and it is possible that the normal hematocrits found for other species are the routine optimals. Little can be done with respect to increasing the oxygen-carrying capacity of red cells, as hemoglobin is packed into the red blood cell to the very limit of its solubility, circa 5 mM · liter^{-1}, so that in fact crystallization is a serious potential problem (Riggs, 1979). The result is that arterial oxygen content is usually fairly constant, close to saturation levels.

Increases in circulating red blood cells with exercise

It has recently been found that during strenous exercise trained racehorses show a transient but marked increase in oxygen capacity (+ 66%), resulting mainly from a release of red blood cells from the spleen (Lykkeboe et al., 1977). Interestingly, a similar short-term increase in hematocrit has also been found in exercised fishes, due again to the spleen releasing and later sequestering red blood cells (Yamamoto et al., 1980), and in Weddell seals a postdive increase in blood hemoglobin concentrations has been attributed to recruitment of red blood cells from venous sinuses (Kooyman et al., 1980). It is very possible that owing to its transient nature this phenomenon is underreported and is more widespread than currently appreciated. Presumably, the "normal" hematocrit is the most efficient for oxygen transport under normal conditions, but during emergencies the prime concern is to maximize oxygen supply by all possible means.

Varying saturation of arterial blood with exercise

During activity, then, the oxygen transport system is working at maximal capacity. In this regard, exercising fish are particularly interesting, as, owing to the high cost of ventilation (estimates range from 10 to 30% of the resting metabolism to 50% of active; White, 1978), cardiovascular adjustments to activity must be especially efficient. Depending on the species, oxygen consumption can increase on maximal exercise as much as 5- to 15-fold over routine levels. In some salmonids this increased tissue delivery is met by an up to 10-fold increase in cardiac output (mainly due to large increases in stroke volume) and a more than doubling of the oxygen extraction from blood (Kiceniuk and Jones, 1977), and in the trout high arterial oxygen pressures (137 Torr; Kiceniuk and Jones, 1977) clearly favor oxygen extraction from blood. Some fish, however, have remarkably low Pa_{O_2} values (e.g., tench, 36 Torr; Eddy, 1974), and Steen and Kruysse (1964) have found that arterial blood in the resting eel is only 50% saturated. We have found similar low arterial oxygen values in the lemon shark, an active predator. The venous oxygen pressures are routinely about 5 Torr, leaving little oxygen reserve in the venous blood (unpubl.). Forced activity in these sharks produces an almost doubling in arterial oxygen content with little change in the venous oxygen content (unpubl.). Rather interestingly, in this way blood oxygen extraction is increased by raising Ca_{O_2}, in contrast to the more conventional lowering of the Cv_{O_2}. It is possible that under routine circumstances there is a minimal blood flow through the respiratory exchange portion of the gills, just sufficient to satisfy demand, with the rest of the blood bypassing the respiratory lamellae. When demand requires, a greater proportion of blood is directed through the lamellae, but at some energy cost.

Although simple respiratory bypass shunts seem to be ruled out by recent anatomical studies, the vascular system of the gills is exceedingly complex with several interconnected circulatory pathways and an array of sphincters under neuronal and hormonal control (Dunel and Laurent, 1980). The microcirculation of the gill is likely to be correspondingly complex. This elaborateness makes some sense when we consider that the gill is the primary interface for exchange of a variety of substances between the fish and its environment. It is not only the site of O_2 and CO_2 transfer, but also of NH_4^+ and HCO_3^- excretion, of acid–base regulation, and of Na, K, Cl, and water flux. All these processes will occur simultaneously, and, according to the circumstances, particular roles will differ in emphasis. Some may even be opposed. For example, in the fresh-water salmon, the processes that allow an increased oxygen uptake during rigorous exercise appear to permit an increase in the passive diffusion of sodium across the gills and thus result in a short-term deficit in total body sodium (Wood and Randall, 1973).

Storage and release of oxygen in diving vertebrates

A contrasting challenge to the oxygen transport system is found in the air-breathing diving reptiles, birds, and mammals. Here the major problem is how to hold or store a maximum supply of oxygen and deliver it in the most effective manner. The marine mammals are by far the best known of aquatic divers, and among them the best divers exhale on diving and use the blood and tissues almost exclusively as oxygen stores (Kooyman, 1973). On diving there are marked circulation changes (Schmidt-Nielsen, 1979), determined to some extent by the anticipated dive duration (Kooyman et al., 1980). In the Weddell seal, for example, a forced dive produces an 80% reduction of cardiac output and a drastically reduced blood flow to all organs except the brain, lung, and adrenals (Zapol et al., 1979).

Achieving large stores of oxygen

For such good divers the respiratory properties of the blood match this oxygen storage function. In the bladdernose seal, for example, the increased blood volume and a high hematocrit of 63% allow a blood oxygen store of 1.7 mM O_2 kg^{-1} body weight (Scholander, 1940), which compares favorably to a calculated store of 0.5 mM O_2 kg^{-1} body weight for man (Farhi and Rahn, 1955). Although no viscosity measurements are available, the hematocrit value in this seal is so high that it is possible that oxygen content may have been developed at the expense, to some extent, of oxygen transport. The increased Bohr effect, characteristic of these animals (killer whale, -0.74; Lenfant et al., 1968) enhances oxygen unloading to the tissues, and higher blood buffering capacities minimize pH shifts (Johansen and Weber, 1976).

A high myoglobin content also allows muscle of these divers to hold considerably more oxygen than in other mammals (0.42 mM O_2 kg^{-1} body weight, bladdernose seal, Scholander, 1940; 0.13 mM O_2 kg^{-1} body weight, man, Farhi and Rahn, 1955). However, this store is quickly depleted, and the basic strategy of the good mammalian diver on a prolonged dive would appear to be to reserve the blood oxygen store for the oxygen-dependent brain, while allowing the muscles to be anaerobic.

But this model does not apply to all marine mammals, particularly those that have more modest diving capabilities. For example, the sea lion, porpoise (Kooyman, 1973), and manatee (Lapennas and Lutz, unpubl.) appear to inspire on diving and may utilize lung oxygen as an important store. The fresh-water turtle, *Pseudemys*, and the green and loggerhead sea turtles also inspire before diving, and it appears that in these animals the lungs hold sufficient oxygen to supply normal routine dives (Burggren et al., 1977; Lutz, unpubl.).

Utilizing oxygen stores in the lung

Interestingly, quite different blood oxygen properties are required if the lung is used as an oxygen store. For this purpose, the blood must continue to pick up oxygen from the lungs as the oxygen is depleted down to low P_{O_2} levels. Because circulation is still an important feature, a high hematocrit with a corresponding increased viscosity may offer no advantage. A large Bohr effect would also be of disadvantage for oxygen binding, as the fall in blood pH that accompanies a dive would result in decreased oxygen affinity. The sea lion and porpoise both have normal mammalian Bohr effects and blood oxygen capacities (Clausen and Ersland, 1968). The manatee has a hematocrit similar to that of man (White et al., 1976), a relatively low blood volume (Scholander and Irving, 1941), and low myoglobin concentrations (Blessing, 1972), yet it is capable of extended dives (15–20 min). The green and loggerhead sea turtles not only have small Bohr effects compared to other reptiles, but, significantly, the Bohr effects decline substantially with oxygen saturation, such that at low P_{O_2} the blood oxygen affinity is unaffected by pH (Lapennas and Lutz, unpubl.). Such a feature would be particularly useful for oxygen uptake at low P_{O_2} when the pH fall would be greatest.

The shape of the oxygen dissociation curve of sea turtle blood is rather unusual. This feature is often expressed in terms of the Hill plot (log $S/(1 - S)$ vs. log P_{O_2}, where S = fractional saturation), and the slope of the plot (n, the Hill coefficient) indicates the degree of sigmoidosity of the curve. Many vertebrates, including man, have rather straightforward, sigmoid-shaped dissociation curves and yield relatively straight-line Hill plots. Except at the extremes of high and low saturation, the n coefficient typically has a constant value of between 2.5 and 3.0. This feature is so common that it is thought to be one of the most stable characteristics of the oxygen dissociation curve (Bartels and Baumann, 1977). However, in the green and loggerhead sea turtles the n value declines sharply with falling oxygen saturation and approaches 1.0 at low P_{O_2} (Lapennas and Lutz, unpubl.). This is due to the curve gradually changing from a sigmoid to hyperbolic shape. An interesting consequence of this feature is that it results in an enhanced blood oxygen content at low P_{O_2} values (Figure 5.1). Perhaps significantly, Farmer et al. (1979) have found that stripped manatee hemoglobin exhibits n values that decline to 1 at less than 30% saturations. If this feature applies to whole blood, and in the sea turtle whole blood and stripped hemoglobin are similar in this respect (Lutz and Lapennas, unpubl.), then as in the sea turtle, it may be an adaptation to assist oxygen loading at low lung P_{O_2}.

If the more modest divers are in fact using the lung as a primary oxygen store, then one might expect other physiological features that facilitate the transport of oxygen from the lung to the tissues while lung P_{O_2} falls, in contrast to the better-known role in the endurance divers of unloading oxygen from a blood store. In this case diving vertebrates can usefully be divided into two functional groups, those that use the lung as the

FIGURE 5.1 Whole blood oxygen dissociation curve of loggerhead sea turtle (pH 7.45, P_{CO_2} = 37 Torr, pH = 7.45) contrasted to curve of same oxygen affinity but constant n. [Lapennas and Lutz, unpubl.]

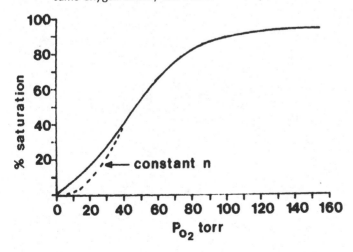

main oxygen store and those that use the tissues, each having quite different sets of physiological adaptations (Lapennas and Lutz, unpubl.).

Birds and diving vertebrates: some similar problems

Curiously, there are some parallels between the operation of the bird lung and that of the lung of this latter class of diving vertebrates. In the bird lung the crosscurrent flow model of Scheid and Piiper (1972) predicts that oxygen is depleted from the air as it passes along the parabronchus (Figure 5.2). It follows then that the oxygen pressures that the afferent venous blood sees, depend on where along the parabronchial axis gas exchange takes place, and the oxygen loading pressures will range from near ambient at the lung entrance (after accounting for CO_2 and water vapor pressure) to venous levels farther down, the latter values determined by the intensity of oxygen demand. The bird oxygen dissociation curve therefore is required to load over a wide range of oxygen pressures. Perhaps significantly, curved Hill plots have been described for the blood of several bird species (Lutz, 1980).

In the typical mammal, of course, the situation is much simpler. Here lung oxygen is maintained at a fairly constant level so that the blood only has to load oxygen over a narrow range of P_{O_2}, and perhaps in consequence the typical mammal has a simpler dissociation curve than many other vertebrates.

FIGURE 5.2 In the crosscurrent model of the bird lung (Scheid and Piiper, 1972) air oxygen pressures will be highest at the entrance to the lung and reach minimal values (determined by blood venous P_{O_2}) at or before the exit. The loading P_{O_2} for venous blood depends on the site of O_2 uptake. [From Schmidt-Nielsen, 1979]

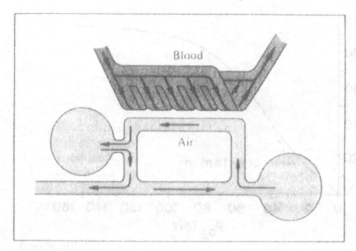

Conclusion

One of the delights of comparative physiology is that of continually turning up the unexpected. It is only by investigating how animals work that we can start to appreciate the detail and variety of functions that physiological systems serve, and thus discover the proper questions to ask to understand their mechanisms.

References

Bartels, H., and Baumann, R. (1977) Respiratory function of hemoglobin. In *International review of physiology – Respiration physiology*, vol. 14 (J. G. Widdicombe, ed.), pp. 107–134. Baltimore: University Park Press.

Blessing, M. H. (1972) Studies on the concentration of myoglobin in the seacow and porpoise. *Comp. Biochem. Physiol.* 41A:475–480.

Burggren, W., Hahn, C. E. W., and Foex, P. (1977) Properties of blood oxygen transport in the turtle *Pseudemys scripta* and the tortoise *Testudo graeca*: effects of temperature, CO_2 and pH. *Respir. Physiol.* 31:39–50.

Clausen, G., and Ersland, A. (1968) The respiratory properties of the blood of two diving rodents, the beaver and the water vole. *Respir. Physiol.* 5:350–359.

Crowell, J. W., and Smith, E. E. (1967) Determinant of the optimal hematocrit. *J. Appl. Physiol.* 22(3):501–504.

Dunel, S., and Laurent, P. (1980) Functional organization of the gill vasculature in different classes of fish. In *Epithelial transport in the lower vertebrates* (Lahlou, B., ed.), pp. 37–58. Cambridge University Press.

Eddy, F. B. (1974) Blood gases of the tench (*Tinca tinca*) in well aerated and oxygen-deficient waters. *J. Exp. Biol.* 60:71–83.

Farhi, L. E., and Rahn, H. (1955) Gas stores of the body and the unsteady state. *J. Appl. Physiol.* 7:472–484.

Farmer, M., Weber, R. E., Bonaventura, J., Best, R. C., and Domning, D. (1979) Functional properties of hemoglobin and whole blood in an aquatic mammal, the Amazonian manatee (*Trichechus inunguis*). *Comp. Biochem. Physiol.* 62A:231–238.

Johansen, K., and Weber, R. E. (1976) On the adaptability of hemoglobin function to environmental conditions. In *Perspectives in experimental biology* (Davis, P. S., ed.), pp. 212–234. New York: Pergamon Press.

Kiceniuk, J. W., and Jones, D. R. (1977) The oxygen transport system in trout (*Salmo gairdneri*) during sustained exercise. *J. Exp. Biol.* 69:247.

Kooyman, G. L. (1973) Respiratory adaptations in marine mammals. *Am. Zool.* 13:457–488.

Kooyman, G. L., Wahrenbrock, E. A., Castellini, M. A., Davis, R. M., and Sennett, E. E. (1980) Aerobic and anaerobic metabolism during voluntary diving in Weddell seals: Evidence of preferred pathways from blood chemistry and behavior. *J. Comp. Physiol.* 138:335–346.

Lenfant, C., Kenny, D. W., and Ducutt, C. (1968) Respiratory function in the killer whale *Orcinus orca* (Linnaeus). *Am. J. Physiol.* 215:1506–1511.

Lutz, P. L. (1980) On the oxygen affinity of bird blood. *Am. Zool.*
20(1):187–198.

Lykkeboe, G., Schougaard, H., and Johansen, K. (1977) Training and exercise
change respiratory properties of blood in race horses. *Respir. Physiol.*
29:315–325.

Riggs, A. (1979) Studies on the hemoglobins of Amazonian fishes: an overview. *Comp. Biochem. Physiol.* 62A:257–272.

Scheid, P., and Piiper, J. (1972) Cross current gas exchange in avian lungs:
effects of reversed parabronchial air flow in ducks. *Respir. Physiol.*
16:304–312.

Schmidt-Nielsen, K. (1979) *Animal physiology*, 2nd ed. Cambridge University
Press.

Scholander, P. F. (1940) Experimental investigations on the respiratory functions in diving mammals and birds. *Hvalradets Skrifter Norske Videnskaps-
Akad., Oslo,* 22. 131 pp.

Scholander, P. F., and Irving, L. (1941) Experimental investigations of the respiration and diving of the Florida manatee. *J. Cell. Comp. Physiol.*
17:169–191.

Steen, J. B., and Kruysse, A. (1964) The respiratory function of teleostean
gills. *Comp. Biochem. Physiol.* 12:127.

White, F. N. (1978) Comparative aspects of vertebrate cardiorespiratory physiology. *Annu. Rev. Physiol.* 40:471–499.

White, I. R., Harkness, D. R., Isaacks, R. E., and Duffield, D. A. (1976)
Some studies on blood of the Florida manatee *Trichechus manatus latrios-
tris. Comp. Biochem. Physiol.* 55A:413–417.

Wood, C. M., and Randall, D. J. (1973) Sodium balance in the rainbow trout
(*Salmo gairdneri*) during extended exercise. *J. Comp. Physiol.* 82:235.

Yamamoto, K. I., Itazawa, Y., and Kobayashi, H. (1980) Supply of erythrocytes into the circulating blood from the spleen of exercised fish. *Comp.
Biochem. Physiol.* 65A:5–11.

Zapol, W. M., Liggins, G. C., Schneider, R. C., Qvist, J., Snider, M. T.,
Creasy, R. K., and Hochachka, P. W. (1979) Regional blood flow during
simulated diving in the conscious Weddell seal. *J. Appl. Physiol.*
47(5):R968–973.

6

Strategies of blood acid–base control in ectothermic vertebrates

DONALD C. JACKSON

In the field of comparative acid–base physiology, the most active and productive area of research in recent years has concerned the effects of body temperature change. The fundamental observation that has emerged is that blood pH varies predictably with body temperature in most ectothermic animals (Figure 6.1). In a variety of species, including reptiles, fish, and amphibians among the vertebrates and annelid worms, crustaceans, and molluscs among the invertebrates, in vivo blood pH has been observed to fall as body temperature rises with a slope of about -0.016 U/°C (dpH/$dT = -0.016$).

Research on this topic actually had its beginnings with isolated observations by such eminent physiologists as Henderson, Austin, and Dill, but it was not until the study of the turtle, *Pseudemys scripta*, by Robin in 1962 that the recent era was launched. Major credit must go to Rahn, Reeves, and their colleagues at Buffalo for seizing upon Robin's observations, extending them to other species, and making important generalizations therefrom. They have also contributed a number of excellent review articles on the subject (e.g., Rahn, 1967; Reeves, 1977; Reeves and Rahn, 1979). Because of the extensive contributions of this group to the development of the pH–temperature story and because of the awkwardness in referring to this important relationship in an unambiguous way, it seems at once fitting and convenient to refer henceforth (at least in this chapter) to the curve depicted in Figure 6.1 as the Buffalo curve.

It is my purpose to discuss the acid–base balance of ectothermic vertebrates in the light of the Buffalo curve with the particular objective of understanding how the relationship is achieved in various animals. Our observation is that an animal at one temperature, say at 20 °C, has a blood pH that is normally close to 7.7, while the same animal at 30 °C has a pH that is close to 7.54. How did this change in pH occur? It is convenient to identify two distinct influences accounting for this change: (1) physical and (2) physiological. The physical influences are those that affect the behavior of blood as a chemical solution and as such can be studied outside the body, in vitro. The physiological effects are due to the vital processes that alter the composition of the blood of a living animal. In contrast to the closed chemical solution, the living animal is an open system that exchanges substances with its environment and between body compartments, and that produces acids as products of its metabolism. Together, the physical and the physiological factors determine the acid–base state of an animal's blood in vivo at each instant.

Physical effects of temperature

If blood is taken from an animal and equilibrated at various temperatures without permitting any gas exchange to occur between the blood and its surroundings, the pH of the blood decreases with temperature by about 0.016 to 0.120 U/°C (dpH/dT = −0.016 to −0.020). This in vitro behavior of blood is sometimes called the *Rosenthal effect* in recognition of the investigator who first studied it systematically (Rosenthal, 1948). The same pattern of change has been reported to occur without exception in blood from many vertebrate and invertebrate species (Reeves, 1977).

The major physical basis for the Rosenthal effect is the temperature dependence of the weak acid equilibria within the blood, which has been thoroughly analyzed by Reeves (1976). The dissociation constants of the weak acids (CO_2, phosphates, and protein) all change with temperature according to their thermodynamic properties. Of particular significance is the observation that the protein pK (the overall pK value of the dissociable groups on hemoglobin and plasma protein) changes in parallel with the blood pH; dpK/dT ~ dpH/dT. As Reeves (1972) has shown, the dominant protein subgroup involved is imidazole, a side chain of the

FIGURE 6.1 The typical relationship between the in vivo blood pH of ectotherms and their body temperature, the Buffalo curve.

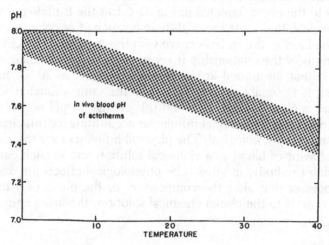

amino acid histidine. It dissociates as follows:

$$HIm^+ = H^+ + Im \tag{1}$$

The equilibrium constant for this reaction, $K(Im)$, is close to the overall dissociation constant for blood protein:

$$K(Im) = ([H^+] \times [Im])/[HIm^+] \tag{2}$$

Expressed as the Henderson–Hasselbalch equation, this becomes:

$$pH = pK(Im) + \log[Im]/[HIm^+] \tag{3}$$

From equation 3, we can see that the fractional dissociation of imidazole, $[Im]/([HIm^+] + [Im])$, what Reeves (1972) has termed alpha-imidazole, remains constant when $pK(Im)$ and pH change in parallel with temperature. This has the important consequence that the net charge state of blood proteins remains constant when blood pH changes with temperature along the Buffalo curve.

The P_{CO_2} is also affected when the temperature of vertebrate blood is changed in vitro. It increases exponentially, and the relationship can be described as: $d(\log P_{CO_2})/dT \sim 0.020$ (Reeves, 1977). A number of blood acid–base variables remain unchanged, however, in the closed in vitro system. The concentrations of the strong ions, such as Na^+, K^+, and Cl^-, do not change, nor, as just discussed, does the protein net charge, $[P^-]$. The concentration of the major ionic form of CO_2 in blood, HCO_3^-, also remains essentially the same, although if one examines the Henderson–Hasselbalch equation for the CO_2 reaction:

$$pH = pK' + \log([HCO_3^-]/S \times P_{CO_2}) \tag{4}$$

it is the case that each of the other factors in the equation, pH, pK' (the apparent dissociation constant of CO_2), S (the CO_2 solubility), and P_{CO_2} are temperature-dependent.

The striking feature of the blood's acid–base response to temperature is the similarity between the dpH/dT in the in vitro Rosenthal system and the change in pH that occurs along the in vivo Buffalo curve. Remember that the in vivo system is also subject to the physiological factors influencing acid–base state. Another compelling aspect of the system is the similarity between the whole blood pH (pH_b) dependence on temperature and that of the pH of pure water (neutral pH or pN). These two variables, pH_b and pN, change in parallel ($dpH_b/dT = dpN/dT$), although pH_b is generally about 0.6 unit above pN. This parallelism has led Rahn (1967) to suggest that the control of acid–base balance at different temperatures represents a maintenance of a constant state of relative alkalinity. The absolute value of pH is thus regarded as less significant than its value with respect to neutrality. Reeves (1972), in his analysis of the significance of the pH–temperature relationship, has focused on the dissociation state of protein (specifically α-imidazole). Because $dpK(\text{Im})/dT = dpH/dT$ in the blood, the ionic state of the blood proteins, a critical variable influencing protein structure and function as well as the distribution of ions and water across cell membranes, remains constant. These two hypotheses, constant relative alkalinity and constant protein net charge state, are not mutually exclusive and are both important insights into the possible functional significance of the Buffalo curve.

Physiological effects of temperature

Because of the identity between the in vitro and the typical in vivo dpH/dT curves, it might be supposed that the Buffalo curve is due strictly to physical effects acting in vivo, and that the physiological activities have no effect. This is by no means the case, however, because living functions have profound effects on acid–base balance. What it does tell us, though, is that in most ectothermic animals, the physiological processes, whether because of control or happenstance, closely preserve the pH changes resulting from the direct physical effects of temperature. The physical effects already described, although studied in vitro, of course occur in vivo as well when the body temperature of an ectotherm changes, but the steady-state response to temperature must depend upon the physiological processes as well. In what ways do physiological processes affect the acid–base balance of the blood?

There are two general mechanisms by which the acid–base state is normally altered in the body: (1) by changes in P_{CO_2}, so-called respiratory changes; and (2) by changes in the relative concentrations of strong cations and strong anions, often called metabolic changes, but more accurately expressed as changes in the strong ion difference (Stewart, 1978). Changes in P_{CO_2}, other than physical effects of temperature discussed above, occur as a result of the diffusion of molecular CO_2 in or out of the blood or by the generation of CO_2 from blood $HCO_3{}^-$. CO_2 diffuses in from the cells where it is produced as the major acid endproduct of aerobic metabolism. Its rate of production is proportional to the overall rate of aerobic metabolism and thus changes with temperature with a Q_{10} normally on the order of 2 or 3. CO_2 diffuses out of the blood as part of the respiratory exchange process by which CO_2 is lost to the environment. In general, the level of P_{CO_2} in the arterial blood is set by the balance between the rate of metabolic CO_2 production and the rate of respiratory CO_2 loss. Control mechanisms concerned with P_{CO_2} regulation, where they exist, act via the respiratory loss of CO_2.

Changes in the strong ion difference (SID) occur by a differential movement of strong base cations (such as Na^+) and strong acid anions (such as Cl^-) in or out of the blood. This exchange may be between the extracellular fluid and intracellular fluid, or it may be between the extracellular fluid and the environment across the epithelial surfaces of the gills (in fish), the skin (in at least some amphibians), or the renal system (in most, or perhaps all, vertebrates). The role of the strong ion difference in determining acid–base balance can be understood in terms of the law of electrical neutrality, which requires that an equality exist in any solution between the positive and negative charges. In the blood, this equality can be summarized as follows:

$$[B^+] - [A^-] = [HCO_3^-] + [P^-] \qquad (5)$$

where $[B^+]$ is the concentration of strong cations, $[A^-]$ is the concentration of strong anions, $[HCO_3^-]$ is the concentration of bicarbonate ion, and $[P^-]$ is the concentration of other weak acid anions (all concentrations in milliequivalents per liter). The concentrations of H^+ and OH^- are omitted because their contribution is negligible in this context. The quantity ($[B^+] - [A^-]$) is the strong ion difference (SID). If SID is abnormally increased, by the addition of a strong cation (or loss of a strong anion), both $[HCO_3^-]$ and $[P^-]$ will increase, and the pH will rise. This type of change is usually called metabolic alkalosis and is commonly identified by the simultaneous rise in pH and $[HCO_3^-]$. However, the rise in $[HCO_3^-]$, like the alkaline shift in pH, is the consequence of the primary SID change, rather than the result of a direct addition of HCO_3^- to the solution. Clearly Na^+ could not have been added without an accompanying anion, but any weak anion (OH^-, HCO_3^-, CO_3^{-2}, etc.) could have accompanied the Na^+ with the same effect. Similarly, a fall in SID, metabolic acidosis, results in a fall in both pH and $[HCO_3^-]$ (and in $[P^-]$). These examples both assume no change in P_{CO_2}, but normally P_{CO_2} will in fact change, either because respiratory processes compensate for the SID effects and tend to restore pH, or because the metabolic (or SID) events overwhelm the respiratory capacity to maintain P_{CO_2}.

The weak acids, CO_2 and HP, and their conjugate bases, HCO_3^- and P^-, are essential elements in the stabilization of pH, however, because they act as sources or sinks for H^+. For example, in the absence of these buffers, the addition of strong acid (such as HCl) to a solution of only strong ions (such as NaCl) will drastically lower the pH because the added H^+ all remains in the ionic form. With the weak acids present, the acid titrates the weak anions, and most of the added H^+ ends up, in effect, as the nondissociated weak acid. But the concentrations of the combined and ionized forms of the weak acids are dependent variables in the system. Their concentrations depend upon the P_{CO_2} (determined by respiratory exchange), the SID (determined by ionic exchange mechanisms), and the parameters of the system, the equilibrium constants and the CO_2 solubility. Temperature has pervasive effects on all aspects of this complex system, both predictable physical effects and far less predictable physiological effects.

Acid–base control at different temperatures

We may now consider how ectothermic vertebrates actually control their in vivo blood pH along the Buffalo curve. From the previous discussion it is clear that the same pH value may be associated with a variety of combinations of P_{CO_2} and SID. In the usual analysis of this system, plasma [HCO_3^-] is evaluated instead of SID because it is more easily measured and, in general, changes in parallel to SID. Confusion arises, however, because [HCO_3^-] also is dependent on P_{CO_2}, so that respiratory changes alter [HCO_3^-] in the absence of any change in SID.

To compare mechanisms of control, I have chosen three representative species: (1) the turtle, *Pseudemys scripta*, a reptile that breathes primarily with its lungs; (2) the rainbow trout, *Salmo gairdneri*, a fish that breathes primarily with its gills, and (3) the hellbender, *Cryptobranchus alleganiensis*, a urodele amphibian that exchanges gas primarily across its skin. Each of these ectotherms regulates its pH along the Buffalo curve, but the values of P_{CO_2} and [HCO_3^-] are notably different, as is the manner in which these variables are affected by temperature (Figure 6.2).

The turtle: a lung breather

It is convenient to begin with the turtle because its acid–base control can serve as a model for comparison with the others. The in vivo blood picture of the turtle (Figure 6.2) matches in all respects the physical in vitro picture discussed earlier (Jackson et al., 1974). Blood P_{CO_2} increases with temperature, whereas plasma [HCO_3^-] is unchanged. This indicates that the turtle is regulating its pH by preserving the physically determined temperature effects, a constant ionic composition but a variable P_{CO_2}. Other lung breathers, including turtles (Howell et al., 1970; Kinney et al., 1977), the alligator (Davies, 1978), lizards (Crawford and Gatz, 1974), and anuran amphibians (Howell et al., 1970; Reeves, 1972), exhibit the same pattern of response.

The physiological adjustment to temperature is revealed by the changes in pulmonary ventilation in the turtle. These changes must be consistent with the observed P_{CO_2} values. Arterial P_{CO_2} (Pa_{CO_2}) is controlled by lung ventilation through exchange of CO_2 between the blood and the alveolar gas. The alveolar P_{CO_2} ($P_{A_{CO_2}}$) is directly determined, under steady-state conditions, by the relationship between metabolic CO_2 production

($\dot{V}_{CO_2}$) and alveolar ventilation ($\dot{V}_A$) as follows:

$$P_{ACO_2} = RT \cdot \dot{V}_{CO_2}/\dot{V}_A$$

where R is the gas constant and T is the temperature in °K. Even in a reptile, with its imperfect separation between arterial and venous blood within the heart, the Pa_{CO_2} is nearly the same as P_{ACO_2} under most circumstances. Therefore, Pa_{CO_2} is also a direct function of the ratio, $\dot{V}_{CO_2}/\dot{V}_A$, or, as commonly expressed, Pa_{CO_2} is an inverse function of $\dot{V}_A/\dot{V}_{CO_2}$. This latter ratio, termed the air convection requirement, represents the volume of air that must ventilate the alveolar portion of the lung in order to eliminate a unit amount of CO_2. For example, if the steady-state value of Pa_{CO_2} doubles from 20 Torr to 40 Torr, then the air convection requirement must have decreased by about 50%.

When a turtle's body temperature rises and an increase in Pa_{CO_2} is physically produced, the air convection requirement falls in order to regulate Pa_{CO_2} at the higher value. Measurements of total ventilation

FIGURE 6.2 Blood acid–base variables of representative ectothermic vertebrates at various body temperatures. [Fish data from Randall and Cameron, 1973; turtle data from Jackson et al., 1974; hellbender data from Moalli et al., 1980]

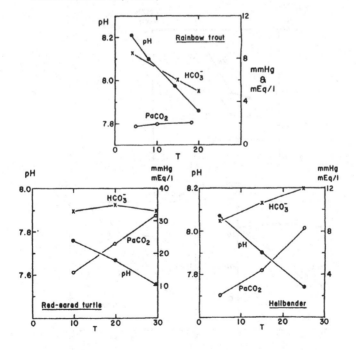

($\dot{V}_E$) and oxygen consumption ($\dot{V}_{O_2}$) of the turtle were made at different temperatures (Jackson, 1971), and the air convection requirement was estimated by the ratio $\dot{V}_E/\dot{V}_{O_2}$ (Figure 6.3). The decrease in the ratio with temperature was not caused by a decrease in the ventilation per se, but rather by a disproportionate increase in $\dot{V}_{O_2}$ compared to $\dot{V}_E$. In the turtle, as with ectotherms generally, metabolic rate ($\dot{V}_{O_2}$ or $\dot{V}_{CO_2}$) increases about 2- to 3-fold ($Q_{10} = 2$–3) for each 10 °C increase in body temperature. Ventilation ($\dot{V}_A$ or $\dot{V}_E$), however, changes less than this in lung breathers, and actually remained unchanged in the turtle between 10 and 30 °C.

The acid–base adjustment to temperature by the turtle, therefore, occurs as follows: Temperature has an immediate effect on cell and body fluid chemistry to induce a change in metabolic rate (Q_{10} effect) and a change to an acid–base state (Rosenthal effect). The ventilation adjusts in accordance with the changed metabolic rate to produce an air con-

FIGURE 6.3 Gas-exchange characteristics of representative ectothermic vertebrates at various body temperatures. See text for details. [Fish data from Randall and Cameron, 1973; turtle data from Jackson et al., 1974; hellbender data from Moalli et al., 1980]

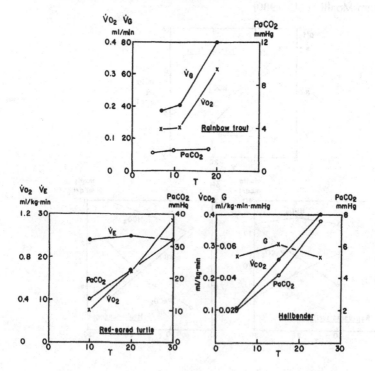

vection requirement that will hold the Pa_{CO_2}, and thereby the blood pH, at their new values. Further changes in metabolic rate at this new temperature, due to varying activity, are matched by equivalent changes in ventilation for continued acid–base control. This pattern of control provides uninterrupted acid–base homeostasis despite rapid fluctuations in body temperature and metabolism, because the physiological process is always matched to the physical effects of temperature.

The rainbow trout: a gill breather

In the rainbow trout, and in most other fish studied, the mechanism of control is radically different from that of the turtle. Blood P_{CO_2} is very low owing to the high rate of gill ventilation, and increases slightly, if at all, with temperature, whereas plasma [HCO_3^-] decreases significantly and largely accounts for the observed fall in blood pH (Figure 6.2). Randall and Cameron (1973) described these changes and attributed them to: (1) a constant relationship between gill ventilation (V_G) and metabolic rate (V_{O_2}) over the temperature range studied, that held P_{CO_2} nearly constant; and (2) an ion exchange mechanism, probably located in the gill epithelium, that produced the change in [HCO_3^-]. Their measurements of ventilation in the trout are shown in Figure 6.3. Similar acid–base changes have been observed in the marine seatrout (Cameron, 1978), the carp (Heisler, 1980), and the juvenile dogfish (Heisler et al., 1976). An interesting exception is the adult dogfish, *Scyliorhinus stellaris*, which with a predominant P_{CO_2} effect more closely resembles the lung breathers than the other fish (Heisler et al., 1976).

Unlike the turtle, the rainbow trout does not conform physiologically to the physically established acid–base pattern. Instead it holds P_{CO_2} constant, thereby changing total CO_2 concentration, and restores the correct pH by the ion exchange mechanism. This means that a period of hours is required following a temperature transition before normal acid–base homeostasis is reestablished. This is an energy-requiring process involving differential transport of Na^+ and Cl^- ions that changes the strong ion difference and thus [HCO_3^-]. Clearly this mechanism is more costly in terms of time and effort than the rapid ventilatory mechanism of the turtle. However, an adjustment in ventilation is not a feasible strategy for the fish because gill ventilation is closely coupled to oxygen uptake requirements, and cannot be adjusted in the interests of acid–base balance without jeopardizing O_2 supply. Compared to air, water has a low O_2 concentration, and a fish must move large volumes of water through its gills to supply adequate O_2. As an adaptation, however, fish possess an efficient countercurrent exchange between blood and water flows in the gills that permits them to extract a large fraction of the incurrent water and thereby minimize water flow. In the trout, about

60% of the O_2 was extracted at all temperatures studied. The ventilation of the turtle, on the other hand, is controlled primarily by acid–base variables, and O_2 supply is satisfied despite large changes in the air convection requirement. At 10 °C, the low end of its temperature range, the turtle's air convection requirement is high, its Pa_{CO_2} is low, and its O_2 extraction is only about 6%. As temperature rises, it can therefore easily afford to lower its ventilation requirement for acid–base purposes without threatening its O_2 supply. At 30 °C, it still extracts only 30% of the inspired O_2, but its air convection requirement has fallen to 25% of the 10 °C value. The fish has much less freedom in altering its water convection requirement without enormously increasing the metabolic cost of ventilation.

The hellbender: a skin breather

The hellbender is a large aquatic salamander that, despite its size, is almost exclusively dependent on its skin for respiratory gas exchange (Guimond and Hutchison, 1976). On the basis of its blood acid–base picture (Figure 6.2), the response of the hellbender to temperature is similar to that of the turtle and to hellbender blood tested in vitro (Moalli et al., 1980).

The increase in plasma $[HCO_3^-]$ was not a significant change statistically, and, just as in the turtle, the increase in P_{CO_2} with temperature represents the effective variable in the pH adjustment. The hellbender, however, unlike the turtle, apparently lacks an effective mechanism for respiratory control. It possesses lungs, but they are simple, saclike structures that are probably unimportant in normal gas exchange. At temperatures between 5 and 25 °C, hellbenders lose more than 90% of their CO_2 through their skin (Guimond and Hutchison, 1976), and, indeed, the blood acid–base response to temperature is the same whether the animals have access to air or not.

Experimental evidence indicates that skin gas exchange in amphibians may be poorly controlled. Skin CO_2 loss ($\dot{V}_{sCO_2}$) occurs by diffusion according to the relationship:

$$\dot{V}_{sCO_2} = G(Pa_{CO_2} - Pm_{CO_2})$$

where $(Pa_{CO_2} - Pm_{CO_2})$ is the transcutaneous P_{CO_2} difference, and G is the total conductance of the skin for CO_2. In order to control Pa_{CO_2}, a skin breather must be able to adjust the conductance, G, in an appropriate manner. This could possibly occur by altering the boundary layer at the skin surface (external convection) or by altering the blood flow to the skin (internal convection). In the bullfrog, *Rana catesbeiana*, an amphibian that utilizes both skin and lungs for gas exchange, no variation in G was observed despite changes in metabolic rate, in Pa_{CO_2}, or in body temperature. Increased skin CO_2 loss was always attributable to a proportionate increase in Pa_{CO_2} (Jackson, 1978). This agrees with the analysis of skin gas exchange in the lungless salamander, *Desmognathus fuscus*, by Piiper et al. (1976), who concluded that gas exchange was primarily limited by the diffusion properties of the skin. Variations in external or internal convection (ventilation and perfusive conductance, respectively) are either lacking or ineffective.

Despite this apparent absence of physiological control, the hellbender nevertheless exhibits the same pH variation with temperature as do ectotherms such as the turtle and the trout which have effective control mechanisms. The explanation appears to be that what appears to be pH control in the hellbender is really only a consequence of the direct effects of temperature on metabolic rate and on P_{CO_2}, as already discussed earlier with respect to the turtle. In the hellbender, an increase in temperature causes immediate and approximately proportionate increases in both CO_2 production (Q_{10} effect) and Pa_{CO_2} (Rosenthal effect). Because skin CO_2 conductance, G, is not significantly affected by temperature, the system automatically assumes a new steady state in which the increased Pa_{CO_2} provides the diffusion head necessary to excrete the increased CO_2 produced (Figure 6.3). This mechanism is essentially passive and has the advantage that it is fast and low in cost. Its disadvantage is that variations in metabolic rate at constant temperature will cause parallel variations in P_{CO_2} and CO_2 concentration (i.e., respiratory acid–base disturbances). In a recent study (Boutilier et al., 1980), hellbenders suffered sustained respiratory (and lactic) acidosis as a result of induced activity at 25 °C. But when an increase in metabolic rate occurs strictly as a consequence of a rise in body temperature, on the other hand, the accompanying elevation in Pa_{CO_2} is not associated with CO_2 retention and is appropriate for the maintenance of normal ectothermic acid–base balance.

Other examples

The three species discussed are in a sense "ideal" because each breathes almost exclusively with a single gas-exchange organ (lung, gill, or skin) over the entire temperature range considered. This is not the case for many other species, for which the control strategies may be different or may be some combination of those discussed. Likely groups to examine in this regard are the Amphibia and the air-breathing fish because simultaneous reliance on two (bimodal) or even three (trimodal) gas exchangers is found in these groups. A change in the respiratory mode may also occur within the same species depending on ambient O_2 availability or temperature. For example, the garfish (*Lepisosteus osseus*) utilizes primarily its gills for O_2 uptake in the winter (10 °C) and its lungs in the summer (25 °C). Its acid–base picture, although following the Buffalo

84 DONALD C. JACKSON

curve, is fishlike in the winter and reptilelike in the summer (Rahn et al., 1971). Even the fresh-water turtle, our example of a lung breather, may switch almost entirely to skin breathing at the very low temperatures (0–5 °C) at which it winters. In a recent study (G. R. Ultsch and D. C. Jackson, unpubl. observations) we observed 3 turtles (*Chrysemys picta belli*), in a group of 10 animals submerged in aerated water with no access to air, go for over 50 days with no significant change in either blood pH or P_{CO_2}, yet with pH values situated comfortably on the Buffalo curve (Figure 6.4).

The Buffalo curve itself, as shown in Figure 6.1, is idealized as well because an examination of the actual pH–temperature relations of many ectotherms reveals considerable diversity both in the position of the curve (with respect to pN) and in the slope of the curve (Figure 6.5). Even more significant, and representing a serious challenge to the unity of ectothermic acid–base physiology, are those species that simply do not follow the Buffalo curve (Figure 6.6). Notable among them are the varanid lizards (Wood et al., 1979; Bennett, 1973), but they include other lizards, *Sauromalus hispidis* (Bennett, 1973), *Dipsosaurus dorsalis* (Withers, 1978), and *Amblyrhinchus cristatus* (Ackerman and White, 1979), over all or part of the temperature range studied. In these lizards, the pH generally decreases with temperature, but the slope is quite low, significantly less than the characteristic −0.016/°C slope of the Buffalo curve.

FIGURE 6.4 In vivo blood pH and P_{CO_2} of three turtles, *Chrysemys picta belli*, during long-term submergence in aerated water at 3 °C.

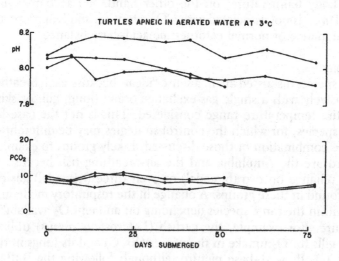

Two interpretations have been offered to account for these deviants: First, Wood et al. (1977) proposed that the high aerobic requirements of the varanid lizards couple their ventilation more closely with O_2 uptake than in less active reptiles and prevent them from making the adjustments in air convection requirement observed in the turtle. Restriction on salt availability in this terrestrial reptile may make the ionic adjustment impossible. The second hypothesis (Withers, 1978) is that the exceptions are all heliothermic reptiles with well-developed behavioral thermoregulatory capacity, and as such are incipient homeotherms. Withers pointed out that the heliothermic lizards resemble the heterothermic mammals which also show little change in the absolute blood pH value between their normothermic (37–78 °C) temperature and the low temperature at which they hibernate (Musacchia and Volkert, 1971) or are torpid (Withers, 1978). Of particular interest in this regard, and it can clearly be seen in the lizard data of Wood et al. (1977) and the ground squirrel data of Musacchia and Volkert (1971), is the fact that there is apparently no loss of control accuracy (i.e., in the variance of the individual values) over the temperature range studied. This suggests that

FIGURE 6.5 Selected Buffalo curves of various species illustrating diversity of slope and position. Note human value at the right. [Carp data from Heisler, 1980; bullfrog and turtle data from Malan et al., 1976; shore crab data from Truchot, 1973; lizard data from Crawford and Gatz, 1974; lugworm data from Toulmond, 1977]

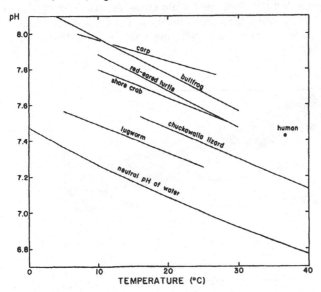

control is as effective in the hibernating squirrel and the cool lizard as in the normothermic squirrel and the warm lizard. Clearly, these animals are not regulating blood protein net charge or relative alkalinity in their extracellular fluids, and it is a challenge in the field to establish what these animals are regulating.

Blood versus cellular pH

Finally, I would like to depart from the exclusive consideration of blood pH as the measure of an animal's acid–base status. Blood has always been the focus of attention in this field for at least two reasons: (1) Blood is readily collected in ample volume from almost all animals, whereas other fluid compartments are comparatively inaccessible; (2) blood composition reflects extracellular fluid composition generally, and it is the extracellular fluid that bathes the cells and is the "milieu interieur" (internal environment) of the cells. The regulation of the internal environment as a condition for the free existence of an organism, first proposed by the French physiologist Claude Bernard, is one of the major paradigms of physiology. New techniques, however, have permitted the attention of investigators to move across the cell membrane into the intracellular compartment so that: "Today the cell is the milieu interieur in the Claude Bernard sense, where the blood system is merely a regulated physical

FIGURE 6.6 Examples of species that control their blood pH at various temperatures, but not along the Buffalo curve. An active lizard, *Varanus exanthematicus*, and a hibernating mammal, *Citellus tridecemlineatus*. [Lizard data from Wood et al., 1979; mammal data from Musacchia and Volkert, 1971]

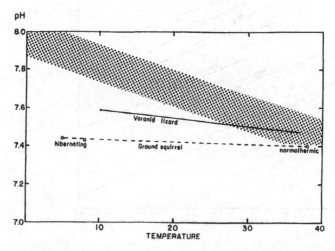

buffer region, the first line of defense, designed to maintain the fixity of the intracellular environment" (Rahn, 1979).

Intracellular pH values, measured by weak acid distribution or by microelectrodes, are typically about 0.6 unit below extracellular pH, which places the cell reaction close to the neutral point of water. In several studies intracellular pH values have been tested in various tissues concurrently with blood measurements on animals acclimated to various temperatures. Although temperature dependence was not uniform in the various tissues, the dpH/dT was generally similar to that of blood, ranging in the turtle, *Pseudemys scripta*, from -0.012 to -0.023 U/°C (Malan et al., 1976), and in the shark, *Scyliorhinus stellaris*, from -0.010 to -0.033 U/°C (Heisler et al., 1976). Individual tissues, however, departed significantly from the condition of constant relative alkalinity in each of these studies. This represents a further unresolved complication in this field, but it may be due to differing intracellular buffer composition in these tissues, and thus differing intrinsic temperature-dependence (Reeves, 1977), or to differing pH–temperature optima in the multienzyme systems operating in these various tissues.

The importance of looking at intracellular pH was dramatically demonstrated recently in studies of CO_2 stress in two aquatic ectotherms. The first of these concerned the transition from water breathing to air breathing in the South American fish *Synbranchus marmoratus* (Heisler, 1980). Over the first 2 or 3 days in air, blood P_{CO_2} rose from 5.6 to 26 Torr and pH fell, an expected result because of the reduction in ventilation that normally accompanies the water–air transition. However, no ionic compensation occurred in blood, so that pH remained more than 0.6 unit below the initial value for at least 5 days. The intracellular pH, in marked contrast, was restored to its initial value by an ionic adjustment of $[HCO_3{}^-]$. A second observation of pH uncoupling between intra- and extracellular fluids was made on the large aquatic salamander, *Siren lacertina*, an inhabitant of water hyacinth–covered lakes in the southeastern United States. Aquatic P_{CO_2} beneath the hyacinth mat can reach 60 Torr (Ultsch, 1976). When experimentally exposed to high ambient P_{CO_2}, these animals suffered a prolonged respiratory acidosis in the blood, but like *Synbranchus*, restored intracellular (muscle) pH to near-normal values (Heisler et al., 1980). These observations illustrate that cells have acid–base control mechanisms independent of the familiar control processes accounting for extracellular control.

These observations further show us that blood values may be misleading as indicators of acid–base regulation. Just as one regulated variable may have higher priority than another in a stress situation, so one area or compartment of the body may be more closely defended than another. In addition to this work on intracellular pH, another recent investigation has shown that the acid–base state of the cerebrospinal fluid of the turtle, *Pseudemys scripta*, is more carefully regulated than the blood during anoxic stress (Hitzig and Nattie, 1979). This priority of the brain environment is already a well-documented observation on mammals. It is now important that these fluid compartments, to which the organisms in their physiological wisdom give high priority, be given equally high priority in future investigations in this field. For example, it is tempting to speculate that intracellular pH may have a different, and more typical, temperature dependence than the blood in heliothermic lizards and hibernating mammals.

Conclusion

The Buffalo curve has proved to be an important general principle in the acid–base physiology of ectothermic animals. It has provided a framework for building a coherent hypothesis for what pH control actually means in these animals and sheds light on the true basis for regulation in homeotherms. For, after all, we ourselves sit without much lateral movement on the high end of the Buffalo curve. The recent questions concerning the precision and ubiquity of the Buffalo curve provide a challenge to the field, but they also serve to remind us of the great majority of ectothermic species whose physiological and biochemical processes have evolved and adapted to keep their blood pH close to this important curve.

References

Ackerman, R. A., and White, F. N. (1979) The effects of temperature on acid–base balance and ventilation of the marine iguana. *Respir. Physiol.* 39:133–147.

Bennett, A. F. (1973) Blood physiology and oxygen transport during activity in two lizards, *Varanus gouldii* and *Sauromalus hispidus*. *Comp. Biochem. Physiol.* 46A:673–690.

Boutilier, R. G., McDonald, D. G., and Toews, D. P. (1980) The effects of enforced activity on ventilation, circulation and blood acid–base balance in the aquatic gill-less urodele, *Cryptobranchus alleganiensis*; a comparison with the semi-terrestrial anuran, *Bufo marinus*. *J. Exp. Biol.* 84:289–302.

Cameron, J. N. (1978) Regulation of blood pH in teleost fish. *Respir. Physiol.* 33:129–144.

Crawford, E. C., Jr., and Gatz, R. N. (1974) Carbon dioxide tension and pH of the blood of the lizard, *Sauromalus obesus* at different temperatures. *Comp. Biochem. Physiol.* 47A:529–534.

Davies, D. G. (1978) Temperature-induced changes in blood acid–base status in the alligator, *Alligator mississipiensis*. *J. Appl. Physiol.* 45:922–926.

Guimond, R. W., and Hutchison, V. H. (1976) Gas exchange of the giant salamanders of North America. In *Respiration of amphibious vertebrates* (G. M. Hughes, ed.), pp. 313–338. London: Academic Press.

Heisler, N. (1980) Regulation of acid–base status in fish. In *The environmental physiology of fishes* (M. A. Ali, ed.), pp. 123–162. New York: Plenum Press.

Heisler, N., Ultsch, G. R., and Anderson, J. F. (1980) Acid–base status in two salamander species, *Siren lacertina* and *Amphiuma*, in response to environmental hypercapnia. *Physiologist* 23:176.

Heisler, N., Weitz, H., and Weitz, A. M. (1976) Extracellular and intracellular pH with changes of temperature in the dogfish *Scyliorhinus stellaris*. *Respir. Physiol.* 26:249–263.

Hitzig, B. M., and Nattie, E. E. (1979) Cerebrospinal fluid acid–base homeostasis during asphyxic diving, hypercapnia, and anoxia in turtles. *Physiologist* 22:56.

Howell, B. J., Baumgardner, F. W., Bondi, K., and Rahn, H. (1970) Acid–base balance in cold-blooded vertebrates as a function of body temperature. *Am. J. Physiol.* 218:600–606.

Jackson, D. C. (1971) The effect of temperature on ventilation in the turtle, *Pseudemys scripta elegans*. *Respir. Physiol.* 12:131–140.

Jackson, D. C. (1978) Respiratory control and CO_2 conductance: Temperature effects in a turtle and a frog. *Respir. Physiol.* 33:103–114.

Jackson, D. C., Palmer, S. E., and Meadow, W. L. (1974) The effects of temperature and carbon dioxide breathing on ventilation and acid–base status of turtles. *Respir. Physiol.* 20:131–146.

Kinney, J. L., Matsuura, D. T., and White, F. N. (1977) Cardiorespiratory effects of temperature in the turtle, *Pseudemys floridana*. *Respir. Physiol.* 31:309–325.

Malan, A., Wilson, T. L., and Reeves, R. B. (1976) Intracellular pH in cold-blooded vertebrates as a function of body temperature. *Respir. Physiol.* 28:29–47.

Moalli, R., Meyers, R. S., Ultsch, G. R., and Jackson, D. C. (1980) Acid–base control in the hellbender, *Cryptobranchus alleganiensis*. *Physiologist* 23:71.

Musacchia, X. J., and Volkert, W. A. (1971) Blood gases in hibernating and active ground squirrels: HbO_2 affinity at 6 and 38 °C. *Am. J. Physiol.* 221:128–130.

Piiper, J., Gatz, R. M., and Crawford, E. C., Jr. (1976) Gas transport characteristics in an exclusively skin-breathing salamander, *Desmognathus fuscus* (*Plethodontidae*). In *Respiration of amphibious vertebrates* (G. M. Hughes, ed.), pp. 339–356. London: Academic Press.

Rahn, H. (1967) Gas transport from the external environment to the cell. In *Development of the lung*, Ciba Foundation Symposium (A. V. S. de Reuck and R. Porter, eds.), pp. 3–23. London: J. & A. Churchill.

Rahn, H. (1979) Acid–base balance and the milieu interieur. In *Claude Bernard and the internal environment. A memorial symposium* (E. D. Robin, ed.), pp. 179–190. New York: Marcel Dekker.

Rahn, H., Rahn, K. B., Howell, B. J., Gans, C., and Tenney, S. M. (1971) Air-breathing of the garfish (*Lepisosteus osseus*). *Respir. Physiol.* 11:285–307.

Randall, D. J., and Cameron, J. N. (1973) Respiratory control of arterial pH as temperature changes in rainbow trout. *Am. J. Physiol.* 225:997–1002.

Reeves, R. B. (1972) An imidazole alphastat hypothesis for vertebrate acid–base regulation: Tissue carbon dioxide content and body temperature in bullfrogs. *Respir. Physiol.* 14:219–236.

Reeves, R. B. (1976) Temperature-induced changes in blood acid–base status: pH and P_{CO_2} in a binary buffer. *J. Appl. Physiol.* 40:752–761.

Reeves, R. B. (1977) The interaction of body temperature and acid–base balance in ectothermic vertebrates. *Annu. Rev. Physiol.* 39:559–586.

Reeves, R. B., and Rahn, H. (1979) Patterns in vertebrate acid–base regulation. In *Evolution of respiratory processes. A comparative approach* (S. C. Wood and C. Lenfant, eds.), pp. 225–252. New York: Marcel Dekker.

Robin, E. D. (1962) Relationship between temperature and plasma pH and carbon dioxide tension in the turtle. *Nature, Lond.* 195:249–251.

Rosenthal, T. B. (1948) The effect of temperature on the pH of the blood and plasma in vitro. *J. Biol. Chem.* 173:25–30.

Stewart, P. A. (1978) Independent and dependent variables of acid–base control. *Respir. Physiol.* 33:9–26.

Toulmond, A. (1977) Temperature-induced variations of blood acid–base status in the lugworm, *Arenicola marina* (L.): II. In vivo study. *Respir. Physiol.* 31:151–160.

Truchot, J. P. (1973) Temperature and acid–base regulation in the shore crab *Carcinus maenas* (L.). *Respir. Physiol.* 17:11–20.

Ultsch, G. R. (1976) Eco-physiological studies of some metabolic and respiratory adaptations of sirenid salamanders. In *Respiration of amphibious vertebrates* (G. M. Hughes, ed.), pp. 287–312. London: Academic Press.

Withers, P. C. (1978) Acid–base regulation as a function of body temperature in ectothermic toads, a heliothermic lizard, and a heterothermic mammal. *J. Thermal Biol.* 3:163–171.

Wood, S. C., Glass, M. L., and Johansen, K. (1977) Effects of temperature on respiration and acid–base balance in a monitor lizard. *J. Comp. Physiol.* 116B:287–296.

Wood, S. C., Johansen, K., Glass, M. L., and Hoyt, R. W. (1979) Acid–base balance during heating and cooling in the lizard, *Varanus exanthematicus*. *Physiologist* 22:135.

7

Blood, circulation, and the rise of air breathing: passes and bypasses

KJELL JOHANSEN

Paleontological evidence suggests that crossopterygian fishes and lung fishes (Dipnoi) were lung breathers in the early Devonian, nearly 400 million years ago (Thomson, 1971). Yet, not until birds appear in the fossil record in the Triassic, 180 million years ago, were the pulmonary and the systemic vascular circuits anatomically separated. In an evolutionary perspective we may ask if this out-of-step development of the lung and a separate pulmonary vascular circuit is the result of random evolutionary processes, or if the incomplete anatomical separation of the principal vascular circuits offers advantages for the exchange and transport of respiratory gases to those vertebrates where it is found. This chapter addresses this question.

Water-breathing fish

Gills of fish are admirably engineered to exchange respiratory gases between water and blood. The exchange system can be modeled as a countercurrent system. Its exchange efficiency exceeds that of other vertebrate gas exchangers, allowing arterial P_{O_2} to approach that of inspired water (Chapter 1).

In water-breathing fish, the heart pumps only mixed venous blood. After passage through the gills, the oxygenated blood flows directly to the arteries of the systemic vascular beds. Energy is imparted to the blood from the heart only once for passage through all the vascular circuits. In this system the gas exchanger receives a higher blood pressure than the systemic circulations, a situation unlike that in all other vertebrates (Figure 7.1). This poses a design dilemma. Sufficient propulsive energy must be left in the blood after passage of the gills to allow effective distribution and perfusion of blood to the various tissues (many of which call for pressures high enough to allow filtration) without the gas-exchange units of the gills (secondary lamellae) becoming fluid filters.

This dilemma is solved by a gill microcirculation permitting efficient gas exchange of the entire cardiac output at very low transbranchial blood pressure gradients, a quality laid down in the design of the sheetlike vascular space of the gill secondary lamellae. Interestingly, active fish, requiring larger cardiac outputs, appear to have smaller branchial vascular resistance than less active species. Once blood has traversed the gills, the capillaries of the systemic vascular beds appear to permit filtration at relatively low capillary pressures. This necessitates an efficient lymph circulation, a subject that so far seems not to have attracted the interest of physiologists.

The presence of alternative routes for blood resulting in nonrespiratory bypass of the secondary lamellae in water-breathing fish has often been reported but later refuted. Anatomical evidence is available, however, showing numerous arteriovenous anastomoses between the efferent lamellar and filament arteries and central venous channels bringing blood directly back to the heart (Vogel et al., 1976; Laurent and Dunel, 1976). Efferent filament arteries also give rise to a nutritive circulation serving the gill tissues and draining into the central venous sinuses. These numerous arteriovenous connections between the efferent gill circulation and the systemic venous drainage of the branchial region appear to entail an efficiency loss because this oxygenated blood will bypass the systemic vascular beds. However, at times when respiratory needs are low this bypass may raise the efficacy of needed ionic, osmotic, and excretory exchange in the gills by rapid recirculation, while lessening the perfusion load on the systemic circulation.

FIGURE 7.1 Perfusion pressures and vascular resistances of the gas-exchange and systemic circulations in representative vertebrates. Vascular resistance is expressed as the width of the bars. [From Johansen, 1972]

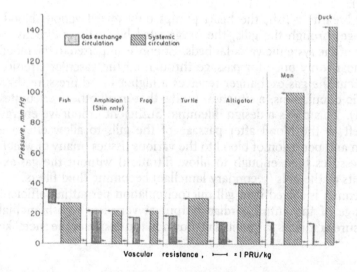

Air-breathing fish

Air breathing is widespread and practiced by members of more than 20
genera of fish, most of them tropical and all of them fresh-water species.
Shortage of O_2 in the water probably was the condition that led to air
breathing, and air breathing in fish must have evolved independently
many times (Johansen, 1970). Air breathing is accessory or supplemental
to water breathing in some species, others are facultative air breathers
and can sustain themselves on either mode of breathing, depending on
the environmental conditions, and some species are obligate air breathers
and drown if prevented direct access to atmospheric air (Johansen, 1970).

Air-breathing organs

In fish, most air-breathing organs are highly vascularized air recep-
tacles that extend from buccal, opercular, or pharyngeal cavities. The
swim-bladder or sections of the gastrointestinal tract act as the air-breathing organ in some species; in the true lungfish (Dipnoi), the air-breathing
organs are true homologues of the tetrapod lung. In obligate air-breathing
fish, there is a reduction in the gill surface area and an increase in the
blood-to-water diffusion distance. In many species, some of the primary
gill arches have lost nearly all their branchial filaments, and the vascular
passages through them become nonrespiratory gill shunts (the South
American and African lungfish, the electric eel (*Electrophorus electricus*),
and the synbranchid fish (*Amphipnous cuchia*) are good examples).

Mixing of oxygenated and deoxygenated blood

In all air-breathing fish, except for the true lungfish, the efferent circulation from the air-breathing organ drains into systemic veins, mixing
its oxygenated blood with deoxygenated systemic venous blood, before
returning to the heart. The hearts of all these fish are undivided and
possess no structural means for selective distribution of blood.

Why have direct vascular channels from the air-breathing organ to the
systemic distributing arteries not evolved? The answer probably lies in
hemodynamic considerations. The circulation through the air-breathing
organs has a high resistance, higher in fact than that through the gills
or the bypasses set up by loss of gill filaments. A large drop in pressure
must occur if significant amounts of blood are to flow through the air-
breathing organ. As a result, the efferent circulation from the aerial gas
exchanger cannot converge with the systemic arteries, but has to return
via low-pressure systemic veins to the heart.

Periodic breathing: alternate storage and utilization of oxygen

Periodic breathing and dependence on lung oxygen stores have been
a major factor in the phylogenetic development of the cardiovascular

system in air-breathing vertebrates. During periodic breathing, the gas-exchange organ is ventilated intermittently by either single breaths or bursts (series) of breaths interspaced by periods of breath holding. This type of breathing is widespread, being utilized not only by air-breathing fish but by most amphibians and reptiles as well.

The importance of periodic breathing as a means of charging O_2 stores for use between breaths can be illustrated by referring to experiments on an obligate air-breathing teleost, *Amphipnous cuchia* (Lomholt and Johansen, 1976). *Amphipnous cuchia* has vestigial gills and periodically ventilates a pair of large air sacs placed laterally as extensions of the buccal chamber (Figure 7.2).

The fourth pair of branchial arteries have been modified to form very large shunts from the ventral to the dorsal aorta. They function as aortic arches, giving rise to the systemic circulation. The blood leaving the air sacs (lungs) empties into the large systemic venous cardinal veins. Similar vascular arrangements are found in many other air-breathing teleosts (e.g., electric eel, *Electrophorus*; Johansen et al., 1968).

This type of circulatory arrangement results in a mixture of oxygenated blood from the gas exchanger with deoxygenated systemic venous blood. To compensate for this, *Amphipnous* has an unusually high O_2 capacity (exceeding 20 vol%) and a high affinity for oxygen. It also has a very high cardiac output (when compared with other fish having the same level of physical activity). When *Amphipnous* is breathing air, more than 75% of the cardiac output is diverted to the air-breathing organs, but between breaths perfusion of the air sacs declines to about 20% of this value (Figure 7.2).

These factors allow *Amphipnous* to store large amounts of oxygen during its periods of breathing. The high blood flow to the lungs coupled with the high blood O_2 affinity enable the animal to achieve better than 90% saturation of all its blood (both arterial and venous) while breathing. Enough oxygen is stored to allow the animal to survive for as long as 30 min without breathing with only a modest reduction in arterial O_2 saturation (from 90 to 60%). These mechanisms for surviving long periods without breathing resemble in some respects those used by diving mammals and birds.

In *Amphipnous*, we can see a crude outline of a functional separation of gas-exchange perfusion and systemic perfusion. The separation is not spatial, but temporal (i.e., a separation in time). Blood flow during periods of air breathing is predominantly through the gas exchanger; it shifts to a dominance through the systemic perfusion when arterial gas exchange is suspended. Continuous O_2 delivery to the tissues depends on tapping the large circulating O_2 stores (stored during periods of breathing) between breaths.

Lungfish: the beginning of an end to shunts

The lungfish (Dipnoi) are phylogenetically the first among vertebrates to possess a separate return of blood from the air-breathing organ directly to the heart. They have a pulmonary vein, and two atrial compartments in their heart that are nearly completely separated by a specialized structure referred to as the pulmonalis fold. The ventricle is also partially divided by a septum. A complicated bulbus structure that forms the outflow tract from the ventricle offers an anatomical means for separating ventricular outflow. Detailed accounts of lungfish cardiovascular morphology have been published by Bugge (1961) and Johansen and Burggren (1981). The lungfish have a cardiovascular system that prevents mixture of blood leaving the air-breathing organ (lung) with systemic venous blood prior to the return to the heart. Admixture, however, still may occur within the heart or in the central outflow vessels to the lung and the systemic vascular beds.

FIGURE 7.2 Left: Schematic drawing of the heart and its major in- and outflow vessels in *Amphipnous cuchia*. Note the paired shunt vessels from the ventral aorta forming the dorsal aorta. The effluent blood from the air-breathing organ converges with systemic venous blood in the jugular veins. Right: Air-breathing increases cardiac output and the fraction of the cardiac output perfusing the air-breathing organ. During breath holding a major fraction of the cardiac output is shunted from the ventral to the dorsal aorta. [Redrawn from Lomholt and Johansen, 1976]

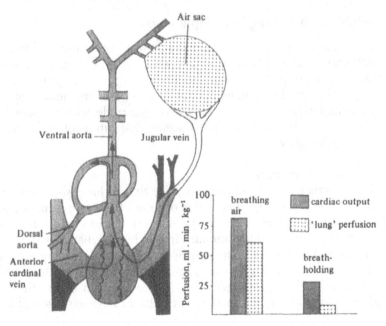

The African and South American lungfish are both obligate air breathers, and species of African lungfish may spend several years in estivation, totally removed from water. The Australian lungfish, *Neoceratodus*, has well-developed gills and employs air breathing only in hypoxic water or during activity.

In the African lungfish, *Protopterus*, the best studied of the lungfish, there exists a clear tendency for separate passage of deoxygenated (systemic venous) and oxygenated (pulmonary venous) blood through the heart and its outflow conduits. Lungfish are periodic breathers, and the extent of selective or preferential circulation depends clearly on the phase of the breathing cycle. During breathing the cardiac output increases, and 95% of the blood returning to the heart via the pulmonary veins may be dispatched selectively through the heart and bulbus to the anterior branchial arteries giving rise to systemic distributing arteries. This fraction declines as the breath-hold period progresses, but never drops below 65%. It seems likely that the clear coupling of the ventilatory and circulatory events is reflexly regulated, and that an inflation reflex involving pulmonary stretch receptors is involved in the response. Significantly, the vascular resistance of the respiratory vascular bed is much lower than the systemic, a situation not met in other air-breathing fish. Lungfish demonstrate that variable shunts and blood flow rates are important for both the filling and the emptying of the O_2 stores used in the periodic breathing pattern they employ.

Amphibians: breathing in multiple ways

Amphibians utilize very unspecialized patterns of multimodal breathing, which may involve gills and lungs as well as skin. Our discussion will be about anuran amphibians. For them, the lungs are the principal organs of O_2 exchange, but cutaneous gas exchange is also important. They have completely separated atria, and a ventricle with no septal compartmentalization. At first sight it seems that intraventricular shunting may be unavoidable.

Cardiovascular arrangements

The bulbus cordis of amphibians, like that of lungfish, plays a significant role in separating ventricular outflow. It has two separate sets of valves at its distal end. The closure of these valves in diastole separates the systemic and pulmocutaneous circuits hemodynamically.

Six arterial arches leave the ventricle: two carotid, two aortic, and two pulmocutaneous. The pulmocutaneous arch supplies the lung and also sends a major branch (arteria cutanea) to the skin of the thorax and abdomen. Thus there is structural specialization for perfusion of a major portion of the skin with blood of the same low oxygen content as that perfusing the lungs. Oxygenated blood draining from the skin, however, is not separated from systemic veins, like that draining the lungs. There are no cutaneous veins connecting with pulmonary veins or with the left atrium. Sections of the skin that are important in respiratory gas exchange are also supplied with blood vessels distributed from the systemic arterial arches. Because of these vascular arrangements, the skin contributes to external gas exchange only by elevating O_2 and reducing CO_2 in the systemic venous return to the heart.

Mixing of oxygenated and deoxygenated blood

To what extent are oxygenated pulmonary venous blood and deoxygenated systemic venous blood separated as they pass through the heart? This question has never been unequivocally answered (de Long, 1962; Johansen and Ditadi, 1966; Tazawa et al., 1979; Meyers et al., 1979) although many attempts have been made.

Unfortunately, none of the studies has given appropriate consideration to the marked effect of periodic breathing on the pattern of cardiac outflow. De Graaf (1957) noted that blood flow in the pulmocutaneous direction of *Xenopus laevis* could greatly exceed the outflow in the systemic direction, and Shelton (1970) and Johansen et al. (1970) noted that the relative flow rates between pulmocutaneous and systemic circulation of frogs depend mainly on the lung-breathing pattern. During periods of frequent breathing the total cardiac outflow increased, and the pulmocutaneous flow exceeded systemic flow. During breath holding, this outflow pattern reversed.

The question of whether or not amphibians effectively separate oxygenated and deoxygenated blood as it flows through the heart can only be answered when patterns of breathing and flow distribution are taken into account.

Blood flow distribution to lungs and skin

Is blood flow to the skin regulated in response to the pattern of pulmonary breathing? Based on microsphere injection, Moalli et al. (1980) claim that during breath holding in frogs (when the pulmonary vascular impedance is increased and flow alleged to decline in both the cutaneous and pulmonary arteries) a compensatory increase in skin blood flow occurs via the systemic vascular supply. In sharp contrast, a recent study by de Saint Aubain and Wingstrand (1979) suggested an alternative mechanism for increased skin blood flow as a compensation for pulmonary apnea. They documented the presence of a hitherto unknown muscular sphincter on the pulmonary artery of the frog *Rana temporaria*, immediately distal to the branching point of the pulmonary artery from the pulmocutaneous arch, as shown in Figure 7.3. The sphincter is richly innervated by vagal fibers, and their stimulation or application of acetylcholine constricts the sphincter and diverts pulmocutaneous flow to the cutaneous artery. Allegedly, deflation of the lungs causes a similar constriction. They reasoned that when the lungs are nonfunctional (e.g., during prolonged diving or overwintering in water), a preferential passage of blood to the cutaneous arteries is of value. These authors suggest that the sphincter is homologous to the pulmonary artery sphincters in chelonian reptiles.

FIGURE 7.3 (a) Comparison of systemic arterial blood flow and cutaneous artery blood flow in the bullfrog, *Rana catesbeiana,* when breathing (air) and during diving. When cutaneous artery flow is reduced during diving, skin perfusion is maintained by an increase in the systemic arterial blood supply to the skin. (b) Schematic drawing of the heart outflow vessels in the frog, *R. temporaria.* The sphincter distal to the bifurcation of the cutaneous artery from the pulmocutaneous artery is important in controlling the balance between cutaneous and pulmonary blood flow. [a, after Moalli et al., 1980; b, after de Saint-Aubain and Wingstrand, 1979]

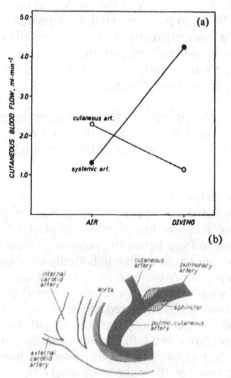

There might be an obvious explanation for the apparent conflict between the above results. If the diving practiced by Moalli and co-workers (1980) on bullfrogs were extended for longer periods than the 10-min dives of their study, there might be an increase in flow in the cutaneous arteries from the pulmocutaneous arches because skin breathing should be made maximal at such times. Clearly the amphibian skin is not a passive gas exchanger because skin blood flow is reciprocally related to the perfusion of the lung.

Reptiles

Reptiles rely almost exclusively on the lungs for respiratory gas exchange. They have higher preferred body temperatures, higher activity levels, and larger metabolic scopes than amphibians or fish. These conditions correlate with a higher systemic arterial blood pressure and higher and more variable blood flow rates.

Cardiovascular arrangements

The heart and main vessels of reptiles are morphologically and functionally organized to accommodate selective blood passage. The two atria are completely separated, but the pulmonary and systemic vascular beds are still perfused in a parallel arrangement from a ventricle complexly divided into three distinct compartments. These ventricular compartments are, however, in anatomical continuity, only partially separated by a longitudinal ventricular septum (Webb et al., 1971). This structural arrangement causes the ventricle to act as a single pressure pump. Because the ventricle releases the same pressure pulse to the two principal arterial trees connected in parallel in the heart, the distribution of the cardiac output to the two circulations is governed entirely by the balance of input impedance between the two circulations (Burggren, 1977).

Periodic breathing

Among vertebrates, reptiles show the largest variations in cardiac output and its distribution between the lung and the systemic circuits. These variations, as in air-breathing fish and amphibians, are related to the phase and pattern of periodic breathing (some reptiles may breath-hold for periods of 50–60 min during normal, undisturbed breathing). As an example, the fresh-water turtle (*Pseudemys scripta*) may show 20- to 30-fold increases in cardiac output associated with the onset of lung ventilation during spontaneous, undisturbed breathing (White and Ross, 1966; Shelton and Burggren, 1976). Such large changes in cardiac outputs, triggered by ventilation, are likely to be regulated for optimal use of the pulmonary O_2 store.

Inequality in flow rate between the two principal vascular circuits implies an operational shunt between them. We know from blood gas analysis that during periods of active ventilation there occurs a high degree of selective circulation and thus minimal admixture of oxygenated and deoxygenated blood. If admixture occurs, it typically involves a left-to-right shunt with recirculation of oxygenated pulmonary venous blood to the lung. During breath hold (apnea) a marked right–left shunt develops, causing recirculation of systemic venous blood and reduced O_2 saturation of the systemic arterial blood.

When the pulmonary O_2 store becomes smaller as breath holding progresses, the usefulness of lung perfusion will also decline. A right–left shunt and partial bypass of the lung then become an asset, and will result in an increased retention of CO_2 in systemic blood during breath holding. This in turn will displace the O_2–Hb dissociation curve to the right, and improve O_2 unloading to the tissues.

The reduction in pulmonary perfusion associated with a right–left shunt will not, of itself, improve the continued use of the remaining lung O_2 store although this has been advocated (White, 1978). Alveolar P_{CO_2} during prolonged breath holding exceeds that in arterial blood, owing to a reduction in the gas-exchange ratio, and hence in the volume of the lung (Lenfant et al., 1970; Glass and Johansen, 1979). The low gas-exchange ratio of the lung during breath holding implies an increased importance of the skin for CO_2 elimination. This elimination would cause blood draining the skin to have a low P_{CO_2}, and when mixed with systemic venous blood, it would reduce systemic venous P_{CO_2}. This reduction would increase the O_2 affinity of the blood supplying the lung. By this means of extrapulmonary CO_2 elimination, O_2 loading from the pulmonary O_2 store during breath holding may be improved.

Another asset of right–left shunting may rest with convection requirements unrelated to O_2 transport. Anaerobic metabolism, so important to many reptiles during hypoxia or diving, will also have a convection requirement for replenishment of blood-borne buffers and substrates and removal of metabolites. Many reptiles (e.g., turtles) may nearly double the systemic perfusion by a large right–left shunt without the need of increasing cardiac output.

The tendency in reptiles of a left–right shunt during the ventilatory periods may be important for increasing the rate of replenishment of the blood O_2 store prior to a subsequent breath hold.

Two exceptional reptiles

We find interesting exceptions to the typical reptilian condition in varanid lizards and crocodilians (White, 1970; Burggren and Johansen, 1981).

Varanid lizards are able to attain two functionally distinct pumps with an incompletely divided ventricle. This is possible because when the ventricular myocardium is shortening during the systolic phase of the cardiac cycle, the partial ventricular septa make contact with the ventricular walls and bring about complete anatomical separation of the pulmonary ventricular pump (cavum pulmonale) and the systemic ventricular pump (cavum arteriosum) (Figure 7.4). During diastole, however, the anatomical separation is incomplete, and admixture of blood filling the ventricle from the two atria is possible.

Figure 7.4 shows schematically the heart and major outflow vessels in a varanid during diastole and systole. Figure 7.5 shows pressure and flow pulses documenting that this heart may function as two distinct pumps. Table 7.1 shows that values of pulmonary and systemic arterial blood pressures may be amazingly similar to those recorded from a similar-size mammal. The table also attests to a great variability in these pressures, suggesting very labile and delicate separating mechanisms (Burggren and Johansen, 1981).

We may say that among ectotherms the varanids are special by having attained near mammal-like cardiovascular characteristics, allowing complete separation of the two circulations, hemodynamically as well as in convective blood gas transport. Yet, the heart retains anatomical shunts, which may become operational, depending on the breathing patterns.

The crocodilian cardiovascular system possesses complete anatomical separation of both the atria and ventricles. Still it retains the possibility

FIGURE 7.4 Schematic drawing of the heart in a varanid lizard during diastole (left) and systole (right). The arrows indicate the channeling of blood during ventricular filling (diastole) and ejection (systole). [After Webb et al., 1971]

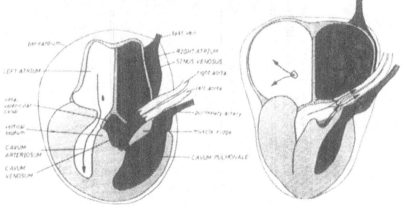

TABLE 7.1 Systemic arterial and pulmonary arterial blood pressures in six varanid lizards. Note the marked difference in pressure in the pulmonary (low) and systemic (high) arteries.

Animal	Body weight (g)	Systemic arterial pressures (cm H_2O)				Pulmonary arterial pressures (cm H_2O)			
		Systolic	Diastolic	Pulse	Mean	Systolic	Diastolic	Pulse	Mean
1	2634	116	95	21	102	65	39	26	48
2	1633	80	59	21	66	80	47	33	58
3	1671	142	96	46	111	70	34	36	46
4	2828	70	46	24	54	47	36	11	40
5	1424	123	90	33	101	32	17	15	22
6	5680	143	80	63	101	46	19	27	20
$\bar{x}$ = 1 SD	2645 ± 1594	112 ± 31	78 ± 21	35 ± 17	89 ± 23	57 ± 18	30 ± 12	25 ± 10	40 ± 13

After Burggren and Johansen, 1981.

of a shunt or bypass of the pulmonary circulation between its aortas. All reptiles have a right and a left aorta. The left aorta originates from the right ventricle together with the pulmonary artery. The right aorta originates from the left ventricle and joins the left aorta. There is a peculiar connection between the two aortas at the base of the heart through a flaplike slit (the foramen Panizzae). During regular breathing the pressures prevailing in the left ventricle and right aorta exceed those in the right ventricle. These pressure conditions will prevent outflow from the right ventricle through the left aorta which now will be filled from the left ventricle through the foramen Panizzae. The entire right ventricular output will go the lungs, and the flow in the two circuits will be balanced. When, however, the breathing pattern is altered as during diving, pulmonary vascular resistance will increase and cause an increase in right ventricular pressure. This situation will result in right ventricular outflow to the left aorta and a variable bypass of the lung.

Conclusion

On the evolutionary time scale, the cardiovascular system shows an out-of-step development between the first appearance of air breathing and the later complete separation of pulmonary and systemic circuits. It is

FIGURE 7.5 Recordings of blood pressure and right aortic flow in a varanid lizard. [After Burggren and Johansen, 1981]

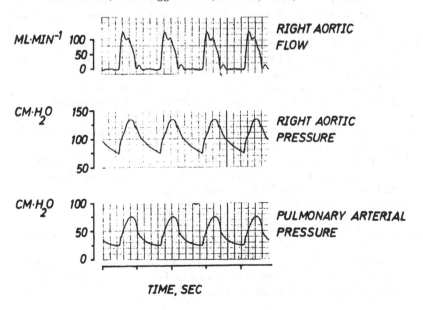

concluded from the preceding discussion that clear advantages can and do accrue to vertebrates that retain the ability to shunt blood in a regulated manner between these two circuits. It is interesting that modern medical technology within a time span of 30 years has started to use artificial vascular circuits as life-sustaining measures for the critically ill. These may be applied in parallel with both the pulmonary and the systemic circuits and may involve supporting pumps as well as diffusion exchangers (e.g., artificial lungs and kidneys). In a way this technology, by introducing temporary vascular bypasses in the cardiovascular system, employs design principles laid down in vertebrate evolution.

References

Bugge, J. (1961) The heart of the African lungfish. *Vidensk. Med. Dansk Naturh. For.* 123:193–210.

Burggren, W. W. (1977) The pulmonary circulation of a chelonian reptile: Morphology, haemodynamics and pharmacology. *J. Comp. Physiol.* 116:303–323.

Burggren, W. W., and Johansen, K. (1981) Hemodynamics of ventricular outflow in the varanid lizard, *Varanus exanthematicus*. *Respir. Physiol.* (In press).

de Graaf, A. R. (1957) Investigations into the distribution of blood in the heart and aortic arches of *Xenopus laevis*. *J. Exp. Biol.* 34:143–172.

de Long, K. T. (1962) Quantitative analysis of blood circulation through the frog heart. *Science* 138:693–694.

de Saint-Aubain, M. L., and Wingstrand, K. G. (1979) A sphincter in the pulmonary artery of the frog *Rana temporaria* and its influence on blood flow in skin and lungs. *Acta Zool. (Stockh.)* 60:163–172.

Glass, M. L., and Johansen, K. (1979) Periodic breathing in the crocodile, *Crocodylus niloticus*: Consequences for the gas exchange ratio and control of breathing. *J. Exp. Zool.* 208(3):319–325.

Johansen, K. (1970) Airbreathing fishes. In *Physiology of fishes II*, vol. IV (W. S. Hoar and D. J. Randall, eds.), pp. 361–411. New York: Academic Press.

Johansen, K. (1972) Heart and circulation in gill, skin and lung-breathing. *Respir. Physiol.* 14:193–210.

Johansen, K., and Burggren, W. W. (1981) Cardiovascular function in the lower vertebrates. In *Hearts and heart-like organs*, vol. I (G. H. Bourne, ed.), pp. 61–117. New York: Academic Press.

Johansen, K., and Ditadi, A. S. F. (1966) Double circulation in the giant toad, *Bufo paracnemis*. *Physiol. Zool.* 39:140–150.

Johansen, K., Lenfant, C., and Hanson, D. (1970) Phylogenetic development of the pulmonary circulation. *Fed. Proc.* 29:1135–1140.

Johansen, K., Lenfant, C., Schmidt-Nielsen, K., and Petersen, J. A. (1968) Gas exchange and control of breathing in the electric eel, *Electrophorus electricus*. *Z. vergl. Physiol.* 61(2):137–163.

Laurent, P., and Dunel, S. (1976) Functional organization of the teleost gill. I. Blood pathways. *Acta Zool. (Stockh.)* 57:189–209.

Lenfant, C., Johansen, K., Petersen, J. A., and Schmidt-Nielsen, K. (1970) Respiration in the fresh water turtle, *Chelys fimbriata. Respir. Physiol.* 8:261–275.

Lomholt, J. P., and Johansen, K. (1976) Gas exchange in the amphibious fish, *Amphipnous cuchia. J. Comp. Physiol.* 107:141–157.

Meyers, R. S., Moalli, R., Jackson, D. C., and Millard, R. W. (1979) Microsphere studies of bullfrog central vascular shunts during diving and breathing in air. *J. Exp. Zool.* 208:423–430.

Moalli, R., Meyers, R. S., Jackson, D. C., and Millard, R. W. (1980) Skin circulation of the frog, *Rana catesbeiana*: Distribution and dynamics. *Respir. Physiol.* 40:137–148.

Shelton, G. (1970) The effect of lung ventilation on blood flow to the lungs and body of the amphibian, *Xenopus laevis. Respir. Physiol.* 9:183–196.

Shelton, G., and Burggren, W. W. (1976) Cardiovascular dynamics of the chelonian during apnea and lung ventilation. *J. Exp. Biol.* 64:323–343.

Tazawa, H., Mochizuki, M., and Piiper, J. (1979) Respiratory gas transport by the incompletely separated double circulation in the bullfrog, *Rana catesbeiana. Respir. Physiol.* 36:77–95.

Thomson, K. S. (1971) The adaptation and evolution of early fishes. *Quart. Rev. Biol.* 46(2):139–166.

Vogel, W., Vogel, V., and Pfautsch, M. (1976) Arterio-venous anastomosis in rainbow trout gill filaments. *Cell Tissue Res.* 167:373–385.

Webb, G., Heatwole, H., and de Bavay, J. (1971) Comparative cardiac anatomy of the reptilia. I. The chambers and septa of the varanid ventricle. *J. Morphol.* 134:335–350.

White, F. N. (1970) Central vascular shunts and their control in reptiles. *Fed. Proc.* 29:1149–1153.

White, F. N. (1978) Circulation: A comparison of reptiles, mammals and birds. In *Respiratory functions in birds, adult and embryonic* (J. Piiper, ed.), pp. 51–61. Berlin: Springer Verlag.

White, F. N., and Ross, G. (1966) Circulatory changes during diving in the turtle. *Am. J. Physiol.* 211:15–18.

PART TWO

Food and energy

Overview

Part Two considers the role of nutrition in the development of animals, how animals convert food into energy, and how the rate of utilization of energy changes with body size.

In Chapter 8 Denis Bellamy discusses the role of nutrition in establishing the timing of developmental events. Low levels of nutrition appear to slow development and extend lifespan. The ability to alter the timing of developmental events provides an experimental tool for studying the programming of development.

In Chapter 9 Hermann Rahn uses allometry (changes in structure and/or function with body size) to examine energy investment and time for embryonic development of birds and mammals. Energetics of embryos and adults are then compared.

In Chapter 10 Peter Hochachka reviews ways organisms have developed to meet their energy needs when oxygen is not available. Substrates, metabolic pathways, and end products of anaerobic metabolism are discussed.

In Chapter 11 Gerry Kooyman considers how mammals meet their energy needs when they dive. He finds that while diving they rely more on oxidative and less on anaerobic metabolism than had previously been thought.

In Chapter 12 C. R. Taylor considers how limits of metabolism (rest, maximum aerobic, and peak) change with body size. The allometric relationships for these limits are used as tools for examining how animals are designed and how their muscles consume energy.

8 Nutrition, growth, and aging: some new ideas

DENIS BELLAMY

Aging and mortality

Aging as a physiological phenomenon has no general relevance to the survival of animals living wild in environments that have not been greatly affected by human activities. The evidence for this comes from mortality statistics which show that death in the wild is not dependent upon the individual's age. Deaths occur through the operation of random factors such as disease, predation, starvation, and fatal accidents. This statement must be qualified in that very few species have been studied – most of the data coming from ringing birds and live-trapping small mammals. Aging as a major mortality factor is only found in animals that now live in environments where they are protected from the random events that originally governed their evolution. In this category we find ourselves and our captive, protected livestock in game parks, zoos, farms, and homes. Deaths in human populations protected from disease and star-vation are markedly age-dependent. Over the last 100 years, demographic data show the trend for improved public-health measures and better nutrition has coincided with the shift of the mortality curve toward a greater age dependence, with most people dying between the ages of 60 and 90 years. Similar-shaped curves are found for laboratory animals such as the mouse. In the wild, the mouse has a mean life expectancy of 100 days (Bellamy et al., 1973) which is increased almost 10-fold in the first laboratory cohort. Causes of death in the laboratory stock are quite different from those in the wild.

Differences in lifespan between species living in protected environ-ments are largely genetic and correlate with differences in species char-acteristics such as body size and rate of development. Within a protected species, differences in lifespan are also under genetic control, but en-vironment may also have a large influence. The most significant envi-ronmental variables are food and temperature, which both appear to exert their action by modifying the rate of development. Slowing devel-opment increases the lifespan of laboratory rodents, provided that the animals are able to assimilate and retain adequate basic nutrients. The aim of this chapter is to delineate some new approaches to the study of

109

connections between development and aging. This new work deals with the role of nutrition in the timing of various physiological characteristics, particularly the clocklike mechanisms governing the gain and loss of fertility and the termination of cellular proliferation. In these approaches aging is viewed as an outcome of programmed development.

Definitions of aging

There are many definitions of aging; the differences between them are connected with differences in viewpoint and the level of working. This approach stresses that aging must be studied in an interdisciplinary context which makes it a difficult field to review and teach. If aging is seen in relation to development, the term *development* must be taken to include all sequential changes, from fertilization to death. The aging organism is therefore viewed against its total life history and the life cycle is the point of reference (Table 8.1). Gerontology is the scientific study of deterioration in those structures and functions that have a definite peak or plateau of development in all members of the species.

Development of an organism is governed by various events that produce physiological change. These events may be classified according to whether they are predictable or not. Predictable events are those that will change physiological function in all members of a species within a narrow age range, at a well-defined rate. Unpredictable physiological events also affect the aging of individuals in a population. Aging is a combination of these predictable and random processes.

For a species in a protected environment, the same age distribution of deaths will always be obtained, spread over a relatively small period of time. The physiological counterpart of the predictable lifespan is the well-documented decline in various physiological functions that in mam-

TABLE 8.1 Scope of research on aging

Objects of study	Processes of interest
Molecules	Chemical deterioration
	Enzyme synthesis
Enzymes	Differential expression of genes
	Growth
Organs	Physiological regulation
	Reproduction
Organisms	Behavioral adaptability
Populations	Heredity
	Natural selection

mals is expressed simply by the formula:

$$Ft = F^0 - F^0 (1 - \alpha)$$

where Ft is function at time t, F^0 is function at time of peak efficiency, and α is a measure of rate of loss of function.

It has been shown experimentally that this type of predictable change, coupled with increased variability in function, could account for the fundamental shape of the laboratory mortality curve (Bellamy, 1970).

Under standard laboratory conditions, the frequency distribution of causes of death will be repeatable. These causes of death will mostly fall into the category of degenerative disease, which is a term used loosely in a medical sense to imply that the underlying morbidity is generated by basic cellular damage and not due to either malnutrition or infectious diseases, which are major factors governing mortality in the wild. On the other hand, it is not possible to predict what particular individuals will die from.

If the genetics or environment is altered, then both lifespan and the pattern of degenerative diseases will change. There are large species differences in lifespans of protected animals, which are greater than those produced by varying the environment, indicating that aging is connected with an evolutionary strategy. This has given rise to the idea that aging is essentially programmed into development. It does not follow from this viewpoint that aging has been selected for in evolution, but that it is an inevitable concomitant of processes that are advantageous in early life. That is, it is difficult to see how aged animals would be advantaged to pass on or help pass on their genes to the next generation. They are usually infertile and nonadaptive; and from wild-type mortality curves, aged individuals make a negligible numerical contribution to field populations.

Turning to the random events of developments, these may have internal causes, through malfunctions in biochemistry, or external causes, through the accidental impact of physicochemical variations in the environment. Theoretically the number of possibilities for basic chemical errors is large, a situation that has given rise to a corresponding number of theories of aging according to which of these various possibilities is taken as the major contributory factor. With regard to external factors of aging, the random impacts of mutagenic agents, both physical and chemical, have played a dominant role in both theorizing and experimentation. On the other hand, attention is being given increasingly to the effects of random nutritional factors, particularly with regard to the origins of cancer and heart disease. The starting point is that human causes of death are clearly linked statistically with environment, with respect to both life style and geography.

Growth and aging at the physiological level

Predictable and unpredictable events

As outlined above, physiological events occurring within the lifetime of a species are either predictable or unpredictable (Table 8.2). Predictable events are governed by the workings of a genetic program that has been selected by evolution. These events occur in all individuals of a species in a definite sequence and over a narrow age range; the overall timed sequence is called development. Unpredictable events, on the other hand, result from random accidents. These unpredictable events may be internal, resulting from the chemical deterioration of key metabolic structures or coming about through errors in the synthesis of macromolecules. External events, which are not predictable, are accidental deaths and the damage arising from accidental hits, which may be physical (e.g., the impact of cosmic rays) or chemical (e.g., the intake of a carcinogen). Individuals within protected populations may die from any one of a range of different degenerative diseases; this is aging due to random events. They all die over a narrow age range, and the timing of these deaths is governed by processes initiated at random earlier. The most likely general timer governing lifespan is probably the mechanism that gradually shifts resources from growth to reproduction. It is becoming clear that surveillance and repair mechanisms operate fully only during the early stages of growth, and that with the shift to the reproductive phase less effort is put into the correction of mistakes.

Programmed aging

The developmental origins of the gradual decline in physiological function are difficult to trace. Viewing the developing body as a whole makes it more difficult. At any time the body is the sum of its independently developing component cellular populations, which, it can be argued, never reach a steady state. At one extreme, the lymphoid organs such as the thymus, although characterized by a high rate of cellular turnover,

TABLE 8.2 Classification of developmental events

Predictable	Unpredictable	
	Internal	External
Genetically timed to occur suddenly or gradually (manifest physiologically in all individuals)	Chemical deterioration Chemical errors in synthesis and transport	Accidental death Accidental hits (physical or chemical)

have a lifespan that is much shorter than that of the body. At the other extreme, postmitotic tissues such as muscle and brain continuously change their makeup, losing cells and gaining extracellular components and intracellular degenerative accumulations.

Changes in cellular balance are the major feature of early life. During this period there are large shifts in the relative growth of organs as the evolutionary program of development unfolds within the various cell populations. Presumably this situation assures the most advantageous partition of the body's total resources in relation to the selection pressures in the wild. At a crude level, this can be seen in the different allometric growth coefficients for various organs and the fact that the brain at first receives the largest fraction of these resources. Evolution works to maximize the efficiency of resource partition at all levels of organization, and at any period of early development the body will probably be in a changing balance of defects as the biological imperative alters from the need to expand cellular populations, and maintain itself virtually error-free, to the stage of reproductive maturation, when a certain level of defects in the life-support systems could be tolerated. Also, because of the high death rate in the wild, reproductive efficiency only needs to be maintained for a relatively short fraction of the artificially long lifespan seen in the laboratory. It may be that these aspects of life in the wild made it advantageous, probabilistically, to gradually withdraw the energy support systems from cellular populations that were no longer giving a prime selection advantage, and that this evolutionary program is expressed in a trajectory of deterioration in these organs. As measured by probability of death, this decline begins at the age of 12 in man when predation, disease, and malnutrition are no longer the main mortality factors.

Nutrition and lifespan

The most dramatic and significant experiments relevant to the concept of aging being determined as a developmental process are those showing that food intake may govern lifespan. Research into the relationship between diet and aging began in the first decades of the twentieth century. This early work established the fact that restricting food intake not only inhibited body weight gain but also extended lifespan in a wide range of animals, from *Bombyx* (Kellogg and Bell, 1903) and *Daphnia* (Ingle et al., 1973) to the rat (Osborne et al., 1917). The rodent studies have been particularly fruitful, the rat model having been developed by Northrop (1917) and McKay et al. (1935). In the 1960s other invertebrate models were examined, using protozoans (Rudzinska, 1962) and rotifers (Fanistil and Barrows, 1965). Extending the rat work to mice not only showed that the same lifespan effect could be obtained but that dietary restriction inhibited the development of spontaneous neoplasia (Saxton et al., 1944).

This antitumor aspect had been discovered in the rat (Tannenbaum, 1940) where it appeared to be part of a general inhibitory action of dietary restriction on degenerative diseases (Saxton and Kimball, 1941). During the late 1960s and 1970s, the emphasis in rodent research shifted toward an examination of the role of different dietary components in the lifespan extension (Ross et al., 1970; Segall, 1979) and its connection with sexual status (Drori and Folman, 1976; Merry and Holehan, 1979).

Comparing the rat results on the basis of similarities and differences among the studies of McCay et al. (1935), Berg and Simms (1960, 1961), Berg et al. (1962), and Merry and Holehan (1979), the following general principles of dietary restriction in relation to lifespan emerge.

The maximum extended lifespan of laboratory rats appears to be about 2000 days with food restriction of the order of 50% of the ad lib intake. This extension amounts to roughly 10 extra days for each 1% reduction in food intake from weaning. However, this relationship is not linear, and the greatest effects are obtained between 25 and 50% restriction. There are also differences in the percentage increase between strains, with a range of between 20 and 60% extension beyond ad lib fed controls. The magnitude of the increase does not appear to be fundamentally sex-linked. It is possible to extend the plateau of fitness without altering the period over which most animals die. Lifespan extension can occur without an inhibition of reproductive activity. Indeed, it is possible to considerably extend the length of the reproductive phase by dietary restriction associated with a large increase in lifespan. Recent work has shown that dietary restriction that delays the onset of puberty by 100 days extends the reproductive phase from 30 to 75% of the lifespan with a decrease in age-linked fetal abnormalities (Merry and Holehan, unpubl.).

We are ignorant of the underlying physiological and biochemical mechanisms that are involved in lifespan extension through dietary restriction. One fundamental is that neither the length of life, percentage increase in the lifespan, nor the extra days added to life correlate with the degree of inhibition of cellular growth. The main correlation is with percentage inhibition of body weight gain, which tends to be slightly less than the percentage increase in lifespan. The only physiological finding common to all studies is that greater extended lifespans occur when a larger fraction of the food consumed is devoted to cellular growth. Very few studies have been made at the biochemical level. Because short-term undernutrition in the rat is coupled with a lowered energy turnover (Lea and Lucia, 1961; Doerr and Hokanson, 1968), reduced turnover at the chemical level could be of significance in relation to lifespan extension.

A final point is that longer lifespans are associated with a later onset of degenerative diseases which follow a different pattern to that in animals fed ad lib. In this connection, there may be a complete inhibition of the development of spontaneous tumors. Despite these differences, the probability of death in dietary-restricted rats matches the probability of onset of lesions.

Models to explain the effects of nutrition on aging

Leaving aside the various "rate of living" theories, which are usually too vague to have experimental value, the dietary-restriction model has been interpreted in terms of a fundamental change or shift in endocrine control. This idea has appeared several times as a "pituitary clock" theory, where aging is considered to be controlled by a centralized pituitary hormone program, possibly governed by a small group of cells in a "master clock" in the brain.

It is interesting that no attempts have been made to fit the effects of dietary restriction into a general model of growth control. Classical growth models deal with body size being controlled through negative feedback of information derived from an ever decreasing proportion of extracellular fluid. This is likely to be too simple a model to explain a process that is more likely controlled by many individual organ "clocks."

Inasfar as growth is a changing state, with the changes varying within narrow limits, an alternative model of the growth process involves feedback of information on the previous growth increment. Comparison with a "desired" value would allow the next increment to be set "high" or "low" according to the "error" in past growth. Recent work on the oscillations of specific growth rates in response to growth inhibitors is in keeping with this idea (Stebbing, 1979). The relevance of this type of model to aging is that dietary restriction would be expected to set new priorities of resource partition. Because the inhibition of cellular growth by restricted diet does not seem to be directly connected with slowing of the aging program, the starting point to fit dietary restriction into a general growth theory must be that there is a new partition of food resources among cellular growth, fat deposition, and turnover. An indication of the kind of "switch" that could operate comes from work on severe dietary restriction followed by refeeding. The rehabilitation period is associated with a four- to fivefold increase in the efficiency of incorporation of digested energy, partly due to a preferentially increased fat deposition but possibly involving a lowered turnover (Williams and Senior, 1979).

A lowered percentage of body fat is a universal finding in dietary-restricted animals. Work on refeeding severely undernourished rats has demonstrated that replication of fat cells may be "turned on" at an age when it would normally have ceased (Kirkland and Harris, 1980). This finding indicates a slowing of the developmental program governing the formation of the fat depots and is linked with an inhibition of DNA turnover. All of these considerations point to a need for more detailed investigations of turnover which may actually be the driving mechanism for the release of the developmental program.

Links between development and aging

In summary, the following links between development and aging emerge from consideration of the dietary-restricted model that gives maximum lifespans of the rat between 1500 and 2000 days. Aging is the outcome of the operation of developmental timing devices that lead to a steady rate of physiological deterioration after a short period of peak fitness. Deaths in old age are largely due to degenerative diseases initiated by random events with rates of progression coupled to the rate of development. Each organ appears to be timed differently, and individual cell populations within organs have their own "clocks." These clocks are likely to be connected with a genetic program that was selected to regulate energy devoted to cell enlargement and maintenance of cell populations. As a size and reproductive output is reached that allows the species to perpetuate itself in the wild, this evolutionary program would be under selection pressure to reduce energy and materials devoted to cell expansion and maintenance. An efficient "balance" of resource utilization would be held for a very short time against a background of a high rate of random and age-independent ecological deaths. In organisms allowed to live beyond their wild life-expectancy, protected from natural selection, the increasing dominance of the energy conservation aspect of development could account for aging. We know virtually nothing about the ways in which the program of development is released. A study of the evolutionary significance of turnover, particularly in dietary-restricted organisms, offers a way to gain new insight into the control of both growth and aging.

Growth and aging at the cellular level

Definitions of cellular aging

Of the several definitions of aging, the one that is the most vividly descriptive of its histological character is that aging is a general involution of the tissues. Cell loss is of major significance in the lymphoid organs and occurs in muscle, kidney, and probably other organs where it is more difficult to quantify.

Death of individual cells may be discussed in the same way as whole-body aging. Cells may die as a result of programmed switches operating to ensure that cellular proliferation is regulated, with or without a morphological differentiation sequence, for the maintenance of a particular mass of cells. They may also die through random biochemical damage. There is also a possible limit on division potential, with cell death as the terminal event that would set a limit on the life of the cell population. This has been proposed as the process that limits the lifespan of the whole organism. The experimental material is the cultured fibroblast which has a finite division potential in vitro that is inversely related to the age of the donor from which the cells were derived. After maturity, aging is expressed grossly at the cellular level in terms of an increase in the proportion of extracellular fluid to cells. This is probably mainly a problem of increased cell death. There is also a rise in the phenotypic variability of cells – expressed morphologically as increased cytological abnormalities and alterations in enzyme patterns. Cellular variability is particularly evident functionally when the organism is called upon to make a physiological response. Such responses of old animals are generally delayed and more variable. This emphasizes that mortality arises from systems failure, and is not due simply to the loss of cells but to failures in the release of appropriate intracellular information. This loss of metabolic accuracy at the cellular level supports the idea of a programmed loss of biochemical control systems which in youth make for precision of response at both the molecular and morphological levels. Various specific degenerative cellular changes have been documented in aging organisms such as chromosome abnormalities, altered enzymes, and accumulated products of antioxidation; but so far none of these are sufficiently widespread with clear connections to any failures to be considered to be of outstanding importance.

Models of cellular aging
Wide use has been made of cultured cells in vitro as models for programmed aging (Hayflick, 1979). Normal vertebrate tissues explanted as primary cultures follow a predictable pattern of growth. After a slow adaptation to the medium there is an exponential phase of proliferation that persists for about 50 generations. During this period, which may last with subculturing for about a year, the cell karyotype appears normal. In the next phase the growth rate declines as cells stop dividing, and the death rate increases despite frequent changes of the culture medium. The culture is said to be committed to a finite lifespan. The programmed aspect of this phenomenon is brought out by the fact that cells of different species show characteristic lifespans often correlated with the average longevity of the species. Also, cells taken from older individuals have a shorter lifespan in culture than those from younger members of the species. Nevertheless there is also a considerable variability between lifespans of cell strains cloned from a culture derived from a given aged

donor, suggesting a high level of flexibility in the setting of controls for the program.

The reasons for the finite lifespan are still unexplained and may be bound up with another characteristic of the development of cultured cells, namely their ability to transform. Occasionally the cultures do not die out but begin to grow at an increasing rate, often exceeding the growth rate of the primary cell strain. These transformed cells, with an abnormal karyotype, become distinguished from the strain that gave rise to them and are then established as a cell line. Cell lines have an indefinite growth period and are often neoplastic.

To explain this phenomenon, a commitment theory of cell aging has been proposed, based on work with fibroblast cultures (Holliday et al., 1977). Starting with a population of uncommitted cells which are potentially immortal, it is assumed there is a given probability that cell division will give rise to fibroblasts which are irreversibly committed to senescence and death. These cells initially multiply normally, but after a given number of cell divisions (the incubating period) all the descendants of the original committed cell die out. If the probability of commitment is reasonably high and the incubation sufficiently long, then the number of uncommitted cells in the population will inevitably be lost from the population. A reduction in the probability of commitment or the length of the incubation period could produce an immortal steady-stage culture consisting of a small subpopulation of uncommitted cells, a majority of committed cells, and a constant fraction of nonviable cells. It was suggested that transformed or permanent lines may be in such a steady state. The process of commitment could be envisaged as the setting of an aging program for the population as a whole. This commitment theory accounts for these observations:

1. Individual fibroblasts produce clones with variable growth potential.
2. Parallel populations of one fibroblast strain grown under the same conditions also vary considerably in their longevity.
3. Longevity is related to population size.
4. When cultures contain equal numbers of two distinguishable types of cell that have stable phenotypes and the same growth rate, one or the other cell type frequently becomes predominant at the end of the lifespan.

It is questionable whether the commitment theory of aging applies to aging of the whole organism because many cell-renewal systems do not show such commitment (e.g., skin, intestinal and germinal epithelium, and stem cells of bone marrow and spleen). Also, much of the cell loss of aging occurs in postmitotic cells of the musculature. However, virtually nothing is known about death in postmitotic differentiated cells. They are difficult to work with, leaving us with models based on committed strains and immortal lines.

Both the characteristics of the fibroblast model in vitro and the phenomenon of increased cell loss in the aging organism indicate that we need to know more about the controls that take cells out of the proliferative pool and determine whether postmitotic cells should await the next call for division or whether they should die. Programmed death is a major feature, not only of the developing embryo, but also of skin, intestinal epithelium, and the hemopoietic tissues; but its controls are still a mystery. The thymus is particularly obvious as a model because not only is the lifespan of its lymphocyte population accurately programmed, probably by a "clock" in its epithelial cells (Bellamy and Hinsull, 1975, 1977), but also because postmitotic aging of individual lymphocytes is part of the controls that govern organ size at any age of the gland. Work on the thymic model has demonstrated the importance of local intraorgan factors in regulating cell division and death (Hinsull and Bellamy, 1974; Hinsull et al., 1977). This work later prompted a study of the effects of manipulating the local extracellular environment within experimental malignant cell lines with the assumption that the commitment theory might have applications in the control of cancer. For example, is it possible to introduce a programmed commitment into an experimental population of tumor cells in vivo and cause the tumor to age and die?

Cellular aging and cancer

To answer this question cancer must be viewed as a problem of anomalous development (Table 8.3). It is well known that some cancers (e.g., papillomas) exhibit complex growth control mechanisms showing that embryonic-type patterns of intercellular coordination have not been lost. This is also true of some solid experimental tumors. For example, the Walker carcinoma shows a definite polarity in its growth as a subcutaneous implant and appears to be responding to positional information rising from the polarity of the host (Auerbach et al., 1978). As this tumor develops, small cell-free areas appear in the periphery, indicative of highly

TABLE 8.3 Cancer as a problem of anomalous development

General principle	All the genetic information of the cancer phenotype is present, but not expressed, in all normal cells.
Criteria	Some cancer properties may be expressed in the absence of neoplasia.
	Proliferation of cancer cells responds to positional information emanating from adjacent cancer cells and normal cells.

120 DENIS BELLAMY

localized cell death, which is consistent with the operation of developmental-type control systems of cell deletion. Although it is conventional to think of centralized necrosis of large solid tumors as being due to failure of the blood supply to match the expanding cell population, some kind of programmed cell death has not been ruled out as a contributing factor. There is also quantitative evidence that mitosis in the Walker tumor is density-dependent, indicating that inhibitory feedback controls are operating at both high and low densities of tumor cells (Bellamy and Hinsull, 1978).

All of this work suggests that the microenvironment of the tumor cells may be an important potential regulator of cellular proliferation and cellular aging. So far the only basic difference that has been found in the intermediary metabolism of tumors, compared with nonneoplastic tissues, is their altered redox state as measured by their higher rate of anaerobic lactic acid production. This is particularly true of the Walker carcinoma, in which the interstitial pH has been recently monitored by radiotelemetry in vivo. The pH of the interior of the developing tumor is about 0.3 unit lower than that of a control area of connective tissues (Bellamy et al., 1980). The establishment of any redox difference may be prevented by coupling the local tumor extracellular environment with that of normal cells elsewhere in the body. This was achieved by means of a simple conducting metal implant and was associated with an inhibition of tumor growth (Bellamy et al., 1979; Hinsull et al., 1980).

The simplest form of coupling employed was two platinum electrodes joined by an electrolyte pathway within a sheath of connective tissue, which was induced to form around an inert silicone tube linking the two metal electrodes. The tumor implant was placed within one of the electrodes, which consisted of a loop of platinum wire. Cancer cells could not survive within the environment of the loop, and some migrated outward to divide in defined positions symmetrically arranged outside the loop and in a fixed relationship to the orientation and geometry of the electrode. In other well-defined areas adjacent to the loop electrode they did not divide. Although the fate of these nondividing tumor cells has not been followed, it is likely that they are held in the postmitotic state by an electrochemical field generated by the implant. Eventually, after several days' delay, the localized populations of dividing cells expanded, and the tumor grew normally. A silver electrode system produced a similar inhibitory effect, indicating that the inhibitory response was not mediated through the formation of specific platinum electrolysis products such as Cis platinum ammonium chloride (Rosenberg et al., 1976). Also, platinum was not detected by electron probe microanalysis at detection limits much lower than the injected levels of Pt required to inhibit tumor growth (Morgan, unpubl.). The fact that thymus transplants were not prevented from developing with the electrode system indicates that the tumor's peculiar redox system supports its proliferation.

These findings point to new ways of investigating cancer cell dynamics by altering the steady-state interstitial environment created by the tumor's own intermediate metabolism. When tumor growth is arrested with the electrode implants, there are three distinct microenvironments, which influence the tumor cells, causing them to die (within the loop electrode) and to divide or remain postmitotic (outside but adjacent to the loop electrode). In themselves, these electrode effects implicate the mitochondria as possible sites for the initiation of the neoplastic proliferation. The task for the future will be to define these environments chemically and follow the life histories of the cells within them. Results of this approach will have an important bearing on the potential controls of proliferation remaining in neoplastic cells and illuminate the mystery as to what proportion of cancer cells die as the result of innate senescence (Cooper, 1971).

References

Auerbach, R., Morrissey, L. W., and Sidky, Y. A. (1978) Gradients in tumour growth. *Nature* 274:697.

Bellamy, D. (1970) Aging and endocrine responses to environmental factors: With particular reference to mammals. *Mem. Soc. Endocrinol.* 18:303.

Bellamy, D., Berry R. J., Jakobson, M. E., Lidicker, W. Z., Morgan, J., and Murphy, H. M. (1973) Aging in an island population of the house mouse. *Age and Aging* 2:235.

Bellamy, D., Colson, R. H., Hinsull, S. M., and Watson, B. W. (1980) Continuous monitoring of tumour extracellular pH in vivo by radiotelemetry. *Proc. Physiol. Soc.* (April), 16 pp.

Bellamy, D., and Hinsull, S. M. (1975) On the role of the reticular–epithelial complex in transplantation of the thymus. *Differentiation* 3:115.

Bellamy, D., and Hinsull, S. M. (1977) Density-dependent cell division after cortisol treatment of rat thymus in relation to age involution. *Virchows Arch. B Cell. Pathol.* 24:251.

Bellamy, D., and Hinsull, S. M. (1978) Density-dependent mitosis in the Walker 256 carcinoma and the influence of host age on growth. *Eur. J. Cancer* 14:747.

Bellamy, D., Hinsull, S. M., Watson, B., and Blache, L. A. (1979) Inhibition of the development of Walker 256 carcinoma with a simple metal–plastic implant. *Eur. J. Cancer* 15:223.

Berg, B. N., and Simms, H. S. (1960) Nutrition and longevity in the rat II. Longevity and onset of disease with different levels of food intake. *J. Nutr.* 71:255.

Berg, B. N., and Simms, H. S. (1961) Nutrition and longevity in the rat III. Food restriction beyond 800 days. *J. Nutr.* 74:23.

122 DENIS BELLAMY

Berg, B. N., Wolf, S., and Simms, H. S. (1962) Nutrition and longevity in the rat IV. Food restriction and the reticuloneuropathy of aging rats. *J. Nutr.* 77:439.

Cooper, E. H. (1973) The biology of cell death in tumours. *Cell Tissue Kinet.* 6:87.

Doerr, H. O., and Hokanson, J. E. (1968) Food deprivation performance and heart rate in the rat. *J. Comp. Physiol. Psychol.* 65:227.

Drori, D., and Folman, J. (1976) Environmental effects on longevity in the male rat: Exercise, mating, castration and restricted feeding. *Exp. Gerontol.* 11:25.

Fanistil, D. D., and Barrows, G. H. (1965) Aging in the rotifer. *J. Gerontol.* 20:462.

Hayflick, L. (1979) The cell biology of aging. *J. Invest. Dermatol.* 73:8.

Hinsull, S. M., and Bellamy, D. (1974) Development and involution of thymus grafts in rats with reference to age and sex. *Differentiation* 2:299.

Hinsull, S. M., Bellamy, D., and Franklin, A. (1977) A quantitative histological assessment of cellular death, in relation to mitosis, in rat thymus during growth and age involution. *Age and Aging* 6:77.

Hinsull, S. M., Bellamy, D., Franklin, A., and Watson, B. W. (1980) The inhibitory influence of a metal–plastic implant on cellular proliferation patterns in an experimental tumour compared with normal tissue. *Eur. J. Cancer* 16:159.

Holliday, R., Huschtscha, L. I., Tarrant, G. M., and Kirkwood, T. B. L. (1977) Testing the commitment theory of cellular aging. *Science* 298:366.

Ingle, L. T., Wood, T. R., and Banta, A. M. (1973) A study of longevity, growth, reproduction and heart rate in *Daphnia longispina* as induced by limitations in the quantity of food. *J. Exp. Zool.* 76:325.

Kellog, V. L., and Bell, R. G. (1903) Variations induced in larval, pupal and imaginal stages of *Bombyx mori* by controlled variations in food supply. *Science* 18:741.

Kirkland, J., and Harris, P. M. (1980) Changes in adipose-tissue of the rat due to early undernutrition followed by rehabilitation 3. Changes in cell replication studied with tritiated thymidine. *Br. J. Nutr.* 43:33.

Lea, M., and Lucia, S. B. (1961) Some relationships between calorie restriction and body weight in the rat.1. Body composition, liver lipids and organ weights. *J. Nutr.* 74:243.

McCay, C. M., Crowell, M. F., and Maynard, L. A. (1935) The effect of retarded growth upon the length of the lifespan and upon ultimate body size. *J. Nutr.* 10:63.

Merry, B. J., and Holehan, A. M. (1979) Onset of puberty and duration of fertility in rats fed a restricted diet. *J. Reprod. Fert.* 57:253.

Northrop, J. (1917) The effect of the prolongation of the period of growth on the total duration of life. *J. Biol. Chem.* 32:123.

Osborne, T. B., Mendel, L. B., and Ferry, E. L. (1917) The effect of retardation of growth on the breeding period and duration of life of rats. *Science* 40:294.

Rosenberg, B., Brace, L. D., and Thomas M. (1976) Tumour cell kinetics with Cis-diamine dichloroplatinum (II) therapy. *Wadley Med. Bull.* 6:3.

Ross, M. H., Brass, G., and Ragbeer, M. S. (1970) Influence of protein and calorie intake upon spontaneous tumour incidence of the anterior pituitary gland of the rat. *J. Nutr.* 100:177.

Rudzinska, M. (1962) The use of a protozoan for studies on aging. *Gerontologia* 6:206.

Saxton, J. A., Boon, M. C. and Furth, J. (1944) Observations on inhibition of development of spontaneous leukemia in mice by underfeeding. *Cancer Res.* 4:401.

Saxton, J. A., and Kimball, G. C. (1941) Relation of nephrosis and other diseases of albino rats to age and modifications of diet. *Arch. Pathol.* 32:951.

Segall, P. E. (1979) Interrelations of dietary and hormonal effects in aging. *Mech. Aging Dev.* 9:515.

Stebbing, A. R. D. (1979) The self-control of growth. *Spectrum* 165:2.

Tannenbaum, A. (1940) The initiation and growth of tumours. Introduction 1. Effects of underfeeding. *Am. J. Cancer Res.* 38:335.

Williams, V. J., and Senior, W. (1979) Changes in body composition and efficiency of food utilization for growth in young adult female rats before, during and after a period of food restriction. *Aust. J. Biol. Sci.* 32:41.

9 Comparison of embryonic development in birds and mammals: birth weight, time, and cost

HERMANN RAHN

I shall ask three questions: How large is the birth weight for mammals and birds, how long does development take, and how much does it cost to produce an offspring?

Partial answers have been given for mammals. Leitch et al. (1959) showed that the birth weight is proportional to maternal weight raised to the 0.83 power and thereby demonstrated that the smaller the adult the larger the relative birth weight, whereas Sacher and Staffeldt (1974) demonstrated a new relationship between gestation time and brain weight at birth. Case (1978) provides a comparative analysis of postnatal growth rates in reptiles, birds, and mammals, with extensive tabular data for mammals. The caloric expenditure above the maintenance level of the pregnant mother was measured by Brody (1945) and will be discussed further. For birds the relationship of egg weight to adult weight was recently established for a large number of species (Rahn et al., 1975), as well as the relationship between egg weight and incubation time (Rahn and Ar, 1974), whereas the energetics of reproduction in birds has been extensively discussed by King (1973) and Ricklefs (1974).

My purpose here is to bring together these data and to compare the development of mammals and birds in order to see common trends or differences. What is impressive is the large differences in strategies that have developed among these homeotherms, which inevitably give rise to a large statistical scatter of the various measurements that have been made. The interpretations, therefore, depend upon how much leeway one is willing to accept. If one takes average trends, they become useful in exposing overall strategies, but they are not necessarily useful in predicting answers for a given species.

Birth weight and adult metabolism

In both mammals and birds birth weight and adult metabolism show similar changes when plotted against adult body weight, suggesting that birth weight is directly proportional to the adult metabolic rate and that therefore the relative energy investment for an offspring by the female is similar regardless of size of the adult. The symbol for birth weight is

124

W (kg), W_L for the litter of mammals, and W_{II} for the single hatchling of birds. B (kg) represents the weight of the adult, and $\dot{M}$ (kcal · day^{-1}), the metabolic rate.

Mammals

In Figure 9.1 are plotted the litter weights against maternal body weights for 140 species of placental mammals, ranging from a 6-g bat to a 79-ton fin whale. These data were taken from the tables provided by Leitch et al. (1959) and Sacher and Staffeldt (1974). The dotted lines indicate ± 2 SEE. The two points that fall below these limits belong to two species of bear noted for their relatively small birth weight. The regression equation is also given in Table 9.1 and has a slope of 0.812; it can be compared with equation 2 for the metabolic rate as a function of body mass (Kleiber, 1965), which has a slope of 0.75. If one assumes that these two slopes are essentially similar, equations 1 and 2 can be combined, eliminating B, so that litter weight is now proportional to adult metabolic rate, or $W_L = 0.0019\ \dot{M}$. This relationship implies that regardless of size of the adult the relative caloric investment of the mother in producing an offspring is the same. In fact the production of a litter may in this sense be likened to milk production of the mother, which Brody (1945) showed from rats to cows to be also proportional to the adult metabolic rate.

FIGURE 9.1 Litter weight (W_L) as a function of maternal weight, B, for 140 placental mammals. See also equation 1 in Table 9.1. [Data from Leitch et al., 1959, and Sacher and Staffeldt, 1974]

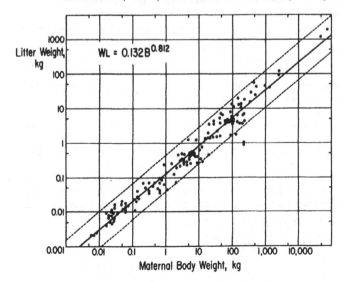

$$WL = 0.132B^{0.812}$$

Birds

In Figure 9.2 are plotted the hatchling weight of single eggs against the adult body weight for 514 species of birds (with the exception of the order Passeriformes). These values were taken from the data of Rahn et al. (1975) in which the birth weight of a hatchling was assumed to be 0.65 of the initial egg mass (Romanoff, 1967). The body weight values range from a 1.7-g hummingbird to the 100-kg ostrich, to which I have added the data for the extinct elephant bird, *Aepyornis* (see circle), whose body mass was estimated to be 457 kg (Amadon, 1947), and whose eggs (n = 23) averaged 9.1 kg (Schönwetter, 1960). The slope of this relationship is 0.703. In this case we are considering the birth weight of a single hatchling rather than that of a clutch because each egg is produced one at a time, spaced one or more days apart.

The slope of 0.703 for hatchlings (see equation 4, Table 9.1) can be compared with the slope of 0.723 for the basal metabolic rate of adult nonpasserine birds (equation 5, Table 9.1) taken from Lasiewski and Dawson (1967). Because these two slopes are not significantly different, these two equations can be combined, eliminating B, and hatchling weight is proportional to the adult metabolic rate, or $W_H = 0.0005\ \dot{M}$.

Mammals and birds

We therefore see a similar pattern for both groups, namely, that the birth weight is proportional to the adult metabolic rate and that the

TABLE 9.1 Allometric equations ($Y = a\ X^b$) for birth mass, development time, and adult metabolism in mammals and birds, where B = adult body mass (kg). SEE of regression (y on x) expressed as antilogarithm, $\dot{X}$ SEE; r = correlation coefficient; n = number of species.

Equation no.	Characteristic	Y	$=$	a	X^b	$\dot{X}$ SEE	r	n
Mammals								
1[a,b]	Litter mass (kg)	W_L	$=$	0.132	$B^{0.812}$	1.85	0.98	140
2[c]	Adult metabolism (kcal·day^{-1})	$\dot{M}$	$=$	69	$B^{0.75}$			
3[b]	Gestation time (days)	G	$=$	130	$W_L^{0.299}$	1.61	0.83	95
Birds								
4[d]	Hatchling mass (kg)	W_H	$=$	0.0371	$B^{0.703}$	1.46	0.94	514
5[e]	Adult metabolism (kcal·day^{-1})	$\dot{M}$	$=$	78	$B^{0.723}$	1.17		72
6[f]	Incubation time (days)	I	$=$	59.1	$W_H^{0.217}$	1.24	0.86	444

[a] Leitch et al. (1959).   [b] Sacher and Staffeldt (1974).
[c] Quoted from Kleiber (1965).   [d] Rahn et al. (1975).
[e] Quoted from Lasiewski and Dawson (1967).
[f] Rahn and Ar (1974).

relative caloric investment within each group producing a litter or a hatchling is similar regardless of the adult size. The fourfold difference in the proportionality constant probably reflects the fact that for the mammals we are considering litter weight, but for birds, the weight of a single hatchling.

Relative birth weight and adult metabolism

As the antilogarithms of exponents of equations 1, 2, 4, and 5 indicate, W and $\dot{M}$ increase only about 5.5 times for every 10-fold increase in body weight of the adult, accounting for the ever decreasing relative birth weight and weight-specific metabolism as the adult body weight increases. Dividing these equations by the adult body mass, B, one obtains the relationships as seen in the semilogarithmic plots of Figures 9.3 and 9.4, showing the parallel changes in relative birth weight and weight-specific metabolism as a function of adult body weight. The typical litter weight of a 10-g mammal (Figure 9.3) is 31% of the adult weight, but it is only 1.5% for a 100-ton whale. In birds (Figure 9.4) the typical hatchling for a 10-g bird is 14% of the adult weight, but it is only 1.5% for a 100-kg ostrich.

The weight-specific metabolic rates of the adult follow in parallel. These relationships are simply another way of expressing the parallel

FIGURE 9.2 Hatchling weight (W_H) as a function of adult weight, B, for 514 species of birds, except for the order Passeriformes. See also equation 4 in Table 9.1. [Calculated from the data of Rahn et al., 1975]

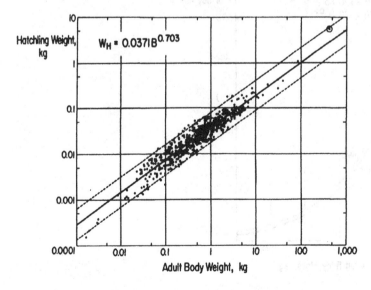

$$W_H = 0.0371B^{0.703}$$

Hatchling Weight, kg

Adult Body Weight, kg

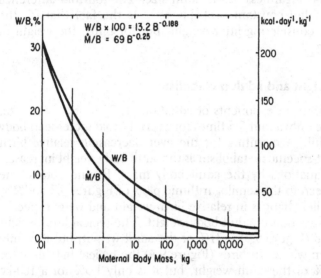

FIGURE 9.3 Relative birth weight (W/B) × 100 and adult weight-specific metabolism, kcal·day^{-1}·kg^{-1}, for mammals, plotted as a function of adult body weight on semilogarithmic coordinates. [Data from Sacher and Staffeldt, 1974]

Within the graph:

$$W/B \times 100 = 13.2\, B^{-0.188}$$
$$\dot{M}/B = 69\, B^{-0.25}$$

FIGURE 9.4 Relative hatchling weight (W/B) × 100 and adult weight-specific metabolism, kcal·day^{-1}·kg^{-1}, for birds, plotted as a function of adult body weight on semilogarithmic coordinates. [Data from Rahn et al., 1975]

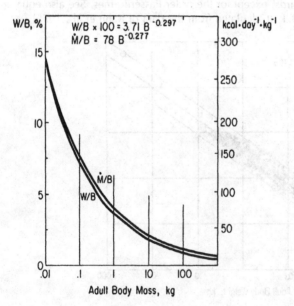

Within the graph:

$$W/B \times 100 = 3.71\, B^{-0.297}$$
$$\dot{M}/B = 78\, B^{-0.277}$$

behavior between birth weight and adult metabolism, indicating that the relative energetic contributions for producing an offspring are the same for small and large animals.

Time of development

Mammals

In Figure 9.5 are plotted the gestation times from the tables of Sacher and Staffeldt (1974) for 95 placental mammals as a function of their birth weight. The shortest period is 16 days for a 10-g litter of a rodent, and the longest is 655 days for the 120-kg newborn elephant. However, the scatter is very large as indicated by the SEE (equation 3, Table 9.1). The average regression slope indicates a 32-day gestation for a 10-g litter and 515 days for a 100-kg offspring. Dividing birth weight by equation 3, we obtain a general expression of an average rate of embryonic growth (birth weight/gestation time) expressed in kilograms per day (kg · day^{-1}), or

$$W_L/G = 0.0077\ W_L^{0.70} \tag{7}$$

This relationship indicates that the mean growth rate (kg · day^{-1}) increases 5 times for every 10-fold increase in birth weight. Thus a 10-g litter accumulates its mass at a mean rate of (10/32) or 0.3 g · day^{-1},

FIGURE 9.5 Gestation time, days, as a function of litter weight in 95 species of placental mammals. See also equation 3, Table 9.1. [Data from tables of Sacher and Staffeldt, 1974]

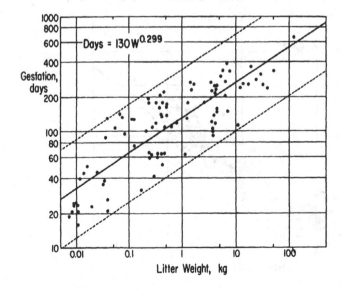

whereas a 1000-g litter accomplishes this at a mean rate of 7.7 g · day^{-1}, or 25 times faster.

Birds

In Figure 9.6 are plotted the incubation times for 440 species of bird as a function of their birth weight (assumed to be equal to 0.65 of the initial egg weight). These values were taken from the data of Rahn and Ar (1974) but exclude the order Procellariiformes (albatrosses, shearwaters, and petrels), whose eggs have unusually long incubation periods that fall outside the indicated confidence limits. The smallest hatchling (hummingbird) weighs less than 0.5 g, whereas the largest hatchling is the ostrich chick at nearly 1000 g. The regression is given by equation 6 (Table 9.1). The average growth rate of the embryo is given by the relation

$$W_H/l = 0.017 \ W_H^{0.78} \tag{8}$$

We may now compare equations 7 and 8. If as a first approximation we assume the exponents to be not significantly different, then the ratio of proportionality constants (0.017/0.0077) is 2.2. This suggests that the mean developmental rate for bird embryos is about twice as fast as in mammals. For example, a 1-kg hatchling develops in 60 days, whereas the 1-kg litter requires 130 days. One might have predicted a reverse ratio

FIGURE 9.6 Incubation time, days, plotted as a function of hatchling weight for 444 species of birds, except for the order Procellariiformes. See also equation 6, Table 9.1. [Data from Rahn and Ar, 1974]

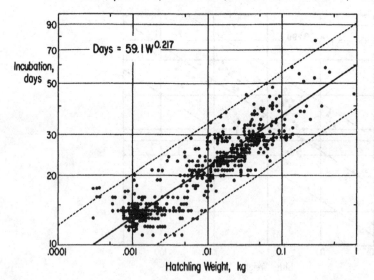

because an average egg temperature (n = 27) is 35.6 °C (Drent, 1975). It is interesting to note that a similar difference in postnatal growth rate between birds and mammals was shown by Case (1978).

Significance of the W/I ratio

The overall development rate, W/I, as defined above, has recently been shown to be an important determinant in the gas exchange of the avian embryo because the oxygen uptake, the water loss, and the diffusing capacity of the shell are all directly proportional to W/I (Rahn and Ar, 1980). An example is shown in Figure 9.7, where $\dot{M}_{O_2}$, the oxygen uptake (ml O_2 · day^{-1}) at comparable stages of development (prepipping stage) for various bird eggs, is plotted as a function of embryonic growth rate, W/I (where in this case W is the initial mass of the egg rather than the hatchling). Below is shown the mean line for G_{O_2}, the O_2 diffusing capacity (ml O_2 · day^{-1} · Torr^{-1}) of egg shells for 91 species. It also has a slope of 1.0 so that the $\dot{M}_{O_2}/G_{O_2}$ represents the mean O_2 difference across the egg shell at the prepipping stage of development and has a value of 42 Torr.

FIGURE 9.7 $\dot{M}_{O_2}$, the O_2 uptake of bird eggs at the prepipping stage of development, plotted against W/I (initial egg mass/incubation time). The slope of the line is 1.0. Below is a parallel line, the O_2 diffusing capacity of the egg shell, G_{O_2}, based upon value for 91 species. $\dot{M}_{O_2}/G_{O_2}$ is equal to 42 Torr and represents the average O_2 difference across the pores of the egg shell at the prepipping stage of development. [From Rahn and Ar, 1980]

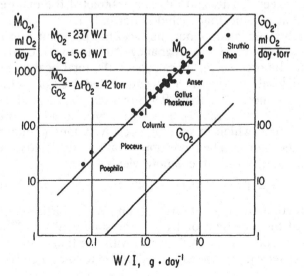

Because $\dot{M}_{O_2}$ at comparable stages of development is proportional to W/I, it is possible to predict that the total amount of O_2 consumed per unit mass of embryo during the embryonic life span is a constant and independent of mass or incubation time. Evidence for this prediction was recently presented by Hoyt and Rahn (1980), who determined a value of about 0.5 kcal $\cdot$ g^{-1} egg or 0.77 kcal $\cdot$ g^{-1} hatchling. This concept, which applies to the embryonic lifespan, is similar to that originally proposed by Max Rubner (1908) for the adult lifespan, namely, that the caloric expenditure per unit body mass during the lifespan for all mammals is a constant (approximately 200 kcal $\cdot$ g^{-1}), which Boddington (1978) has aptly named the animal's *absolute metabolic scope*. Thus it will be of interest, when commensurate data on oxygen uptake of the mammalian embryo become available, to see whether it will be proportional to W/G and whether one can speak of an absolute metabolic scope during the gestation period, and to compare its value with that for avian embryos.

The cost of producing an offspring

Some estimates can be made of the energy requirements to produce an offspring by measuring or calculating the energy requirements above the maintenance level of the adult that are necessary either to produce an egg or to maintain the growing embryo.

Mammals

Brody (1945) measured the metabolism throughout pregnancy in horses, cows, swine, sheep, goats, and rats and subtracted the maintenance cost to arrive at a value that included not only the oxygen requirements of the fetus and membranes, but also the pregnant uterus and all other tissues stimulated by the pregnancy. Adding up these extra demands throughout the pregnancy, he calculated the total heat increment of gestation in kilocalories (Figure 9.8a) and plotted this against the birth or litter weight as shown in Figure 9.9. For example, to produce a 75-kg foal required 600 000 kcal. To this figure he also added two estimates derived for man, which in spite of a relatively long gestation fall in line with the other values. The regression equation calculated from Brody's values (Table 14.6) for 21 observations yields

$$E_{tot} = 4094 \, W_L^{1.24}, \qquad r = 0.99, \quad \bar{X}\,SEE = 1.42 \qquad (9)$$

where E_{tot} = heat increment of gestation (kcal) and W_L = birth weight (kg). Brody explained the exponent of 1.24 by the statement that "the heat increment per kg young necessarily varies with the time required to develop it." This statement can be supported by the following consid-

FIGURE 9.8 (a) Total metabolism during pregnancy and maintenance metabolism during pregnancy are plotted schematically as a function of gestation time. Subtracting maintenance metabolism from total gives the heat increment of gestation (shaded area). (b) Same plot as in Figure 9.8a, except an average daily value for the heat increment ($\bar{E}$) has been calculated and is shaded in on the graph.

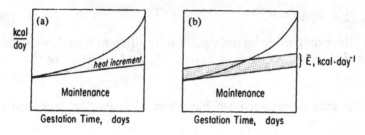

FIGURE 9.9 Cost of producing a litter, in kilocalories, as a function of litter weight. Inset: cost of producing a precocial egg, in kilocalories, as a function of hatchling weight. The dotted line expresses the total caloric expenditure of the embryo during development and is equal to one-fifth of the initial caloric investment. [Mammal data plotted from Brody, 1945]

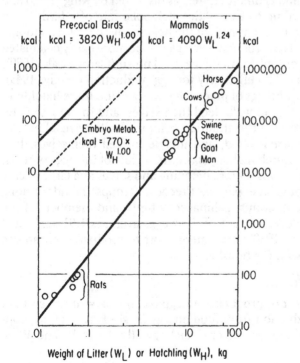

eration. Because we know the mean gestation time as a function of birth weight, we can divide equation 9 by equation 3 (Table 9.1) and obtain a mean daily heat increment of gestation, $\overline{E}$, as shown in Figure 9.8b. Thus

$$\overline{E} = E_{tot}/G = 32\ W_L^{0.94} \tag{10}$$

Assuming the exponent to be not significantly different from 1, we obtain a constant for mammals, namely $\overline{E}/W_L = 32\ \text{kcal} \cdot \text{day}^{-1} \cdot \text{kg}^{-1}$.

Birds

The total cost of producing a hatchling, E_{tot}, can be described as follows:

$$E_{tot} = E_{egg} + E_{incub} \tag{11}$$

where E_{egg} is the energy that has gone into producing the egg, and E_{incub} is the heat increment above the maintenance cost to keep the egg warm as well as to maintain the proper humidity and ventilation of the nest's microclimate. The extra energy required for incubation of the egg is difficult to measure, and estimates differ, as discussed by King (1973) and Ricklefs (1974). According to King, "the metabolic rate of a bird at rest can supply a large fraction, if not all, of the heat required for incubation."

The energy required for producing an egg can be more easily calculated because the caloric content for many eggs is known (Carey et al., 1980), and one may assume a 77% efficiency for egg production (Brody, 1945). Here I shall consider only eggs of precocial birds because these hatchlings emerge from the egg able to walk, feed themselves, and thermoregulate, and thus require relatively little attention compared with altricial hatchlings, which emerge naked, blind, and unable to feed themselves. Precocial eggs also have a much higher yolk content (40% of the egg content), part of which is absorbed just before hatching and serves as a large energy reserve during the postnatal period. Precocial groups include among others the ducks, geese, swans, gallinaceous birds, and members of the ostrichlike group. Caloric content of their eggs averages 1.91 kcal $\cdot$ g^{-1} ($n = 20$) (Carey et al., 1980). The energy cost to produce such an egg is given by the following relationship:

$$E = 1.3 \times 1.91 \times W \tag{12}$$

where E = energy cost of producing an egg (kcal), 1.3 = inverse of net efficiency of egg production from food energy (kcal $\cdot$ kcal^{-1} of egg content), 1.91 = caloric content of precocial eggs (kcal $\cdot$ g^{-1}), and W = weight of egg (g).

Thus E (kcal) = 2.4 W. Because we want to consider the weight of a hatchling instead of egg weight, W must be factored by 1/0.65 or 1.54, and converting to kilograms we have $E = 3820\,W_H$. This value is plotted in Figure 9.9 and can be compared with E value for mammal $= 4094\,W_L^{1.24}$. Because the proportionality constants are similar, the E_{tot} for a birth weight of 1 kg is similar for birds and mammals. However, the exponents are not, possibly because it was here assumed that the caloric content of precocial birds was independent of incubation time. The energy content may eventually be shown to increase with longer incubation because Ar and Yom-Tov (1978) have recently demonstrated that among precocial birds the yolk content increases with egg size.

Caloric expenditure of bird embryo

The average oxygen uptake during development is now known for 24 precocial species and is equal to about 500 kcal $\cdot$ kg^{-1} egg mass (Hoyt and Rahn, 1980). Converted to hatchling mass this is equal to 770 kcal $\cdot$ kg^{-1}. Thus 770/3820 or about 20% of E, the energy cost to produce the hatchling, is dissipated during its development (see Figure 9.9). Whether comparable values will be found for the relative energy expenditure of the mammalian embryo and fetus needs to be explored.

Similarity of metabolism between fetus and mother

It is of interest to inquire about the metabolism of the fetus shortly before birth. In spite of the large weight differences between fetus and mother, the weight-specific metabolic rates appear to be the same. This observation was first made by Christian Bohr (1900) on the guinea pig. By ligation of the umbilical cord he discovered that the CO_2 production per unit weight of the embryo was the same as that of the mother. Kleiber et al. (1943) measured the oxygen uptake of 13-day-old rat fetuses in vitro and showed that this uptake per unit weight was the same as for maternal tissue, suggesting that embryos behave metabolically as an integral part of the maternal body. Similar evidence is cited by him for sheep and man (Kleiber, 1965), suggestive of a maternal regulation that is severed with birth, at which time the newborn quickly attains a metabolic rate comparable to its size.

Similar evidence was recently presented for a large number of precocial birds in which the oxygen uptake had been measured just before internal pipping took place. These weight-specific metabolic rates were essentially the same as those calculated for their parents (Hoyt and Rahn, 1980). These observations demonstrate that no direct maternal regulation is required to "suppress" the metabolic rate while the bird is confined to the egg shell. However, postnatally the precocial hatchling doubles its metabolic rate very quickly (Dawson et al., 1976) as it establishes its own thermoregulation and mobility to seek food and shelter. In fact, if such a doubling of metabolism were to take place before the internal pipping event, it is easy to calculate that the shell conductance would be too small to provide the necessary oxygen. Thus it is likely that all embryos, mammalian or bird, have compromised to develop at a relatively low metabolic rate, which in the birds does not exceed the capabilities of the shell conductance and in mammals does not compromise the mother by requiring too large an oxygen delivery to the uterus. As Kleiber (1965) writes: "If the eight fetuses in a rat at the end of pregnancy had a rate of oxygen consumption equal to that of a one- to two-day-old infant rat, the metabolic rate of the pregnant rat would be forty-seven times her basal metabolic rate!"

Summary

Comparing the embryonic development of mammals and birds, there are overall similarities in relation of birth weight to adult weight, growth rates, and metabolic costs of producing offspring.

Birth weight (in terms of litter or single hatchling) is proportional to the metabolic rate of the adult and accounts for the general trend that the smaller the adult mammal and bird, the larger is the relative birth weight.

Developmental time in mammals and birds is approximately proportional to the fourth root of birth weight. The mean growth rate in bird embryos (birth weight/incubation time) is twice that of mammals (birth weight/gestation time) and directly proportional to their metabolism, water loss, and shell conductance.

In mammals the total caloric investment to produce offspring, in kilocalories, is equal to about $4000 \, W_L^{1.24}$ (Brody, 1945), whereas estimates of caloric expenditure for egg production in precocial birds are about $4000 \, W_H^{1.00}$, setting aside the possible additional parental heat increment of incubation. Of the total energy input into an egg about one-fifth is expended by the embryo.

Evidence is reviewed that suggests that in spite of the large size difference between fetus and mother and bird embryo and incubating parent, their weight-specific metabolisms are similar. This relationship is quickly severed at birth or hatching when the newborn quickly attains a metabolic rate commensurate with its size.

References

Amadon, D. (1947) An estimated weight of the largest known bird. *Condor* 49:159–164.

Ar, A., and Yom-Tov, Y. (1978) The evolution of parental care in birds. *Evolution* 32:655–669.

Boddington, M. J. (1978) An absolute metabolic scope for activity. *J. Theor. Biol. 75*:443–449.

Bohr, C. (1900) Der respiratorische Stoffwechsel des Säugethierembryo. *Skand. Arch. Physiol. 10*:413–424.

Brody, S. (1945). *Bioenergetics and growth.* New York: Hafner Press. (Reprinted 1974)

Carey, C., Rahn, H., and Parisi, P. (1980) Calories, water, lipid, and yolk in avian eggs. *Condor 82*:335–343.

Case, T. J. (1978) On the evolution and adaptive significance of postnatal growth rates in the terrestrial vertebrate. *Quart. Rev. Biol. 53*:243–282.

Dawson, W. R., Bennett, A. F., and Hudson, J. W. (1976) Metabolism and thermoregulation in hatchling ring-billed gulls. *Condor 78*:49–60.

Drent, R. (1975). Incubation. In *Avian biology,* vol. 5 (D. S. Farner and J. R. King, eds.), pp. 334–420. New York: Academic Press.

Hoyt, D. F., and Rahn, H. (1980) Respiration of avian embryos – A comparative analysis. *Respir. Physiol. 39*:255–264.

King, J. R. (1973) Energetics of reproduction in birds. In *Breeding biology of birds* (D. S. Farner, ed.), pp. 78–107. Washington, D.C.: National Academy of Sciences.

Kleiber, M. (1965) Respiratory exchange and metabolic rate. In *Handbook of respiration,* sect. III, vol. II (W. O. Fenn and H. Rahn, eds.), pp. 927–938. Washington, D.C.: American Physiological Society.

Kleiber, M., Cole, H. H., and Smith, A. H. (1943) Metabolic rate of rat fetuses in vitro. *J. Cell. Comp. Physiol. 22*:170–176.

Lasiewski, R. C., and Dawson, W. R. (1967) A re-examination of the relation between standard metabolic rate and body weights in birds. *Condor 69*:13–23.

Leitch, I., Hytten, F. E., and Billewicz, W. Z. (1959) The maternal and neonatal weights of some mammalia. *Proc. Zool. Soc. Lond. 133*:11–28.

Rahn, H., and Ar, A. (1974) The avian egg: Incubation time and water loss. *Condor 76*:147–152.

Rahn, H., and Ar, A. (1980) Gas exchange of the avian egg: Time, structure and function. *Am. Zool. 20*:477–484.

Rahn, H., Ar, A., and Paganelli, C. V. (1975) Relation of avian egg weight to body weight. *Auk 92*:750–756.

Ricklefs, R. E. (1974) Energetics of reproduction in birds. *Publication of the Nuttall Ornithological Club 15*:152–297. Cambridge, Mass.

Romanoff, A. L. (1967) *Biochemistry of the avian embryo,* p. 325. New York: Wiley.

Rubner, M. (1908) *Das Problem der Lebensdauer und seine Beziehungen zu Wachstum und Ernährung.* Berlin: R. Oldenburg.

Sacher, G. A., and Staffeldt, E. F. (1974) Relation of gestation time to brain weight for placental mammals: Implications for the theory of vertebrate growth. *Am. Naturalist 108*:593–615.

Schönwetter, M. (1960) *Handbuch der Oologie.* (W. Meise, ed.). Berlin: Akademie Verlag.

10 Anaerobic metabolism: living without oxygen

P. W. HOCHACHKA

All animals can withstand short periods of total O_2 lack, but some species are exceptionally tolerant of anoxia. The most outstanding anaerobic abilities are found in certain invertebrate groups (parasitic helminths, burrowing annelids, intertidal bivalves), but even among the vertebrates there are species with suprising tolerance to anoxia. The common gold-fish can survive several days of total O_2 lack (Walker and Johanson, 1977), and some fresh-water turtles at low temperature are able to withstand up to months of anoxia (Chapter 6); furthermore, the Weddell seal is capable of breath-hold diving for about 1.2 h, of which only about 0.3 h can be sustained on stored O_2 (Chapter 11). When an organism becomes cut off from its normal sources of O_2, three critical needs must be satisfied. First, provision must be made for suitable storage substrate(s) and for their regulated utilization. Second, provision must be made to store, recycle, or minimize the production of noxious waste products. And third, some provision must be made for reestablishing metabolic homeostasis in recovery. In this paper, I shall briefly review the means that have been developed by various organisms to satisfy these needs for sustaining periods of O_2 lack.

Matching substrate availability with energy needs

Glycogen in a central depot

Glycogen is the primary fuel for anaerobic metabolism in all animals. In vertebrates most of the glycogen is stored in the liver, which supplies the bulk of body needs for glycogen and can be viewed as a central depot of this storage substrate. The primary carbon and energy source utilized during periods of severe O_2 limitation in goldfish (Walker and Johanson, 1977), the turtle (Daw et al., 1967) and diving animals such as seals (Murphy et al., 1980) is undoubtedly liver glycogen, which is stored at very high levels and is mobilized as blood glucose. Invertebrates that have been studied show a similar central depot for glycogen storage. In *Mytilus* the hepatopancreas and the mantle can accumulate glycogen to 50% of dry weight during energy-rich conditions (De Zwann and

Zandee, 1972a). As nutritive conditions change through the year, the glycogen content of these central depots is used.

Vertebrates derive little blood glucose from glycogen stored in the liver under normal oxygen-rich conditions, but the liver becomes a major source of glycogen under hypoxic conditions. In diving mammals physiological reflexes (Zapol et al., 1979) are activated during dives that partition blood O_2 stores between O_2-needy tissues and anoxia-tolerant tissues. Similar physiological reflexes probably occur in lower vertebrates such as molluscs (Jorgensen et al., 1980). This partitioning of O_2 also leads to an automatic partitioning of blood glucose supplies. The O_2-needy tissues require relatively little glucose because oxidative metabolism is efficient in producing energy, and the mass of the tissues involved is small. The anoxic tissues require large amounts of glucose because anaerobic metabolism is inefficient, and a large mass of tissues is involved (Murphy et al., 1980). In the complete absence of O_2; glucose consumption rates by all tissues would rise 18-fold if demand for ATP were the same and the classical glucose → lactate fermentation were used. Something must change, however, because rapid depletion of central depot supplies and declining blood glucose levels are not observed during anoxia. For example, not all of the liver glycogen stores are utilized by goldfish after 4 days of anoxia (Shoubridge and Hochachka, 1979), and glycogen depots are only partially depleted after extreme anoxia in mussels and oysters (De Zwann and Zandee, 1972b). After near-maximum-duration diving excursions in the Weddell seal, blood glucose levels are near normal (Chapter 11). Massive and rapid glucose depletion of central-depot glycogen reserves is avoided during anoxic stress by at least three mechanisms: storage of endogenous glycogen in many tissues; utilization of more efficient fermentations; and reductions in metabolic rates. Each of these strategies deserves separate consideration because they vary in importance in different species.

Peripheral stores of glycogen

Perhaps the simplest way of reducing dependence upon central-depot glycogen is to maintain elevated levels of glycogen in many tissues. These depots could be mobilized during anoxia independently or concurrently with blood glucose utilization. Not surprisingly, glycogen levels are high in most tissues of animals that are tolerant of anoxia. Gill and muscle in *Mytilus* can store up to about 200 μmol glucosyl–glycogen g^{-1} tissue (De Zwann and Zandee, 1972a). Turtle and Weddell seal hearts contain higher levels of glycogen than are found in any other species of vertebrates thus far studied (Daw et al., 1967; Kerem et al., 1973). Glycogen levels in white and red muscles of the goldfish are also high, $\frac{1}{10}$ and $\frac{1}{3}$, respectively, of the astronomical levels found in the liver (Shoubridge, pers. commun.). In vertebrates, when glycogen levels are low, glycogen is stored as β-particles; when glycogen levels are greatly elevated, as in the Weddell seal heart, lungfish muscle, or goldfish liver, glycogen is stored as β-particles, as large (1000 Å diameter) α-particles, and as glycogen bodies (Hochachka and Hulbert, 1978). It seems safe to conclude that this pattern of peripheral storage of glycogen is general in most animals that survive well under anoxic conditions.

Energetically improved fermentations

We find the best examples of alternative fermentation pathways to the classical glucose → lactate fermentation (yielding 2 mol ATP/mol glucose) among the champion invertebrate anaerobes, the helminths and the marine bivalves. These have been reviewed several times in recent years (De Zwann, 1976, 1977), so only a brief summary will be considered here. They can be viewed as a series of linear, and probably linked, pathways (Hochachka, 1980). The most important are:

1. Glucose → octopine, alanopine, or strombine, ATP yield 2 mol ATP/ mol glucose
2. Glucose → succinate, energy yield 6 mol ATP/mol glucose
3. Glucose → propionate, energy yield 8 mol ATP/mol glucose
4. Aspartate → succinate, energy yield 1 mol ATP/mol aspartate
5. Aspartate → propionate, energy yield 2 mol ATP/mol glucose
6. Glutamate → succinate, energy yield 1 mol ATP/mol glutamate
7. Glutamate → propionate, energy yield 2 mol ATP/mol glutamate
8. Branched chain amino acids → volatile fatty acids, energy yield 1 mol ATP/mol substrate
9. Glucose → acetate, energy yield 6 mol ATP/mol glucose

Pathways 1 through 4 are known in various bivalve molluscs; 2, 3, 8, and 9 are often utilized by helminths; 5, 6, and 7, although theoretically possible in bivalve molluscs, do not appear to be utilized to any significant extent. There is little argument about the pathways to most of these end products. However, the reaction path from succinate to propionate is only now being established in bivalves (Schroff and Wienhausen, 1979); in helminths, current evidence (Tkachuck et al., 1977; Barrett et al., 1978) favors

$$\text{succinate} \rightarrow \text{succinylCoA} \xrightarrow{\quad} \text{methylmalonylCoA} \xrightarrow{CO_2} \text{propionylCoA} \rightarrow \text{propionate}$$

Any animal anaerobes utilizing such fermentations can increase their yield of ATP from glucose by two to four times.

There is good evidence that vertebrates can utilize schemes 4 and 6. More than a decade ago Penney and Cascarano (1970) emphasized that the anoxic rat heart works better and longer if supplemented with a source of fumarate (for the fumarate reductase reaction) and a source of succinylCoA (for the succinic thiokinase reaction). Aspartate, malate, or fumarate could satisfy the former needs, whereas glutamate or 2-ketoglutarate could satisfy the latter. Both reactions are energy-yielding, with succinate clearly serving as a carbon sink for the two arms of the Krebs cycle feeding into it. Useful energy is formed in the mitochondria, which may be important for sustaining mitochondrial integrity during anoxic episodes (Schertzer and Cascarano, 1979). Succinate accumulation has been observed in the blood of diving animals (Hochachka et al., 1975), in various hypoxic or anoxic fish tissues (Johnston, 1975), and in working mammalian heart preparations (Taegtmeyer, 1978; Sanborn et al., 1979). However, the yield of ATP by these succinate-forming reactions is not high enough to contribute significantly to the overall anaerobic energy production. They appear to serve a local function of maintaining mitochondria and do not aid in minimizing the animal's anaerobic demands for glucose.

Reducing anaerobic demands for glucose by depressing metabolism

Among vertebrate anaerobes, the goldfish and diving turtle are the best studied, and both reduce metabolism during prolonged anoxia. It is well known that the goldfish is capable of surviving several days of total anoxia. The actual duration seems to vary depending upon the condition of the fish and the season. In our laboratory, winter-acclimatized goldfish at 4 °C are able to survive 4–5 days of total anoxia (Shoubridge and Hochachka, 1979), whereas others have observed survival for 2 to several weeks (see Walker and Johanson, 1977). Liver glycogen levels for winter-acclimatized goldfish are sufficient to maintain the resting metabolic rate of a goldfish at 4 °C (Fry and Hochachka, 1970) for only 2.2 days, but if the resting normoxic metabolic rate were depressed fivefold, this glucose supply could keep the goldfish going for about 11 days. This reduction in metabolic rate and duration of anoxia tolerance are within published estimates for this species (Andersen, 1975).

Jackson and his colleagues (pers. commun.) have maintained turtles under anoxic conditions for up to 6 months. In the turtle, liver glycogen levels are sufficient to maintain the resting metabolic rate at 3 °C (Jackson, 1971) for only about 2.9 days. Other body stores of glycogen may extend this time period, but in tissues such as the heart, glycogen stores are largely depleted after a few hours of anoxia (Daw et al., 1967). In short-duration diving at 24 °C, the metabolic rate is known to drop to about 15% of predive levels (Jackson, 1968). If these results are extrapolated to the situation at 3 °C, then the anoxia tolerance of the turtle could be extended six- to sevenfold (i.e., to only about 19 days). It is unlikely that the turtle augments anaerobic glycolysis with amino acid fermentation because no changes in free amino acid profiles are observed during 3- to 4-week diving periods; nor do succinate and alanine, two anaerobic end products that would be expected from such a metabolism, accumulate to any large extent during diving (B. Emmett and P. W. Hochachka, unpubl. data). Thus, to account for the outstanding duration of anoxia the diving turtle can withstand, it is necessary to assume a depression of metabolism to about $\frac{1}{80}$ of the resting normoxic rate. The same estimate is obtained if the calculation is made on the basis of lactate accumulation during prolonged anoxia. Thus it appears that the anoxic turtle operates primarily the classical glucose → lactate fermentation during prolonged anoxia.

A similar strategy of reducing anaerobic demands for glucose by reducing metabolic rate is also utilized by bivalve molluscs. For example, *Mytilus* could sustain a normoxic metabolic rate during anoxia using glycogen stored in mantle and liver for only 3 days if glucose were fermented to lactate. If we assume glucose is fermented to succinate, the ATP yield doubles, and anoxia could be sustained for about 6 days. From measurements of succinate accumulation, phosphoarginine depletion, and ATP depletion, De Zwann and Wijsman (1976) estimate that the rate of ATP turnover during anoxia is only $\frac{1}{20}$ the normoxic metabolic rate. This means that even at the usual glycolytic efficiency level, *Mytilus* can be sustained by mantle and hepatopancreas glycogen for about 60 days. If all glucose were channeled into succinate, it could survive 120 days, whereas if all glucose flowed to propionate, it could survive 150 days, assuming nothing else was limiting.

The "end products" problem

Nature of the problem
The accumulation of anaerobic end products to very high levels can create pH problems, osmotic problems, and/or metabolic disruptions through inhibitory effects of end products. Animals have used at least four strategies for dealing with the end products problem. They have (1) minimized accumulation by metabolic depression; (2) tolerated high accumulations by maintaining high buffering capacities of blood and tissues; (3) "detoxified" anaerobic end products; and (4) formed alternative, less toxic end products.

Reducing end product formation by metabolic depression

By depressing metabolism during anoxia, an organism not only reduces the need for pouring glucose into inefficient fermentative reactions, it also automatically reduces the accumulation of anaerobic end products. Clearly, both outcomes are advantageous. Of course, when a diving turtle ferments 7600 mol of glucose it will generate 15 200 mol of lactate, whether the process occurs in 3 days or 3 months.

Tolerating end product accumulation

A simple means for dealing with periodic massive accumulations of an acid end product, such as lactic acid, is to improve buffering capacity. Diving turtles do this with high bicarbonate reserves (Jackson, 1969). A similarly improved buffering capacity is also utilized by bivalves (Wijsman, 1975) and intestinal parasites (Podesta et al., 1976).

Another way to tolerate high accumulations of end product is to design enzymes whose function is either actually favored, or at least unaffected, by the accumulations. In diving mammals and turtles, key regulatory enzymes such as pyruvate kinase (see Hochachka and Murphy, 1979) display distinctly acid pH optima compared to homologous enzymes in terrestrial animals. This means that in diving animals, conditions become more favorable for pyruvate kinase function at a metabolically appropriate time: when O_2 is lacking and the organism is depending more and more upon anaerobic glycolysis. Lactate dehydrogenase in diving turtles supplies another example of an enzyme that has been adjusted to function adequately when lactate levels rise (see Hochachka, 1980).

"Detoxifying" lactate by its further metabolism

An effective strategy for dealing with end products is to convert them into less harmful derivatives (i.e., to "detoxify" them). The anoxic goldfish utilizes this strategy. The goldfish, like most vertebrates, ferments glycogen (glucose) to lactate during anoxia. All its tissues display high activities of LDH and have the capacity to form lactate. However, little is accumulated during anoxia. This phenomenon has intrigued numerous workers over the last 15 years (see Hochachka, 1980), and recently has been partially explained. The key observation (Shoubridge and Hochachka, 1979, 1980) is that ^{14}C-lactate is partially oxidized in anoxic goldfish. The reaction path generating CO_2 from lactate appears to be

$$lactate \longrightarrow pyruvate \xrightarrow{\text{CoASH}} acetylCoA + CO_2$$
$$NAD^+ \; NADH \quad NAD^+ \; NADH$$

and the subsequent fate of acetylCoA is reduction to ethanol, which is largely excreted or diffused out into the surrounding water. The probable sequence in this segment is

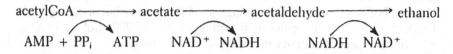

acetylCoA —————→ acetate —————→ acetaldehyde —————→ ethanol

AMP + PP$_i$ ATP   NAD$^+$ NADH   NADH   NAD$^+$

The main tissue site at which ethanol formation occurs is skeletal muscle. In winter-acclimatized goldfish, ethanol dehydrogenase occurs in high concentrations in white and red muscles, but does not occur in any other tissues. Concentrations of ethanol never build up above about 4 μmol g^{-1} in tissues (4 μmol ml^{-1} in blood) because it is so readily removed to the outside water. This mechanism is wasteful of carbon because much of the 2-carbon compound formed is lost to the medium, but the loss seems to be a reasonable trade-off for survival through anoxic periods imposed on goldfish by the wintertime freeze-up.

It is not known whether the goldfish mechanism is general among anoxia-tolerant fish. To data, ethanol production during anoxia has been observed in only one other animal species, namely *Chironomus* during larval stages (Wilps and Zebe, 1976). However, numerous hypoxia-adapted teleosts are known to maintain unusually high RQ values with rates of CO_2 production exceeding rates of O_2 uptake under hypoxic conditions by 1.5–2 times (Randall, pers. commun.; Kutty, 1968). Similarly, true metabolic CO_2 production under anoxic conditions is known for other species, especially carp (see Hochachka, 1980). Thus, it may well turn out that comparable mechanisms are utilized by other fish as well.

Another strategy appears to be utilized during experimental diving in the Weddell seal. During experimental diving, the seal relies almost exclusively upon glycogen (glucose) to sustain its various functions. Under these conditions many hypoperfused tissues and organs ferment glucose and slowly release lactate into the blood. Two of the central organs (heart and lung) maintain oxidative metabolism and utilize this lactate in preference to glucose (Murphy et al., 1980). This oxidation of lactate reduces its accumulation in blood and also spares glucose supplies for more glucose-needy tissues. It is possible that other tissues, such as the liver and kidney (Keith et al., 1980), utilize blood lactate during prolonged diving.

Utilization of nontoxic anaerobic end products

In many invertebrate anaerobes (parasitic helminths, polychaetes, sipunculoids, bivalves, cephalopods, gastropods, and probably many other groups as well), PEP functions at a distinct metabolic branchpoint:

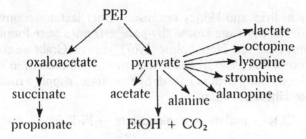

PEP carboxykinase (PEPCK) catalyzes carboxylation of PEP to form oxaloacetate, from which ultimately succinate and propionate can be formed. Under other circumstances (particularly in molluscs), PEP is converted to pyruvate, which can be converted into several kinds of end products such as octopine, alanopine, strombine, and lysopine. These end products are formally analogous to LDH and serve the same function as do lactate and LDH, that is, as hydrogen and carbon sinks to maintain redox balance during anaerobic metabolism (Fields et al., 1980; Gade, 1980).

None of the above end products is as acidic as lactic acid; some, like octopine, have isoelectric points close to neutrality. Some, like acetate and propionate, although acidic, are readily volatized. Of the entire series, succinate is probably most like lactate, but even it has a higher pK than lactate. Furthermore, in some of the most capable intertidal anaerobes (*Mytilus*, *Littorina*, etc.) more than one terminal dehydrogenase can function in any given tissue. Under anoxic stress, these tissues can generate more than one anaerobic end product in the process of maintaining an oxidizing potential, and this in itself reduces the chance of any one compound building up to unacceptably high levels.

Reestablishing metabolic homeostasis

Nature of the problem
The third major metabolic problem that must be solved by animal facultative anaerobes is the rapid reestablishment of metabolic homeostasis when the period of anoxia ends. The lactate or other anaerobic end products formed during anoxia must be cleared, and the storage depots of glycogen must be recharged. In invertebrate anaerobes, this problem has just begun to be studied; so the discussion will be limited to vertebrates.

Fate of lactate formed in peripheral tissues (muscle)
Skeletal muscle is the largest tissue in the vertebrate body, and during anoxic stress it is usually the largest source of lactate. Krebs and Woodford (1965) concluded that lactate reconversion to glucose (then glycogen)

occurs only in liver and kidney because neither lactate reconversion to glycogen nor the pyruvate kinase "by-pass" enzymes were found in muscle. However, Opie and Newsholme (1967) (see also Crabtree et al., 1972) have now found that PEPCK and malic enzyme are present in vertebrate muscles, which, along with malate dehydrogenase, supply a route around the pyruvate kinase reaction:

$$\text{pyruvate} + CO_2 \rightarrow \text{malate} \rightarrow \text{oxaloacetate} \rightarrow \text{PEP}$$

Newsholme and his colleague also confirmed the occurrence of fructose diphosphatase, an indicator of glycogenic capacity, in both frog muscle and mixed mammalian muscles. These findings served as points of departure for Bendall and Taylor (1970), who found that respiring frog sartorius can rapidly convert lactate to glycogen. Their data for mammalian muscle, however, were not so convincing. More recently, several other studies of the problem have appeared, that by McLane and Holloszy (1979) probably being the most convincing. From all of these studies it is now clear that much (about half) of the lactate formed in fast-twitch white and fast-twitch red muscles during anaerobic glycolysis is reconverted to glycogen in situ. Slow-twitch red fibers in mammals are unable to make this conversion, probably because of the absence of key gluconeogenic enzymes, fructose biphosphatase in particular. In fish, recent studies show that the bulk of lactate formed in white muscle does not equilibrate with blood (Wardle, 1978), and it is a reasonable guess that much of it is reconverted to glycogen in situ.

A significant fraction of lactate is washed out of muscle into the blood and is thus metabolized elsewhere. For years it has been known that an important fate of this lactate is reconversion in liver and kidney via the Cori cycle to glucose, which in turn may be used to replenish glycogen depots everywhere in the body. In recent years, the rate of the Cori cycle has been estimated in various metabolic states (Kusaka and Ui, 1977). High rates of cycling lactate and glucose carbon through the Cori cycle seem to be found in diving marine mammals (Keith et al., 1980), but these data are in need of being repeated in different species and under a greater number of metabolic conditions.

It should also be emphasized that lactate is an excellent substrate for various organs and tissues in the vertebrate body, and therefore an additionally significant fate of lactate during recovery from anoxia is complete oxidation. At elevated blood concentrations, skeletal muscle (Jorfeldt, 1970; Bilinski and Jonas, 1972), lung (Wolfe et al., 1979), heart (Keul et al., 1972), and even brain (Siesjo and Nordstrom, 1977) are all fully capable of utilizing lactate as a carbon and energy source, often in preference to glucose. Special significance is attached to these capacities during recovery from long-duration diving in the Weddell seal and presumably other diving animals as well because they speed up the rate of lactate clearance (Hochachka and Murphy, 1979). Even so, lactate clearance (which metabolically represents the sum of lactate reconversion to glycogen in situ, lactate entry into the Cori cycle, and lactate entry into other metabolic pathways, particularly into the Krebs cycle for complete oxidation) remains a sluggish process in marine mammals. For example, in the Weddell seal, the lactate accumulated in a 46-min dive takes twice as long to clear during recovery (Hochachka et al., 1977).

Summary

This chapter has concentrated upon how animals survive prolonged periods without oxygen. Glycogen is the main source of energy during anoxia. The most effective animal anaerobes (*Mytilus*, goldfish, turtles) seem capable of very significant depressions of metabolism during anoxia. This strategy reduces both the need for glycogen and the accumulation of anaerobic end products. In the anoxic goldfish (and probably other animals as well), lactate is "detoxified" by being metabolized to ethanol and CO_2; ethanol is readily removed to the outside medium and never accumulates to high levels. Invertebrate anaerobes also utilize energetically more efficient fermentations, yielding up to four times as much ATP from a mole of glucose as the classical glucose → lactate fermentation.

Lactate accumulated during anoxia is reconverted to glycogen in the liver, kidney, and muscle. In most tissues, it serves as an excellent substrate, often being utilized in preference to glucose. Clearance of lactate remains extremely sluggish. In one of the most capable of diving marine mammals, the Weddell seal, the lactate formed during a maximum-duration (1.2-h) dive takes at least that long, and usually longer, to clear.

References

Anderson, J. R. (1975) The anaerobic resistance of *Carassius aratus*. Ph.D. thesis, Australian National University, Canberra.

Barrett, J., Coles, G. C., and Simpkin, K. G. (1978) Pathways of acetate and propionate production in adult *Fasciola hepatica*. *Intl. J. Parasitol.* 8:117–123.

Bendall, J. R., and Taylor, A. A. (1970) The Meyerhof quotient and the synthesis of glycogen from lactate in frog and rabbit muscle. A reinvestigation. *Biochem. J.* 118:887–893.

Bilinski, E., and Jonas, R. E. E. (1972) Oxidation of lactate to CO_2 by rainbow trout tissues. *J. Fish. Res. Bd. Can.* 29:1467–1471.

Crabtree, B., Higgins, S. J., and Newsholme, E. A. (1972) The activities of pyruvate carboxylase, phosphoenolpyruvate carboxykinase, and fructose diphosphatase in muscle from vertebrates and invertebrates. *Biochem. J.* 130:391–396.

Daw, J. C., Wenger, D. P., and Berne, R. M. (1967) Relationship between cardiac glycogen and tolerance to anoxia in the western painted turtle, *Chrysemys picta bellii. Comp. Biochem. Physiol. 22:69–73.*

De Zwaan, A., and Wijsman, T. C. M. (1976) Anaerobic metabolism in Bivalvia (Mollusca). *Comp. Biochem. Physiol. 54B:313–324.*

De Zwaan, A., and Zandee, D. I. (1972a) Body distribution and seasonal changes in the glycogen content of the common sea mussel *Mytilus edulis. Comp. Biochem. Physiol. 43A:53–58.*

De Zwaan, A., and Zandee, D. I. (1972b) The utilization of glycogen and the accumulation of some intermediates during anaerobiosis in *Mytilus edulis. Comp. Biochem. Physiol. 43B:47–54.*

Fields, J. H. A., Eng, A. K., Ramsden, W. D., Hochachka, P. W., and Weinstein, J. A. (1980) Alanopine and strombine are novel imino acids produced by a dehydrogenase found in the adductor muscle of the oyster, *Crassostrea gigas. Arch. Biochem. Biophys. 201:110–114.*

Fry, F. E. J., and Hochachka, P. W. (1970) Fish. In *Comparative physiology of thermoregulation* vol. 1 (G. C. Whittow, ed.), pp. 79–134. London: Academic Press.

Gade, G. (1980) Biological role of octopine formation in marine molluscs. *Marine Biol. Lett. 1:121–135.*

Hochachka, P. W. (1980) *Living without oxygen*, pp. 1–173. Cambridge, Mass.: Harvard University Press.

Hochachka, P. W., and Hulbert, W. C. (1978) Glycogen seas, glycogen bodies, and glycogen granules in heart and skeletal muscles of two air-breathing burrowing fishes. *Can. J. Zool. 56:774–786.*

Hochachka, P. W., and Murphy, B. (1979) Metabolic status during diving and recovery in marine mammals. In *Intl. rev. physiol. environ. physiol. III,* vol. 20, pp. 253–287. Baltimore: University Park Press.

Hochachka, P. W., Owen, T. G., Allen, J. F., and Whittow, G. C. (1975) Multiple end products of anaerobiosis in diving vertebrates. *Comp. Biochem. Physiol. 50:17–22.*

Hochachka, P. W., Liggins, G. C., Qvist, J., Snider, M., Schneider, R., Wonders, T., and Zapol, W. M. (1977) Pulmonary metabolism during diving: Conditioning blood for the brain. *Science 198:831–834.*

Jackson, D. C. (1968) Metabolic depression and oxygen depletion in the diving turtle. *J. Appl. Physiol. 24:503–509.*

Jackson, D. C. (1969) The response of the body fluids of the turtle to imposed acid–base disturbance. *Comp. Biochem. Physiol. 29:1105–1110.*

Jackson, D. C. (1971) The effect of temperature on ventilation in the turtle, *Pseudemys scripta elegans. Respir. Physiol. 12:131–140.*

Johnston, I. A. (1975) Anaerobic metabolism in the carp (*Carassius carassius* L.) *Comp. Biochem. Physiol. 51B:235–241.*

Jorfeldt, L. (1970) Metabolism of L(+)lactate in human skeletal muscle during exercise. *Acta Physiol. Scand. Suppl. No. 338:1–67.*

Jorgensen, D. D., Ware, S. K., and Redmond, J. R. (1980) The reverse dive reflex in an intertidal mollusc. *Fed. Proc. 39:4161.*

Keith, E. O., Pernia, S. D., and Oritz, C. L. (1980) Intermediary metabolism and bioenergetics of northern elephant seals during starvation. *Fed. Proc.* 39:3295.

Kerem, D., Hammond, D. D., and Elsner, R. (1973) Tissue glycogen levels in the Weddell seal, *Leptonychotes weddelli*: A possible adaptation to asphyxial hypoxia. *Comp. Biochem. Physiol.* 45A:731–737.

Keul, J., Doll, E., and Keppler, D. (1972) *Energy metabolism of human muscle*, pp. 104–125. Baltimore: University Park Press.

Krebs, H. A., and Woodford, M. (1965) Fructose 1,6-diphosphatase in striated muscle. *Biochem. J.* 94:436–445.

Kusaka, M. and Ui, M. (1977) Activation of the Cori cycle by epinephrine. *Am. J. Physiol.* 232:E145–E155.

Kutty, M. N. (1968) Respiratory quotients in goldfish and rainbow trout. *J. Fish. Res. Bd. Can.* 25:1689–1728.

McLane, J. A., and Holloszy, J. O. (1979) Glycogen synthesis from lactate in the three types of skeletal muscle. *J. Biol. Chem.* 254:6548–6553.

Murphy, B., Zapol, W. M., and Hochachka, P. W. (1980) Metabolic activities of heart, lung, and brain during diving and recovery in the Weddell seal. *J. Appl. Physiol.* 48:596–605.

Opie, L. H., and Newsholme, E. A. (1967) The activities of fructose 1,6-diphosphatase, phosphofructokinase, and PEP carboxykinase in white muscle and red muscle. *Biochem. J.* 103:391–399.

Penney, D. G., and Cascarano, J. (1970) Anaerobic rat heart: Effects of glucose and tricarboxylic acid–cycle metabolites on metabolism and physiological performance. *Biochem. J.* 118:221–227.

Podesta, R. B., Mustafa, T., Moon, T. W., Hulbert, W. C., and Mettrick, D. F. (1976) Anaerobes in an aerobic environment: Role of CO_2 in energy metabolism of *Hymenolepis diminuta*. In *Biochemistry of parasites and host–parasite relationships* (H. Van den Bossche, ed.), pp. 81–88. New York: North-Holland Press.

Sanborn, T., Gavin, W., Berkowitz, S., Perille, T., and Lesch, M. (1979) Augmented conversion of aspartate and glutamate to succinate during anoxia in rabbit heart. *Am. J. Physiol.* 237:H535–H541.

Schertzer, H. G., and Cascarano, J. (1979) Anaerobic rat heart: Mitochondrial role in calcium uptake and contractility. *J. Exp. Zool.* 207:337–350.

Schroff, G., and Wienhausen, G. (1979) Formation of acetate and propionate by isolated mitochondria of *Arenicola marina* (Polychaeta). In *Proc. Intl. Symp. Physiol. Euryoxic Animals* (D. A. Holwerda, ed.), pp. 73–74.

Shoubridge, E. A., and Hochachka, P. W. (1979) Lactate oxidation in the anoxic goldfish. *Intl. Cong. Biochem. Abstracts* 13(1):R123.

Shoubridge, E. A., and Hochachka, P. W. (1980) Ethanol: A novel endproduct of vertebrate anaerobic metabolism. *Science* 209:308–309.

Siesjo, B. K., and Nordstrom, C. H. (1977) Brain metabolism in relation to oxygen supply. In *Oxygen and physiological function* (F. F. Jobsis, ed.), pp. 459–475. Dallas, Tex.: Professional Information Library.

Taegtmeyer, H. (1978) Metabolic responses to cardiac hypoxia: Increased production of succinate by rabbit papillary muscles. *Circ. Res.* 45:808–815.

Tkachuck, R. D., Saz, H. J., Weinstein, P. P., Finnegan, K., and Mueller, J. F. (1977) The presence and possible function of methylmalonylCoA mutase and propionylCoA carboxylase in *Spirometra mansonoides*. *J. Parasitol.* 63:769–774.

Walker, R. M., and Johanson, P. H. (1977) Anaerobic metabolism in goldfish *Carassius auratus*. *Can. J. Zool.* 55:1304–1311.

Wardle, C. S. (1978) Non-release of lactic acid from anaerobic swimming muscle of plaice *Pleuronectes platessa* L.: A stress reaction. *J. Exp. Biol.* 77:141–156.

Wijsman, T. C. M. (1975) pH fluctuations in *Mytilus edulis* L. in relation to shell movements under aerobic and anaerobic conditions. *Proc. 9th Eur. Marine Biol. Symp.* (H. Barnes, ed.), pp. 139–149. Aberdeen: University Press.

Wilps, H., and Zebe, E. (1976) The end-products of anaerobic carbohydrate metabolism in the larvae of *Chironomus thummi thummi*. *J. Comp. Physiol.* 112:263–272.

Wolfe, R. R., Hochachka, P. W., Trelstad, R. L., and Burke, J. F. (1979) Lactate oxidation in perfused rat lung. *Am. J. Physiol.* 236:E276–E282.

Zapol, W. M., Liggins, G. C., Schneider, R. C., Qvist, J., Snider, M. T., Creasy, R. K., and Hochachka, P. W. (1979) Regional blood flow during simulated diving in the conscious Weddell seal. *J. Appl. Physiol.* 47:R968–973.

11 How marine mammals dive

G. L. KOOYMAN

It was pointed out in Knut Schmidt-Nielsen's text, *Animal physiology* (1979), that all aquatic mammals and birds have retained lungs and breathe air. Lengthy dives involve long periods without breathing and require effective utilization of a limited oxygen supply. Most diving animals utilize several physiological methods for prolonging the duration of dives: large oxygen stores; changes in circulation (slowing of the heart and redistribution of the blood flow away from muscles and abdominal organs and to heart and brain); a decreased metabolic rate; and a reliance of peripheral tissues on anaerobic metabolism (evidenced by a buildup of lactate). The sudden drops in heart rate (bradycardia) and redistribution of blood flow away from peripheral tissues are so characteristic of the forced dives studied by physiologists in the laboratory that this phenomenon has acquired the name "the diving reflex."

Schmidt-Nielsen cautions that the diving reflex observed during forced submersion of an animal may not always occur when animals dive in nature. In Weddell seals the degree of bradycardia is related to the duration of the dive, with the longer dives having the slowest heart rates (Kooyman and Campbell, 1972). Also, the seal trained to dive on command for short periods exhibits less slowing of heart rate than one forced to dive (Elsner, 1965). These observations indicate that the dive reflex is rather variable in marine mammals. This chapter considers the physiological mechanisms marine mammals use as they dive in nature and specifically how often they normally use the diving reflex.

How long can marine mammals hold their breath?

Oxygen stores
When one thinks of the oxygen stores necessary for an extended dive, the first source that comes to mind is oxygen in the lungs. However, the lungs of marine mammals are no larger than those of other mammals (Kooyman, 1973). Furthermore, many marine mammals do not dive on a full lung volume and, therefore, do not take complete advantage of this potential store. For example, we know that seals, sea lions, and sea otters

151

dive with only 40 to 60% of total lung capacity (Kooyman et al., 1971; Kooyman and Sinnett, unpubl. observ.).

Many marine mammals have exceptional oxygen-carrying capacities in the hemoglobin of their blood and the myoglobin of their muscles. Of these two, the blood oxygen-carrying capacity is perhaps the more important because it is a transporter of oxygen. The blood oxygen stores of seals amount to about 30 to 40 ml $\cdot$ kg^{-1} of body weight compared with 10 to 15 ml $\cdot$ kg^{-1} of body weight in some small cetacea, a value that is similar to that of man (Lenfant et al., 1970). However, most cetacea and sea lions are probably intermediate between man and seals (Ridgway and Johnston, 1966; Ridgway, 1972). Thus, from this observation alone, phocids ought to be able to breath-hold longer than most sea lions or small whales. Furthermore, the myoglobin concentrations of most diving marine birds and mammals exceed those of terrestrial forms (Lenfant et al., 1970; Castellini, 1980). During a breath hold this attribute makes possible a longer aerobic period in muscle of marine birds and mammals than in terrestrial mammals.

Breath holds during sleep

When most marine mammals sleep, they exhibit a marked arrythmic respiration with long periods between breaths. In some species several minutes may elapse between breaths. During these periods, heart rate slows (about 20%). This slowing appears to be accompanied by a small reduction in blood flow to some organs and a fall in arterial oxygen tension (Pa_{O_2}), which decreases because the organs continue to consume blood O_2 at a normal, resting rate. In sleeping Weddell seals, *Leptonychotes weddelli*, blood Pa_{O_2} reaches a minimum of about 25 Torr (Figure 11.1). This occurs after about 8 min without breathing. The duration of the periods between breaths appears to be correlated with metabolic rate (Kooyman et al., 1973); that is, the lower the metabolism, the longer the breath hold. Unfortunately, there are no comparable Pa_{O_2} measurements from Weddell seals during forced dives. However, in the harbor seal, *Phoca vitulina*, after 8 min of breath holding during a forced dive, the Pa_{O_2} is about 50 Torr (Kerem and Elsner, 1973), or about twice the tension found in the sleeping Weddell seal when it holds its breath for the same period. It is now well established that little flow occurs to most organs during forced dives (Elsner et al., 1978; Zapol et al., 1979), and the lower Pa_{O_2} of the sleeping Weddell seal most likely reflects a greater utilization of arterial O_2 by peripheral organs.

Breath holds during diving

It has been estimated that the Weddell seal might have a maximum breath hold of 1.1 h (66 min) (Table 11.1, Figure 11.2; Kooyman, 1975)

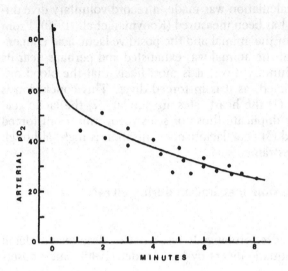

FIGURE 11.1 Arterial gas tensions in resting Weddell seals. The plot is a pooled sample from several resting apneuses. The longest apneusis was 8 min. [Data from Kooyman et al., 1980]

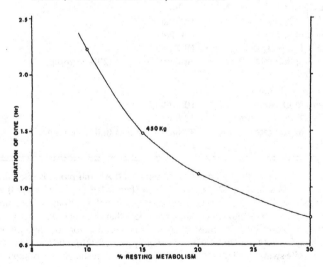

FIGURE 11.2 The calculated breath-holding limit of a 450-kg Weddell seal based on the oxygen stores of the body and the oxygen consumption rate during the dive. See Table 11.1 for the bases of the calculations. [Modified from Kooyman 1975]

by assuming that only the brain, heart, and lung have access to blood O_2 stores. Since that calculation was made, a record voluntary dive duration of 1.2h (72 min) has been measured (Kooyman et al., 1980). From the recovery behavior of the animal and the postdive lactic acid concentration, it appeared that the animal was exhausted and perhaps near its limit. In such long voluntary dives, it is most likely that the blood distribution is quite restricted, as it is in forced dives. This conclusion is based on findings that (1) the heart rates are similar, (2) the lactic acid recovery curves nearly duplicate those of seals recovering from forced dives (Figure 11.3), and (3) peak lactate concentration is high, although not as high as in the restrained seal.

How important is anaerobic metabolism during dives?

Forced dives

The lactic acid pulse after prolonged dives was first described in detail for several species of aquatic divers by Scholander (1940). Such obser-

TABLE 11.1 Basic values used to estimate an adult Weddell seal's maximum breath-hold capacity during severe restriction in blood flow (A) and when the muscle mass is perfused (B).

A. During restriction of blood flow	
Seal mass	450 kg
Total blood volume	68 liters
Available blood O_2 store	18.3 liters
Available lung O_2 store	1.4 liters
Total available body O_2 store	19.7 liters
Oxygen consumption rate	18 liters $O_2 \cdot h^{-1}$ (20% of resting)
Breath-hold limit	1.1 h
B. During perfusion of muscle mass	
Available muscle O_2 store	10.3 liters
Total available body O_2 store	30 liters
Oxygen consumption rate	90 liters $O_2 \cdot h^{-1}$ (full resting rate)
Breath-hold limit	20 min

Note: Blood values are based upon the following assumptions: (1) Arterial blood is one-third of the blood volume and 95% saturated; (2) O_2 can be extracted to 10% of saturation; (3) in A, muscle perfusion is zero so O_2 store of muscle is not considered in the consummable store; (4) $\dot{V}_{O_2}$ of 20% of resting is quite arbitrary, and this considers that the heart and brain are consuming most of the O_2 store; (5) in B, muscle O_2 store is available for consumption in addition to blood and lung O_2 store, and muscle is 30% of body mass; (6) $\dot{V}_{O_2}$ under conditions of perfused muscle and while swimming is estimated to be equivalent to resting O_2 consumption rate.

Source: Estimates derived from calculations of Kooyman (1975) and Kooyman et al. (1980).

vations are a clear indication that some areas of the body have been starved for oxygen and have had to function anaerobically. This blood lactic acid pulse after forced dives is so distinctive that it has been called "the Hallmark" of diving in aquatic vertebrates (Hochachka and Murphy, 1979). It reflects clearly the important involvement of anaerobic metabolism in this kind of dive. The major source of this lactic acid is skeletal muscle, but, surprisingly, there has been only one experiment in which muscle biopsies were taken from a diving seal to measure directly the concentration of lactic acid at its source (Scholander et al., 1942). In this study, Scholander showed a relationship between muscle oxygen depletion and the onset and increase in lactic acid accumulation.

Voluntary dives

During voluntary dives in Weddell seals, there is no such hallmark of diving (see preceding paragraph) until dives exceed 20 to 30 min duration (Figure 11.4). Dives of shorter duration show very little or no increase in blood lactate concentrations. These observations suggest a much broader circulation pattern in which muscle is supplied with blood oxygen during the dives. The myoglobin concentration of Weddell seal muscle is similar to that of harbor seal muscle (Castellini, 1980), and it was calculated that a depletion of this O_2 store would occur in about 5 to 10 min. This seems to be an overestimate considering that the Weddell seals were swimming during their dives.

FIGURE 11.3 Arterial lactic acid concentration changes during recovery from a voluntary dive of 43 min (curve A), and a forced dive of 46 min (curve B). [Modified from Hochachka et al., 1977, and Kooyman et al., 1980]

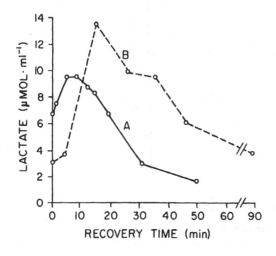

Taking into account that much more of the body is utilizing O_2 during the dive than just the brain, heart, and lung, and incorporating the myoglobin O_2 stores in the calculation, it was estimated that the Weddell seal could dive for about 20 min before its O_2 store was depleted (Table 11.1; Kooyman et al., 1980). Obviously no marine mammal is going to completely exhaust its O_2 store during a dive and survive, but this calculation yields a coarse estimate of the aerobic diving limit of this species of seal. Significantly, fewer than 3% of 4600 free-ranging dives of Weddell seals exceeded 26 min (Figure 11.5).

When are the different types of dives used?

Short "aerobic" dives

Most dives made by Weddell seals are so short that there would be no lactate accumulation (Figure 11.5). Groups of repetitive deep dives of <20 min duration often lasted for 8 or more hours (Figure 11.6). The majority of these dives were usually to about the same depth, in this case 400 m. Presumably, the record illustrated indicates that the seal had found a good source of food and was exploiting it. As a result, he spent

FIGURE 11.4 The peak arterial lactic acid concentrations obtained during recovery from a variety of dive durations in three different adult Weddell seals. [Modified from Kooyman et al., 1980]

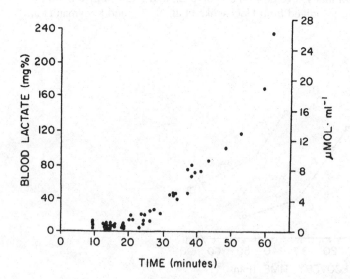

much of his time at depth and little time descending and ascending from the resource.

The diving episodes of fur seals have some similar features to those of Weddell seals (Figure 11.6). The durations were similar, but in most instances, little time was spent at depth. Instead, the dive was mainly a rapid descent and ascent over a 1- to 3-min period. This probably reflects different strategies in type and method of prey capture. Also, the body O_2 stores of fur seals are smaller and therefore would be able to maintain aerobic metabolism for less time. In both species, one can see that another common event was rapid exposure to high hydrostatic pressure. Although there is no space to consider it here, it should be kept in mind that such a powerful variable must have some marked physiological effects.

Let us consider one last aspect of these diving periods. The useful time during a hunting episode is the dive time. The surface time serves to restore O_2 stores, restore acid–base balance, and process any excess metabolites in preparation for the next dive. Thus, the shorter the surface time, the more efficient the total feeding period. If the recovery times for a short and long dive are considered, the advantage of a short, aerobic dive over a prolonged dive is clear. It takes 3 to 4 min for a Weddell seal to recover from a dive of <20 min in which it replenishes depleted O_2

FIGURE 11.5 A summary of the length of dives recorded for 22 free-ranging Weddell seals. Each bar represents the percentage of the total number of dives measured. The total was 4600 dives; 123 dives or 2.7% exceeded 26 min. [Modified from Kooyman et al., 1980]

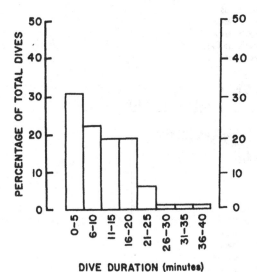

DIVE DURATION (minutes)

stores. If the dives are so long that anaerobic metabolites must be proc-
essed and acid–base balance restored, then the recovery becomes substantially longer. From Figure 11.3, we see that about 60 min is required
to recover from a 45-min dive, with regard to blood lactate concentrations. Therefore, about 43% of the time is spent diving, compared with

FIGURE 11.6 A single day's diving activities for a Weddell seal (upper)
and an Afro-Australian fur seal (*Arctocephalus pusillus*) (lower).
Time is presented relative to a 24-h clock in the area where the measurements were made. [Weddell seal data from Kooyman and Castellini,
unpubl. observ.; fur seal data from Kooyman and Gentry, unpubl.
observ.]

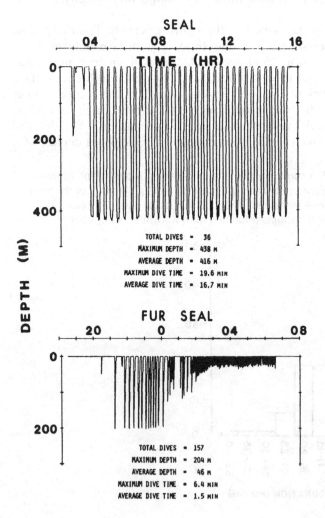

SEAL

TIME (HR)

TOTAL DIVES = 36
MAXIMUM DEPTH = 438 M
AVERAGE DEPTH = 416 M
MAXIMUM DIVE TIME = 19.6 MIN
AVERAGE DIVE TIME = 16.7 MIN

DEPTH (M)

FUR SEAL

TOTAL DIVES = 157
MAXIMUM DEPTH = 204 M
AVERAGE DEPTH = 46 M
MAXIMUM DIVE TIME = 6.4 MIN
AVERAGE DIVE TIME = 1.5 MIN

79% for 5 dives of 15 min each in which there is a 4-min recovery following each dive. No such blood lactate information is available for fur seals, but it is likely that the problems of recovery from high blood and tissue lactate levels would be the same and therefore would affect the time spent diving in a similar way.

Prolonged "anaerobic" dives

Perhaps it is appropriate to think of prolonged and short dives in terms of track and field sports. The prolonged dive is like a sprint. Regard for endurance over the long haul is sacrificed to achieve an immediate end, such as escaping from a predator or finding a new breathing hole, so one long submergence in which there will be a long recovery later is appropriate. In contrast, the short dive is a marathon event in which energy supplies are utilized aerobically to last for a long period of vigorous exercise.

Conclusions

Several conclusions can be drawn about how seals, and very likely other vertebrates, dive.

In natural or voluntary dives there is a correlation between the dive duration and blood distribution. During short dives blood circulation is more broadly distributed than in long dives, and, therefore, blood O_2 stores are more widely used.

Seals depend primarily upon oxidative metabolism for most of their energy needs during the great majority of their dives.

The continuous diving effort of seals over many hours indicates that O_2 *is replenished rapidly between dives.*

Prolonged "anaerobic" dives in which the diving reflex is observed are used infrequently, but are probably very important for escaping predators and/or finding breathing holes.

This review was supported by USPHS Grant No. HL 17731 and NSF, Division of Polar Programs, Grant No. 78-22999.

References

Castellini, M. A. (1980) Enzyme biochemistry in the tissues of marine mammals. Ph.D. thesis, University of California, San Diego, La Jolla. 123 pp.

Elsner, R. (1965) Heart rate response in forced versus trained experimental dives in pinnipeds. *Hvalradets Skrifter* 48:24–29.

Elsner, R., Blix, A. S., and Kjekshus, J. (1978) Tissue perfusion and ischemia in diving seals. *Physiologist* 21:33.

Hochachka, P. W., Liggins, G. C., Qvist, J., Schneider, R., Snider, M. Y., Wonders, T. R., and Zapol, W. M. (1977) Pulmonary metabolism during diving: Conditioning blood for the brain. *Science* 198:831–834.

Hochachka, P. W., and Murphy, B. (1979) Metabolic status during diving and recovery in marine mammals. In *Intl. rev. physiol., environ. physiol. III*, vol. 20 (D. Robertshaw, ed.), pp. 253–287. Baltimore: University Park Press.

Kerem, D., and Elsner, R. (1973) Cerebral tolerance to asphyxial hypoxia in the harbor seal. *Respir. Physiol.* 19:188–200.

Kooyman, G. L. (1973) Respiratory adaptations in marine mammals. *Am. Zool.* 13:457–468.

Kooyman, G. L. (1975) Physiology of freely diving Weddell seals. In *Biology of the seal*, Rapports et Proces-verbaus des Reunions, vol. 169, (K. Ronald and A. W. Mansfield, eds.), pp. 441–444. Charlottenlund Slot, Denmark: Conseil International pour L'Exploration de la Mer.

Kooyman, G. L., and Campbell, W. B. (1972) Heart rates in freely diving Weddell seals, *Leptonychotes weddelli*. *Comp. Biochem. Physiol.* 43A:31–36.

Kooyman, G. L., Kerem, D. H., Campbell, W. B., and Wright J. J. (1971) Pulmonary function in freely diving Weddell seals, *Leptonychotes weddelli*. *Respir. Physiol.* 12:271–282.

Kooyman, G. L., Kerem, D. H., Campbell, W. B., and Wright, J. J. (1973) Pulmonary gas exchange in freely diving Weddell seals, *Leptonychotes weddelli*. *Respir. Physiol.* 17:283–290.

Kooyman, G. L., Wahrenbrock, E. A., Castellini, M. A., Davis, R. W., and Sinnett, E. E. (1980) Aerobic and anaerobic metabolism during voluntary diving in Weddell seals: Evidence of preferred pathways from blood chemistry and behavior. *J. Comp. Physiol.* 138:335–346.

Lenfant, C., Johansen, K., and Torrance, J. D. (1970) Gas transport and oxygen storage capacity in some pinnipeds and the sea otter. *Respir. Physiol.* 9:277–286.

Ridgway, S. H. (1972) Homeostasis in the aquatic environment. In *Mammals of the sea* (S. H. Ridgway, ed.), pp. 590–747, Springfield, Ill.: C. C. Thomas.

Ridgway, S. H., and Johnston, D. G. (1966) Blood oxygen and ecology of porpoises of three genera. *Science* 151:456–458.

Schmidt-Nielsen, K. (1979) *Animal physiology*, 2nd ed. Cambridge University Press. 560 pp.

Scholander, P. F. (1940) Experimental investigations on the respiratory function in diving mammals and birds. *Hvalradets Skrifter Norske Videnskaps-Akad., Oslo*, 22. 131 pp.

Scholander, P. F., Irving, L., and Grinnell, S. W. (1942) Aerobic and anaerobic changes in seal muscles during diving. *J. Biol. Chem.* 142:431–440.

Zapol, W. M., Liggins, G. C., Schneider, R. C., Qvist, J., Snider, M. T., Creasy, R. K., and Hochachka, P. W. (1979) Regional blood flow during simulated diving in the conscious Weddell seal. *J. Appl. Physiol. Respir. Environ. Exercise Physiol.* 47:968–973.

12 Scaling limits of metabolism to body size: implications for animal design

C. RICHARD TAYLOR

The metabolic rate of organisms changes in a regular manner with body size. This phenomenon is so consistent that Knut Schmidt-Nielsen in his text, *Animal physiology* (1979), suggests it represents a general biological rule. He has pointed out that at rest each gram of shrew tissue consumes energy at a rate some 100-fold that of a gram of elephant tissue. These large differences in mass-specific metabolic rate have important implications for how animals are built and what they can do. In this paper I shall examine several relationships between metabolic rate and body size in adult mammals and the implications of these relationships for structure and function.

Scaling as a tool for studying structure and function

The design of structures should change dramatically in order to accommodate the 100-fold changes in mass-specific metabolic rates that occur within mammals. The study of how structures and/or functions scale with body mass is called allometry, and it provides a powerful tool for understanding the design constraints under which animals are built. In allometry, one calculates the power function that describes how any structural or functional parameter, Y, changes with body mass, M_b:

$$Y = a \cdot M_b{}^b$$

where the exponent b is called the scaling factor. In order to apply statistics to the allometric equation it is convenient to use the logarithmic transformation:

$$\log Y = \log a + b \cdot \log M_b$$

where b becomes the slope of the linear regression plot. This transformation enables one to calculate regression coefficients and confidence limits.

Two conditions must be met in order for allometry to yield meaningful results. First, the range of body mass and number of species must be great enough to discriminate between variation in a parameter that occurs as a result of adaptations to different environments and/or differences in behavior among animals of the same size, and variations that occur as a result of changing body mass. The 95% confidence limits of the scaling factor enable one to decide whether or not the range of M_b has been large enough. The second condition is that errors in measurement that change systematically with changing size must be avoided. These types of errors can easily creep in because the techniques one uses to make a measurement on a 10-g shrew may be very different from those used to measure the same parameter on a 10 000-kg elephant.

What metabolic rates does one compare?

The metabolic rate of a cheetah or a gazelle sprinting at top speed has been estimated to be 50 to 100 times greater than its resting metabolic rate (Taylor, 1974). Thus, the change in metabolic rate that occurs within a gram of tissue may be as great as the differences in metabolic rate that occur with changing size in mammals. Obviously it is important to compare metabolic rates of animals under similar conditions when studying how rates change with body size. The lower and upper limits should be the most important metabolic parameters setting constraints on animal design. This chapter defines three metabolic limits that should be useful for comparisons: a lower limit, resting metabolism ($\dot{V}_{O_2std}$); an upper limit for oxidative metabolism ($\dot{V}_{O_2max}$); and a peak metabolism that includes maximal simultaneous rates for both oxidative and anaerobic metabolism ($\dot{E}_{peak}$).

The lower limit: resting metabolism

By definition, basal metabolism is the lower limit to metabolism, but what one measures is not really basal. Resting metabolism ($\dot{V}_{O_2std}$) can easily vary two- to threefold without any visible outward signs. Max Kleiber (1961) dealt with this problem by defining a *standard metabolic rate* as the "rate of fasting katabolism." Measuring this parameter is complicated by the fact that different animals require different amounts of time after a meal to reach fasting katabolism. Kleiber (1961) states that fasting rates are achieved "12 hours after last feeding for rats, 24 hours for rabbits, and 4 days for cows. If the measurement is carried out long enough – 3 hours in our trials with rats . . . then the metabolic rate represents reasonably well an average condition." It should be noted that large animals take longer after a meal to reach a fasting rate of katabolism than small animals. Thus one could introduce a systematic error into allometric studies of $\dot{V}_{O_2std}$ by using the same period for fasting (e.g., 2 h after the last meal) for animals over the entire size range.

The upper limit for aerobic metabolism

Two procedures have commonly been used for measuring the upper limit for aerobic metabolism ($\dot{V}_{O_2max}$): exercise or exposure to very cold environments (see Lechner, 1978).

The cold-exposure technique involves measuring $\dot{V}_{O_2}$ when an animal is placed in an environment that is so cold that aerobic metabolism cannot be increased sufficiently to maintain a constant body temperature. A modification of this method involves substituting a helium–oxygen mixture for air (Rosenmann and Morrison, 1974). This mixture has a much higher thermal conductivity than air and effectively creates a much "colder environment," decreasing the insulation of the fur by increasing the thermal conductivity of the gas trapped in the fur. A major disadvantage of this procedure is that it gives significantly lower values (16–23%) for $\dot{V}_{O_2max}$ than those obtained during exercise (Seeherman et al., 1981). Also, it can only be used with relatively small mammals (approx. 1 kg) because it is difficult to create cold enough environments to elicit $\dot{V}_{O_2max}$ for larger animals. Using the cold-exposure procedure for small animals and the exercise procedure for large animals would result in an error in the allometric relationship yielding an incorrect scaling factor.

The exercise technique for measuring $\dot{V}_{O_2max}$ was developed for humans by Margaria and his colleagues (Margaria et al., 1933; Margaria, 1976). Their subjects ran on a treadmill (level and/or inclined), and the rate of oxygen consumption ($\dot{V}_{O_2}$) was measured for sustained exercise at a constant speed as a function of treadspeed. $\dot{V}_{O_2}$ increased with increasing speed up to a maximal rate of $\dot{V}_{O_2}$ ($\dot{V}_{O_2max}$). Further increases in speed resulted in an increased rate of utilization of metabolic energy by the muscles without an increase in $\dot{V}_{O_2}$. The additional energy was supplied to the muscles by anaerobic glycolysis, resulting in an accumulation of lactic acid and R values exceeding 1.0. Exercise could be sustained at these intensities until lactic acid concentration reached 20 to 30 mM $\cdot$ kg^{-1}. It has been shown by many investigators that $\dot{V}_{O_2max}$ for an individual human subject is approximately the same for a variety of exercises and independent of the ambient temperature at which the exercise is performed (e.g., running, bicycling, rowing, cross-country skiing, swimming, climbing stairs, etc.) (Åstrand and Rodahl, 1977).

Seeherman et al. (1981) found that it was possible to obtain $\dot{V}_{O_2max}$ during treadmill exercise for mammals ranging in body mass from 7 g to over 100 kg using this treadmill procedure (Figure 12.1). An important advantage of this procedure over the cold-exposure method is that it can be used over a wide range of body sizes. The major disadvantage is that it is often difficult and time-consuming to train animals to perform at exercise intensities where $\dot{V}_{O_2max}$ is reached.

The peak limit for metabolism, aerobic and anaerobic

Techniques have not yet been devised for measuring the simultaneous peak rates of aerobic and anaerobic metabolism ($\dot{E}_{peak}$). The problems associated with such measurements are difficult. Peak metabolism will

FIGURE 12.1 Mass-specific oxygen consumption $\dot{V}_{O_2}/M_b$ (solid circles and solid lines) and rate of increase in blood lactic acid concentration averaged over the duration of a run (open circles and dashed lines) plotted as a function of running speed for nine species of mammals. $\dot{V}_{O_2}$ increased linearly up to a maximum rate ($\dot{V}_{O_2max}$) and then remained unchanged as the additional energy utilized for further increases in speed was provided by anaerobic glycolysis ($\dot{L}actate_{(exercise-rest)}$). The scales are drawn so that the rates of ATP provided by the rate of oxygen consumption on the left-hand side approximately equal rates of ATP provided by the rate of increase in lactate concentration on the right by assuming a P/O_2 ratio of 6 and a $P/lactate$ ratio of 1.5 (i.e., 1 ml $O_2 \cdot s^{-1} \cdot kg^{-1}$ approximately equals 11 mM $lactate \cdot kg^{-1} \cdot min^{-1}$). The onset of significant rates of anaerobic glycolysis occurred only after $\dot{V}_{O_2max}$ had been reached. [From Seeherman et al., 1981]

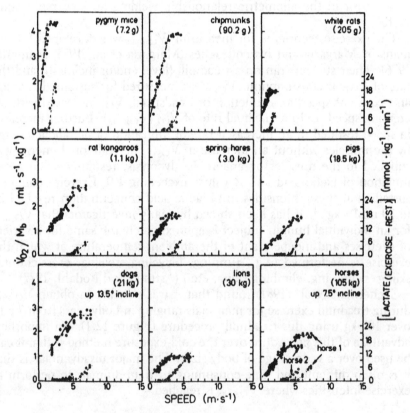

only be reached for short periods, and duration will change with body size. During these short periods, a large amount of the oxygen for aerobic metabolism may come from oxygen stored in hemoglobin and myoglobin and dissolved in the tissues. Creatine phosphate levels in the muscles will fall, yielding energy anaerobically. The lactate produced by anaerobic glycolysis is released slowly from the cells and will be used for aerobic metabolism after the peak activity stops. These problems represent a major challenge for innovative scientists as they develop techniques for measuring $\dot{E}_{peak}$.

How do metabolic limits scale with body size?

The lower limit

Knut Schmidt-Nielsen discusses the scaling of the lower limit ($\dot{V}_{O_2std}$) to body size in a most thoughtful and readable manner on pp. 183–190 of his text, *Animal physiology* (Schmidt-Nielsen, 1979). I shall deal with it very briefly, and I suggest that the reader review his discussion.

It was almost 50 years ago that Max Kleiber (1932) found that the rate of oxygen consumption measured under standardized resting conditions was proportional to $M_b^{0.75}$, instead of to body surface area or $M_b^{0.66}$ Kleiber's relationship has been substantiated by measurements on hundreds of species of mammals during the last 50 years (see Schmidt-Nielsen, 1979). Nonetheless, a simple rational explanation for this scaling factor of 0.75 has eluded physiologists.

Kleiber's equation for standardized resting oxygen consumption is:

$$\dot{V}_{O_2std} = 0.188 \cdot M_b^{0.75} \tag{1}$$

where $\dot{V}_{O_2}$ has the units milliliters of O_2 per second and M_b is body mass in kilograms.

The upper limit for aerobic metabolism

It has long been assumed that the upper limit ($\dot{V}_{O_2max}$) is some simple multiple of $\dot{V}_{O_2std}$, and therefore also proportional to $M_b^{0.75}$ (Wilkie, 1959; Hemmingsen, 1960; McMahon, 1975). However, the data base is not as well established as for $\dot{V}_{O_2std}$. A variety of methods have been used (e.g., maximal metabolic response to cold, exercise, or exercise in cold) (Pasquis et al., 1970; Lechner, 1978), and most of the measurements have been made on small mammals. There seemed good reason to question the scaling factor of 0.75 for $\dot{V}_{O_2max}$ because Ewald Weibel and his colleagues had found that the morphometrically determined diffusing capacity of lungs is scaled proportionally to $M_b^{1.0}$ (see Chapter 3). Ewald Weibel and I believe that the respiratory system of mammals is designed optimally, and that this applies to each step in the system from the oxygen store in environmental air to the oxygen sink in the mitochondria (Taylor and Weibel, 1981). It therefore seems that $\dot{V}_{O_2max}$ should be scaled proportionally to $M_b^{1.0}$ rather than to $M_b^{0.75}$. If the two scaling factors differ by 0.25, then large mammals would have much more diffusing capacity per unit oxygen flow at $\dot{V}_{O_2max}$ than small mammals, and this seems unreasonable.

We (Taylor and Weibel, 1981) decided to determine a reliable allometric relationship for $\dot{V}_{O_2max}$ and to measure diffusing capacity of the lungs on the same individual animals in which we measured $\dot{V}_{O_2max}$.

We measured $\dot{V}_{O_2max}$ on 22 wild and domestic species of mammals ranging in size from 7 to 263 kg. Our data (Taylor et al., 1981) confirmed that maximum rate of oxygen consumption, $\dot{V}_{O_2max}$, was a nearly constant multiple of 10 times resting metabolism, $\dot{V}_{O_2std}$, and scaled approximately proportionally to $M_b^{0.75}$ (Figure 12.2). The equation was:

$$\dot{V}_{O_2max} = 1.92\, M_b^{0.809} \tag{2}$$

where $\dot{V}_{O_2}$ has the units milliliters of O_2 per second, and M_b is in kilograms. The 95% confidence limits for the scaling factor were from 0.747 to 0.870. A comparison of equation 2 with Kleiber's equation for $\dot{V}_{O_2std}$ given in equation 1 shows no significant difference in the exponent, but approximately a 10-fold difference in the numerical constants.

Another important finding to emerge from our study was the enormous variability in $\dot{V}_{O_2max}$ among different species of the same body size. For example, we found that $\dot{V}_{O_2max}$ for horses and dogs was about 3.5 times greater than that for cattle and goats or sheep of the same M_b, respectively. These great differences in $\dot{V}_{O_2max}$ between animals of the same body size provide another important tool for the study of structure–function relationships (see Chapter 3).

Implications for animal design

Oxygen delivery system

It seems reasonable to suggest that the design of the oxygen delivery system at each level of organization from lung to mitochondria will be closely matched to the maximal rates of oxygen consumption. Thus if mass-specific rates of oxygen consumption are 100 times as great in a small mammal as in a large one, the mitochondria should be capable of using the oxygen at 100 times the rate, the circulatory system should be capable of delivering the oxygen from lung to tissue at 100 times the rate, and the lung should be capable of supplying the oxygen to the blood at 100 times the rate. Ewald Weibel (see Chapter 3) considers the data available for testing this hypothesis. He concludes that the close quantitative relationships one might reasonably expect between maximal flow of oxygen and structural parameters is more complex than we had anticipated. Further experiments are still needed to demonstrate a close link, and these are outlined by Weibel.

Locomotory system

It is generally assumed that the energetics and mechanics of the locomotory system are closely linked (i.e., maximum rates of energy consumption and maximal stresses in bones, tendons, and muscles both occur when an animal runs at top speed). If one also makes the reasonable assumption that the muscles transform energy consumed into mechanical work as the animal runs, and that this conversion takes place at close to some optimal efficiency, then one can develop allometric equations for the dimensions, forces, and energetics of mammalian locomotion.

FIGURE 12.2 Average values of $\dot{V}_{O_2}$ for 14 species of wild mammals (circles) and 7 species of laboratory/domestic animals (triangles) are plotted as a function of body mass on logarithmic coordinates. Two average values have been used for cattle because their M_b varied twofold. The allometric regression line drawn on the graph has been calculated for all animals using the least squares regression analysis and has a slope of 0.809. [From Taylor et al., 1981]

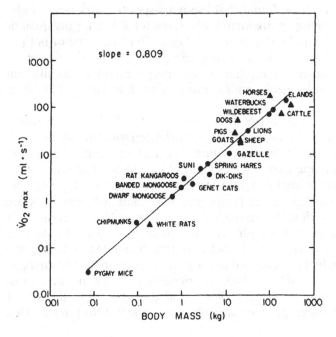

A. V. Hill (1950), T. A. McMahon (1975), and R. McN. Alexander (1977) have developed such allometric equations for describing animal loco-motion. Alexander reviews these allometric relationships in Chapter 21 of this volume. In this section I shall examine the assumption that en-ergetics and mechanics of locomotion are indeed closely linked.

Are the rates at which muscles of running animals consume energy and perform mechanical work closely related? It has been more than 10 years since Knut Schmidt-Nielsen, Jake Raab, and I observed that the energetic cost of locomotion increased linearly with speed and varied as a regular function of body mass (Taylor et al., 1970). The data base has expanded greatly in the last 10 years and has substantiated the initial observation. The mass-specific rate of energy expenditure for animals running at the same speed is proportional to $M_b^{-0.33}$ (Taylor, 1980). Thus each gram of muscle of a 30-g mouse is consuming energy at about 20 times the rate of a gram of muscle of a 300 000-g horse when both animals run at the same speed. Are the mouse muscles also performing me-chanical work at 20 times the rate?

Norman Heglund, Giovanni Cavagna, Michael Fedak, and I decided we needed to measure mechanical work rate directly to find out. We used a combination of high-speed film and force-plate analysis to deter-mine the increments in energy that occur within an animal during a stride. These increments have to be supplied by the muscle–tendon units (either by muscular work or by release of energy stored elastically). Much to our surprise, we have found that the mass-specific work rate is inde-pendent of size, being approximately the same for a 30-g pygmy quail or kangaroo rat and a 150 000-g ostrich or horse. Evidently one or both of our "reasonable" assumptions linking energetics and mechanics of lo-comotion are incorrect (i.e., the major energetic cost of locomotion occurs when muscles work, and/or muscles work at close to maximal efficiency).

How do muscles use energy during locomotion? If the rate at which muscles perform mechanical work is not correlated with the rate at which they utilize metabolic energy when animals run at a constant speed, what are the muscles doing that requires energy? During level running, mus-cles of a running animal generate force and consume energy not only while they are shortening and performing work, but also while they are stretched (e.g., to slow the animal's fall and reverse the direction of its limbs), and while their length remains unchanged (e.g., to stabilize joints). Work is performed on the active muscle when it is stretched, and no work is done by the active muscle when its length does not change.

We (Taylor et al., 1980) studied the energetic cost of generating mus-cular force of running animals. We found that the energetic cost of running at a particular speed was directly proportional to the force exerted by the muscles, and that the cost of generating force by a gram of muscle, like the cost of running, was proportional to $M_b^{-0.33}$. We have proposed that the higher energy cost of generating force in smaller animals is due to a faster intrinsic velocity of muscles (i.e., faster rate of cross-bridge cycling).

Muscular force must be generated and decay more rapidly as an animal increases its speed. This is accomplished by recruiting faster muscle fibers within a muscle, which have more rapid rates of cycling of their cross-bridges between actin and myosin. Each breaking of a cross-bridge consumes a unit of energy. Thus the linear increase in energy cost of locomotion with speed could be the result of recruiting faster fibers with faster cycling time.

Muscular force must also be generated and decay more rapidly in small animals than in large ones when they run at the same speed, because the small animal takes more steps per unit time to move at the same speed. Equivalent muscles of small animals have faster fibers with more rapid cross-bridge cycling rates than those of large animals. This decrease in rate of cross-bridge cycling with increasing body size could account for the scaling factor of -0.33 for energy cost of running. These ideas are currently being tested in several laboratories.

Conclusions

Four conclusions can be drawn from this chapter:

Both $\dot{V}_{O_2std}$ and $\dot{V}_{O_2max}$ of mammals scale about proportionally to the $\frac{3}{4}$ power of body mass, $\dot{V}_{O_2max}$ being approximately 10 times $\dot{V}_{O_2std}$. However, enormous interspecies variability was observed in $\dot{V}_{O_2max}$.

One would expect that at $\dot{V}_{O_2max}$ the systems for delivery of oxygen at each organizational level from lung to mitochondria should be operating close to their limits and therefore have the same scaling factors. Ewald Weibel discusses this hypothesis in Chapter 3.

Entirely different scaling factors apply for energetic cost and the mechanical work of running. Therefore our assumptions linking the mechanics and energetics of locomotion appear to be incorrect.

Finally, allometry is seen to be a powerful tool for testing assumptions about the design of animals.

References

Alexander, R. McN. (1977) Terrestrial locomotion. In *Mechanics and energetics of terrestrial locomotion* (R. McN. Alexander and G. Goldspink, eds.), pp. 168–203. London: Chapman and Hall.

Åstrand, P., and Rodahl, K. (1977). *Textbook of work physiology*. New York: McGraw Hill. 681 pp.

Hemmingsen, A. M. (1960) Energy metabolism as related to body size and respiratory surfaces, and its evolution. *Rep. Steno Mem. Hosp.* 4:1–110.

Hill, A. V. (1950) The dimensions of animals and their muscular dynamics. *Sci. Prog.* 38:209–230.

Kleiber, M. (1932) Body size and metabolism. *Hilgardia* 6:315–353.

Kleiber, M. (1961) *The fire of life: An introduction to animal energetics.* New York: Wiley. 454 pp.

Lechner, A. J. (1978) The scaling of maximal oxygen consumption and pulmonary dimensions in small mammals. *Respir. Physiol.* 34:29–44.

McMahon, T. A. (1975) Using body size to understand the structural design of animals: Quadrupedal locomotion. *J. Appl. Physiol.* 39:619–627.

Margaria, R. (1976) *Biomechanics and energetics of muscular exercise.* Oxford: Clarendon Press. 146 pp.

Margaria, R., Edwards, H. T., and Dill, D. B. (1933) The possible mechanism of contracting and paying the oxygen debt and the role of lactic acid in muscular contraction. *Am. J. Physiol.* 106:689–715.

Pasquis, P., Lacaisse, A., and Dejours, P. (1970) Maximal oxygen uptake in four species of small mammals. *Respir. Physiol.* 9:298–309.

Rosenmann, M., and Morrison, P. (1974) Maximum oxygen consumption and heat loss facilitation in small homeotherms by He–O_2. *Am. J. Physiol.* 226:490–495.

Schmidt-Nielsen, K. (1979) *Animal physiology: Adaptations and environment.* Cambridge University Press. 560 pp.

Seeherman, H. J., Taylor, C. R., Maloiy, G. M. O., and Armstrong, R. B. (1981) Design of the mammalian respiratory system: Measuring maximum aerobic capacity. *Respir. Physiol.* 44:11–24.

Taylor, C. R. (1974) Exercise and thermoregulation. In *MTP international review of science. Environmental physiology*, series one, vol. 7 (D. Robertshaw ed.), pp. 163–184. London: Butterworths.

Taylor, C. R. (1980) Running machines: Muscles in living animals. In *Proc. Intl. Union Physiol. Sci.*, XXVIII Intl. Cong., Budapest. 14:24–42. Budapest: Akadémiai Kiadó.

Taylor, C. R., Heglund, N. C., McMahon, T. A., and Looney, T. R. (1980) Energetic cost of generating muscular force during running: A comparison of large and small animals. *J. Exp. Biol.* 86:9–18.

Taylor, C. R., Maloiy, G. M. O., Weibel, E. R., Langman, V. A., Kamau, J. M. Z., Seeherman, H. J., and Heglund, N. C. (1981) Design of the mammalian respiratory system: Scaling maximum aerobic capacity to body mass – wild and domestic mammals. *Respir. Physiol.* 44:25–38.

Taylor, C. R., Schmidt-Nielsen, K., and Raab, J. L. (1970) Scaling of energetic cost of running to body size in mammals. *Am. J. Physiol.* 219:1104–1107.

Taylor, C. R., and Weibel, E. R. (1981) Design of the mammalian respiratory system: Problem and strategy. *Respir. Physiol.* 44:1–10.

Wilkie, D. R. (1959) The work output of animals: Flight by birds and by manpower. *Nature* 183:1515–1516.

PART THREE

Temperature

Overview

Part Three considers temperature regulation in vertebrates.

In Chapter 13, Jürgen Aschoff looks at how the 24-hr cycles in body temperature of birds and mammals change with body size. He finds that the cycles change in a regular manner with body size. He then considers how heat production and heat loss vary during the day to produce the observed cycles in body temperature.

In Chapter 14, Marvin Bernstein discusses temperature regulation of exercising birds. He finds that high body temperature, separate control of brain temperature, and heat loss from extremities play important roles in enabling birds to dissipate the 10-fold increase in their heat production during exercise.

In Chapter 15, Eugene Crawford considers the interaction of behavioral and physiological mechanisms used by terrestrial ectotherms to regulate their body temperature. He uses this information to speculate about the evolution of temperature regulation in vertebrates.

In Chapter 16, Frank Carey discusses temperature regulation in fish. He considers how and why fish elevate their body temperature above that of the surrounding water. Records of body temperature of fish in nature obtained by telemetry provide important insights into the phenomenon of "warm-blooded" fish.

13 The circadian rhythm of body temperature as a function of body size

The body temperature of homeothermic animals is characterized by a stabile long-term mean and by regular 24-h (circadian) variations around that mean. Owing to these properties, body temperature is one of the typical examples of two major principles in the makeup of living systems: homeostasis and circadian temporal organization. Many of the processes that operate in support of homeostatic mechanisms show a strong dependence on body size, as is especially well documented for the production of heat (Benedict, 1938; Kleiber, 1947). The question is, To what extent and how do body temperature and its circadian pattern depend on body size?

The circadian rhythm of body temperature has been measured in more than 100 avian and mammalian species, since its first documentation in man by Gierse in 1843 and in pigeons by Chossat in 1844 (for references, see Aschoff, 1955). A similar amount of data is available on the circadian variation of oxygen uptake. Hence it is possible to make a comparative analysis of the rhythms in body temperature and in heat production. In this analysis, emphasis is placed on the range of oscillation in these rhythms, measured as the difference between maximal and minimal values within one circadian period. It will be shown that the range of oscillation has a similar negative correlation with body weight in both these functions. Heat transfer from the body to the environment, as measured by thermal conductance, is examined to interpret this dependence, and to shed light on the interrelationship between heat production and body temperature.

Data collection and problems of a comparative analysis

Figure 13.1 shows circadian temperature rhythms of birds and mammals from three and four weight classes, respectively. Two trends can be seen: (1) The range of oscillation decreases with increasing body weight; (2) the maximal values are at about the same level in all weight classes, but the minimal values increase from small to large species. Consequently, the 24-h mean of temperature seems to be positively correlated with body weight.

173

The examples of Figure 13.1 were deliberately selected to indicate the supposed dependence on body weight. Although several published curves do not agree with this picture, it corresponds to a statistical trend for which evidence will be given in this chapter. All data are taken from publications, and animals were studied under dissimilar conditions and with different techniques. This renders a comparative analysis difficult, because experimental conditions and measuring devices can have drastic effects on the circadian pattern. The following are a few of the factors that especially influence the range of oscillation.

1. The range of oscillation is usually larger in animals exposed to a light–dark cycle (LD) as compared to animals kept in continuous light (LL) or continuous darkness (DD) (see Pohl, 1970; Aschoff et al., 1973; Spencer et al., 1976; Fuller et al., 1978).
2. The range of oscillation depends on ambient temperature. For the rhythm of oxygen uptake in the evening grosbeak, *Hesperiphona vespertina*, a range of 2.65 ml O_2 g^{-1} h^{-1} has been found at 10 °C, but only of 0.98 ml O_2 g^{-1} h^{-1} at 35 °C (West and Hart, 1966). More

FIGURE 13.1 Circadian rhythms of body temperature in birds and mammals from different classes of body weight. [For references see Aschoff 1981a (birds) and Aschoff 1981b (mammals).] Inset: temperature profiles in the rectum. [For references see Figure 5 in Aschoff, 1981b.]

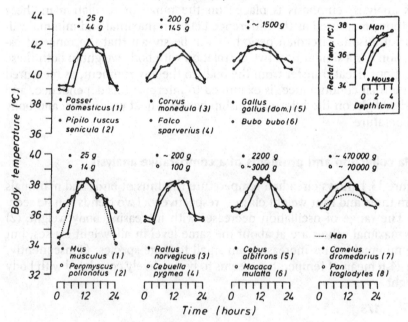

examples of this kind are provided elsewhere (see Aschoff, 1981a, for birds; Aschoff, 1981b, for mammals).

3. The state of nutrition, the number of animals in the cage, and whether nesting material is provided or not, can influence the range of oscillation (see Figures 3 and 4 in Aschoff, 1981b).

4. Absolute values and also ranges of oscillation depend on the site (the depth in the rectum) where the temperature is taken, because within the shell of a homeothermic organism, temperature increases from the surface to the core of the body in a somewhat parabolic fashion (Figure 13.1, inset).

5. Whenever measurements are made by hand, it is likely that the temperature will increase, owing to the excitement of the animal. Such effects are most pronounced when the animal is at rest, and the temperature is at a low level (Miles, 1962). Because of this dependence on circadian phase, increments in temperature based on handling result in a decrease of the range of oscillation (see Figure 6 in Aschoff, 1981b). The smaller the animal is (and possibly more excitable), the more this "flattening" of the temperature curve must be taken into account.

In view of the many interfering factors listed, one might doubt that a correlation could be established between the range of oscillation and body weight on the basis of data obtained in different laboratories under a variety of conditions. Thus the results described in the following section are all the more noteworthy. The analysis has been restricted to eutherian animals, with the exclusion of Marsupialia and Chiroptera as well as of data obtained during states of evident torpor.

Results

Body temperature

In Figure 13.2, ranges of oscillation in the rhythm of body temperature are graphed as a function of body weight with logarithmic scales on both coordinates. For the mammals, a negative correlation is clearly indicated within a range from about 10 to 10 000 g. In the higher weight classes, the range either remains unchanged or increases again. The nonpasserine birds show a dependence similar to that of the mammals. The dependence is only marginally significant in the passerine birds ($p < 0.05$), but the ranges of oscillation as shown in the diagram may not reflect the "true" ranges because of handling effects. Passerine birds may simply be more excitable than nonpasserine birds (see the discussion in the foregoing section, list item no. 5). Up to a weight of 10 000 g the data can be approximated by regression lines that are represented by the following equations:

Nonprimates: $\qquad \log R_T = \log\ 4.762 - 0.197 \log W \qquad$ (1)

Primates: $\qquad \log R_T = \log 16.333 - 0.289 \log W \qquad$ (2)

Nonpasserines: $\qquad \log R_T = \log 10.856 - 0.396 \log W \qquad$ (3)

Passerines: $\qquad \log R_T = \log\ 4.182 - 0.125 \log W \qquad$ (4)

In these equations, R_T represents the range of oscillation in body temperature (°C) and W the body weight (g). The numbers inscribed in Figure 13.2 refer (as in the following figures) to the two coefficients in

FIGURE 13.2 Range of oscillation in the rhythm of body temperature in birds and mammals, drawn as a function of body weight. LD: light–dark cycle; DD: continuous darkness. a and b: coefficients in the regression equation $\log y = \log b - a \cdot \log x$. [For references see Aschoff, 1981a (birds), and Aschoff, 1981b (mammals).]

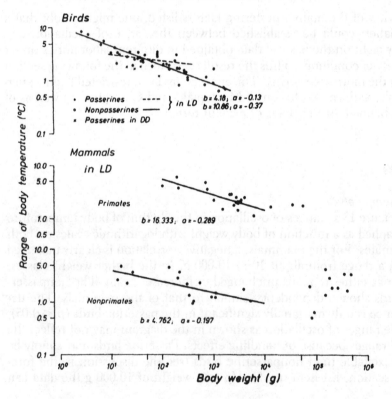

the general regression equation $\log y = \log b - a \cdot \log x$, in which the coefficient a indicates the slope of the regression line.

It should be noticed that all data used for computing regression equations were obtained from animals kept in LD. The few data taken from passerine birds in DD (crosses) are lower than the LD data, in accordance with the expectation (see the foregoing section, list item no. 1).

In summary, the LD data show that, apart from the passerine birds, the range of oscillation in body temperature is 2.5 to 5.5 times larger in an animal of 10 g weight than in a 1000-g animal.

Figure 13.3 shows the temperatures measured either during the activity time α or the rest time ρ of the animals. In agreement with Figure 13.1, the α-temperatures are at about the same level in the nonpasserine birds and the mammals (nonprimates) of all weight classes; the ρ-temperatures increase from small to large species. In the passerine birds the indicated negative correlation between α-temperatures and body weight is not significant. The overall mean of α-temperatures is 38.36 °C for the non-

FIGURE 13.3 Body temperature in birds and nonprimate mammals, measured during the activity time α and the rest time ρ, drawn as a function of body weight. [References as in Figure 13.2]

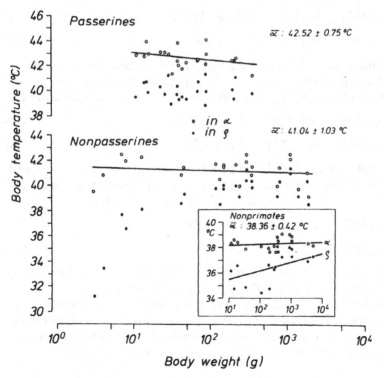

primates (primates: 38.61 °C), 41.04 °C for the nonpasserines, and 42.52 °C for the passerines.

Oxygen uptake

Oxygen consumption per unit of body weight of mammals is negatively correlated with body weight. This correlation can be described by the equation

$$\log M/W = \log 5.26 - 0.29 \log W \tag{5}$$

in which M is oxygen uptake (ml h^{-1}) and W body weight (g) (Kleiber, 1947). In this general formulation no reference is made to the fact that oxygen uptake has a circadian rhythm. Different results might be expected when measurements are made either during the daytime or at night (Aschoff and Pohl, 1970a). Hildwein (1972) has recorded the oxygen consumption of starved and resting mammals over 24 h in dim LL. His data can be grouped according to whether measurements have been made during the animal's activity time, α, or its rest time, ρ. If these data are plotted on a graph with logarithmic scales on both coordinates (Figure 13.4C, left), two regression lines can be drawn, which are represented by the following equations:

$$\alpha: \quad \log M/W = \log 8.32 - 0.34 \log W \tag{6}$$

$$\rho: \quad \log M/W = \log 7.38 - 0.40 \log W \tag{7}$$

The α-values are 50% larger than the ρ-values. The difference amounts to 80% for data taken from mammals (nonprimates) that were not starved and were kept in LD (Figure 13.4B, left). The regression equations are:

$$\alpha: \quad \log M/W = \log 20.00 - 0.47 \log W \tag{8}$$

$$\rho: \quad \log M/W = \log 13.34 - 0.51 \log W \tag{9}$$

Finally, data are available from primates that were measured over 24 h in dim light except one measurement in LD (Figure 13.4A, left). The regression lines are represented by the following equations:

$$\alpha: \quad \log M/W = \log 14.20 - 0.49 \log W \tag{10}$$

$$\rho: \quad \log M/W = \log 7.88 - 0.44 \log W \tag{11}$$

As in the case of the nonprimates kept in LL (diagram C), the α-values of the primates are about 50% larger than the ρ-values.

The differences between paired α- and ρ-values can be considered an approximation of the range of oscillation. As a consequence of the parallel course of the two regression lines in each of the three diagrams, the range decreases steadily with increasing body weight (Figure 13.4, right three diagrams), as indicated by the following equations:

Primates: $\log R_{M/W} = \log 7.80 - 0.61 \log W$ (12)

Nonprimates in LD: $\log R_{M/W} = \log 5.69 - 0.40 \log W$ (13)

Nonprimates in LL: $\log R_{M/W} = \log 2.84 - 0.34 \log W$ (14)

According to these equations, in which R_M represents the range of oscillation in oxygen uptake (ml h^{-1}) and W body weight (g), a mammal of 10 g weight has a 4.5 to 6.5 times larger range in oxygen uptake than a 1000-g mammal.

The dependence of oxygen consumption on body weight in birds can also be described by two parallel regression lines representing measure-

FIGURE 13.4 Weight-specific oxygen consumption drawn as a function of body weight. Left: oxygen uptake measured during the activity time α and the rest time ρ. Right: range of oscillation. LD: light–dark cycle; LL: continuous dim illumination. a and b: coefficients in the regression equation $\log y = \log b - a \cdot \log x$. [For references see Aschoff, 1981b]

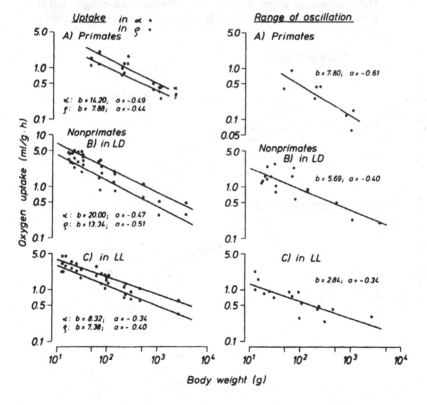

ments made in a, α, and ρ (Aschoff and Pohl, 1970b). Ranges of oscillation computed from those data are plotted in Figure 13.5. On the average, passerine birds have a 75% larger range than nonpasserine birds. Hence, the regression equations differ in the coefficient b, but they have the same coefficient a (i.e., the same slope):

Passerines: $\log R_{M/W} = \log 3.105 - 0.385 \log W$ (15)

Nonpasserines: $\log R_{M/W} = \log 1.721 - 0.385 \log W$ (16)

According to these equations, a bird of 10 g weight has a 5.5 times larger range in oxygen uptake than a 1000-g bird.

In summary it can be stated that in both birds and mammals, the range of oscillation in body temperature and in oxygen uptake depend in a similar way on body weight. This observation could mean that the rhythm in oxygen uptake (i.e., in heat production) causes the rhythm in body

FIGURE 13.5 Range of oscillation in the circadian rhythm of oxygen uptake in birds, drawn as a function of body weight. LD: light–dark cycle; DD: continuous darkness. a and b: coefficients in the regression equation $\log y = \log b - a \cdot \log x$. [For references see Aschoff, 1981a]

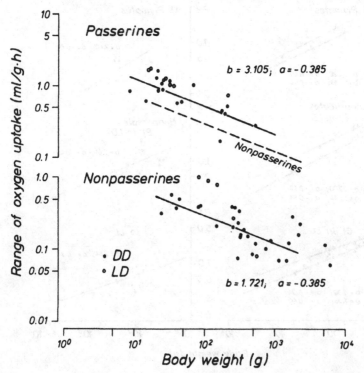

temperature. Such a conclusion, however, is never justified if based only on the argument that two functions follow a similar circadian pattern. Furthermore, the temperature at a given site in the body results from two factors: the heat produced in (or transported into) that tissue, and the conditions for heat transfer from that tissue to the environment. Because nearly all functions within the organism are programmed in a circadian fashion, one has to assume that the conditions for heat transfer also show circadian variations and hence participate in determining the circadian rhythm of body temperature.

Conductance

The heat produced in an organism flows along a temperature profile from the core of the body to the environment. If the amount of heat dissipated per unit of time is divided by the difference between core and ambient temperature, one gets thermal conductance, usually expressed in watts per degree Celsius (W/°C). Thermal conductance is a measure of the ease by which heat is transferred to the environment; it comprises all channels of heat dissipation, including the respiratory tract and evaporative heat loss. Often conductance is expressed in units of oxygen uptake; the dimension then is milliliters O_2 per hour per degree Celsius (ml O_2/h · °C), or for weight-specific conductance milliters O_2 per gram per hour per degree Celsius (ml O_2/g · h · °C). The organism has several means to change conductance, including physiological functions (e.g., circulation and respiration) as well as behavioral patterns (e.g., body posture or ruffling of feathers).

In a warm environment, conductance has large values to facilitate heat loss. If the organism is exposed to a stepwise lowering of ambient temperature, conductance decreases and eventually reaches a minimal value (= maximal insulating power), which, in the ideal case, remains unchanged when the ambient temperature is further lowered. (For the principles involved, see Scholander et al., 1950.) *Minimal* conductance has been determined for many mammalian and avian species. To allocate those data to either the activity time α or the rest time ρ of the animal, one has to find out from the publication whether the measurement has been made during daytime or at night, and has to take into account whether the species tested is diurnal or nocturnal. Data extracted in this way from the literature are summarized in Figures 13.4 and 13.7. The diagrams document for mammals and birds a negative correlation between minimal conductance and body weight. This dependence has been described by several authors (Herreid and Kessel, 1967; Lasiewski et al., 1967; McNab, 1970, 1979; Bradley and Deavers, 1980), but they do not refer to the fact that minimal conductance is subject to circadian variation (Aschoff and Pohl, 1970a). Figures 13.6 and 13.7, show the α-values of

FIGURE 13.6 Minimal conductance of mammals measured during the activity time α and the rest time ρ, drawn as a function of body weight. a and b: coefficients in the regression equation log y = log b − a · log x. [For references see Aschoff, 1981c]

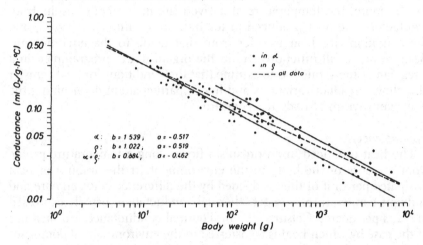

FIGURE 13.7 Minimal conductance of birds measured during the activity time α and the rest time ρ, drawn as a function of body weight. Inset: α- and ρ-values measured in the same individuals. a and b: coefficients in the regression equation log y = log b − a · log x. [For references see Aschoff, 1981c]

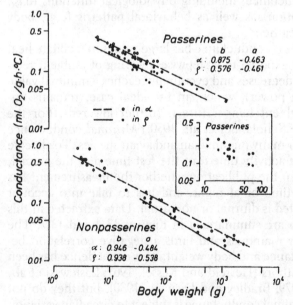

conductance are about 50% larger than the ρ-values. In other words, the circadian rhythm in conductance has a range of oscillation equal to that in heat production. The three sets of data can be described by the following regression equations, in which R_C represents the range of oscillation in minimal conductance (ml $O_2/h \cdot {}^\circ C$) and W body weight (g):

Mammals:
$\quad\quad\quad\quad$ α: $\quad$ $\log R_{C/W} = \log 1.539 - 0.517 \log W$ $\quad$ (17)

$\quad\quad\quad\quad\quad\quad\quad$ ρ: $\quad$ $\log R_{C/W} = \log 1.022 - 0.519 \log W$ $\quad$ (18)

Passerine birds:
$\quad\quad\quad$ α: $\quad$ $\log R_{C/W} = \log 0.857 - 0.463 \log W$ $\quad$ (19)

$\quad\quad\quad\quad\quad\quad\quad$ ρ: $\quad$ $\log R_{C/W} = \log 0.576 - 0.461 \log W$ $\quad$ (20)

Nonpasserine birds:
$\quad$ α: $\quad$ $\log R_{C/W} = \log 0.946 - 0.484 \log W$ $\quad$ (21)

$\quad\quad\quad\quad\quad\quad\quad$ ρ: $\quad$ $\log R_{C/W} = \log 0.947 - 0.583 \log W$ $\quad$ (22)

The slope of the curve representing the ρ-values of nonpasserine birds is least reliable because of the scarcity of data at the upper end of the weight scale. Otherwise, the regression equations may be considered representative.

In Figure 13.6, a third (dashed) line is drawn. It represents the dependence of minimal conductance on body weight if the computation is made for all data together, without a distinction between α- and ρ-values. Such a procedure results in slope for the regression line that differs substantially from the "true" slope. The reason is that α- and ρ-values (or nocturnal and diurnal species, respectively) are not equally distributed over all weight classes. (For a more detailed discussion of this problem, see Aschoff 1981c).

Several conclusions can be drawn from Figures 13.6 and 13.7:

1. The dependence of minimal conductance on body weight is somewhat stronger in mammals (average slope -0.587) than in birds (average slope -0.47 if the ρ-data of the nonpasserines are discarded).
2. On the average, minimal conductance is about 35% larger in mammals than in birds, a finding that confirms a conclusion drawn earlier by McNab (1966). The difference in conductance between mammals and birds decreases with increasing body weight (compare the different slopes of the regression lines).
3. In mammals and in birds, minimal conductance has a circadian rhythm that facilitates heat loss during the activity time and supports heat conservation during the rest time of the animal.

It should be noted that within the zone of metabolic thermoneutrality where conductance increases with increasing ambient temperatures, the difference between α- and ρ-values becomes smaller but still remains

significant (see Figure 8 in Aschoff 1981a). Hence, a circadian variation
in the conditions for heat transfer has to be taken into account in warm
as well as in cold environments. The conclusion then is that the circadian
rhythm of body temperature can only be explained as a combined effect
of the rhythm in heat production and the rhythm in heat loss which in
itself is to a large extent determined by the rhythm in conductance.

The interplay between the components: a theoretical approach

Transport of heat needs time. This time factor, together with the heat
capacity of the tissue, results in a time lag between heat production and
heat loss. In circadian terms this means that the rhythms of heat production and heat loss are out of phase with each other. The phase relationship between these rhythms determines, among other factors,
phase and range of oscillation of the rhythm in body temperature. The
temperature rises as long as more heat is produced than lost, and the
temperature starts to decrease when heat production falls below heat
loss. The situation is schematically illustrated in Figure 13.8. Models II
to IV show increasing phase lags between heat loss (HL) and heat production (HP), indicated by the phase-angle difference $\psi_{HL/HP}$. In all three

FIGURE 13.8 Hypothetical schemes of the interaction between the
rhythms of heat production (HP), heat loss (HL), body temperature (BT),
and conductance (C). ψ: phase angle difference between various
rhythms, given in angular degrees. W = watt.

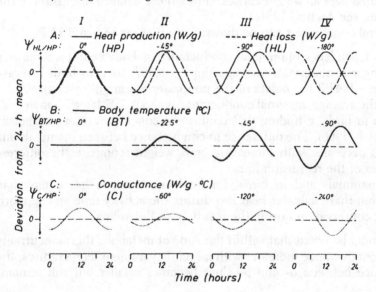

models, body temperature (BT) reaches its daily maximum when the descending slope of the HP curve crosses through the ascending slope of the HL curve. The range of oscillation in BT increases as $\psi_{HL/HP}$ gets larger. Conductance can be computed from the amount of heat lost at a given time, and the corresponding difference between BT and ambient temperature. As shown in the third row of Figure 13.8, the phase-angle difference between the rhythms of conductance and heat production, $\psi_{C/HP}$, differs from $\psi_{HL/HP}$ as well as from $\psi_{BT/HP}$. Notice that $\psi_{HL/HP}$ and $\psi_{C/HP}$ keep a constant ratio to each other; in all three models, $\psi_{C/HP}$ is 33% larger than $\psi_{HL/HP}$. Model I is unrealistic insofar as it is based on the assumptions that no time lag exists between HL and HP; in this case, body temperature remains constant throughout 24 h, which is only possible if conductance has a rhythm in phase with that of the two other rhythms. Another theoretically possible but also unrealistic case would be a model in which conductance remains constant throughout 24 h; in such a model, the rhythm of body temperature must be in phase with that of heat loss, both rhythms being out of phase with the rhythm of heat production by $-60°$ (see Figure 9 in Aschoff, 1981a).

It must be emphasized that the simplistic models drawn in Figure 13.8 are based on several assumptions that cannot be discussed here in detail. They serve to illustrate a few principles in the interaction of the various rhythmic functions, and they refer to one given body mass only. Hence, the effects of changes in the phase-angle differences on body temperature cannot be applied to organisms of different body size. To give one example: If an increase in body size results in an increase of $\psi_{HL/HP}$, as one might expect, the range of oscillation in body temperature should according to Figure 13.8 also increase, in strong contrast to the findings summarized in Figure 13.2. This apparent contradiction has to be explained by interfering effects of the rhythm in conductance. Of this rhythm we know that its range of oscillation decreases with increasing body weight (Figures 13.6 and 13.7), and it does so to a larger extent than the range of heat production (Figures 13.4 and 13.5); on the other hand, we do not know its exact phase relationship to other rhythms. Without such additional knowledge, the dependence on weight of the rhythm of body temperature cannot readily be explained. The assumption, however, seems justified that the increases in the range of body temperature seen in the very high weight classes (see Figure 13.2, nonprimates) is at least partly due to an increase in $\psi_{HL/HP}$.

To test whether the principles outlined in Figure 13.8 apply to a true homeothermic organism, it is necessary to measure continuously and simultaneously heat production, heat loss, and body temperature. Such data are not available from animals. However, records of that kind were obtained from eight human subjects, exposed in the nude to an ambient temperature of 28 °C for 24 h (Schmidt, 1972; Aschoff et al., 1974). The means of the hourly data, plotted on a 24-h abscissa, can be approximated by sine functions. The resulting curves are drawn in Figure 13.9. As indicated by the upper dashed line, the rhythm of rectal temperature reaches its maximum when the curves of heat production and heat loss cross through each other. It is further shown that heat loss lags behind heat production by −21° (see the two solid arrows). Finally, the rhythm in conductance has a phase-angle difference of −33° from the rhythm of heat production. These ψ-values are only about half as large as the corresponding ψ-values of model II in Figure 13.8, which is not of much relevance; what matters is that they maintain the same ratio to each

FIGURE 13.9 Circadian rhythms of rectal temperature, heat production (HP) and heat loss (HL), and conductance (C), measured in human subjects at an ambient temperature of 28 °C. The curves represent the best-fitting sine functions adapted to the mean values from eight subjects. ψ: phase-angle difference between various rhythms, given in angular degrees. [Data from Schmidt, 1972]

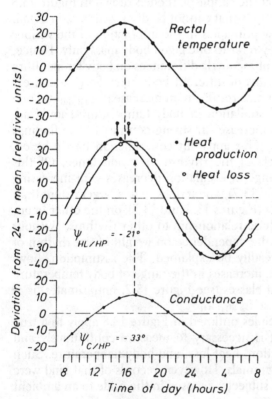

other. In principle, then, these findings are in agreement with the theoretical concept developed in Figure 13.8. It remains to be seen whether by the simultaneous use of direct and indirect calorimetry, similar patterns can be documented for animals. The results hopefully will answer the question of to what extent the rhythms of heat production and of conductance contribute to the rhythm of body temperature, and in what way the mutual relationship between the rhythms depends on body weight.

References

Aschoff, Ch., Aschoff, J., and Saint Paul, U. v. (1973) Circadian rhythms of chicken brain temperature. *J. Physiol.* 230:103–113.

Aschoff, J. (1955) Der Tagesgang der Körpertemperatur beim Menschen. *Klin. Wschr.* 33:545–551.

Aschoff, J. (1981a) Der Tagesgang der Körpertemperatur von Vögeln als Funktion des Körpergewichtes. *J. Ornithol.* (In press.)

Aschoff, J. (1981b) Der Tagesgang der Körpertemperatur und des Energieumsatzes bei Säugetieren als Funktion des Körpergewichtes. *Z. Säugetierk.* (In press.)

Aschoff, J. (1981c) Thermal conductance in mammals and birds: its dependence on body size and circadian phase. *J. Comp. Physiol.* (In press.)

Aschoff, J., Biebach, H., Heise, A., and Schmidt, T. (1974) Day night variation in heat balance. In *Heat loss from animals and man* (J. L. Monteith and L. F. Mount, eds.) pp. 147–172. London: Butterworths.

Aschoff, J., and Pohl, H. (1970a) Rhythmic variations in energy metabolism. *Fed. Proc.* 29:1541–1552.

Aschoff, J., and Pohl, H. (1970b) Der Ruheumsatz von Vögeln als Funktion der Tageszeit und der Körpergröße. *J. Ornithol.* 111:38–47.

Benedict, F. G. (1938) *Vital energetics. A study in comparative basal metabolism.* Washington, D.C.: Carnegie Inst. of Washington. 215 pp.

Bradley, S. R., and Deavers, D. R. (1980) A re-examination of the relationship between thermal conductance and body weight in mammals. *Comp. Biochem. Physiol.* 65:465–476.

Fuller, C. A., Sulzman, F. M., and Moore-Ede, M. C. (1978) Thermoregulation is impaired in an environment without circadian time cues. *Science* 199:794–796.

Herreid, C. F., and Kessel, B. (1967) Thermal conductance in birds and mammals. *Comp. Biochem. Physiol.* 21:405–414.

Hildwein, G. (1972) Métabolism énergétique de quelques mammifères et oiseaux de la foret équatoriale. II. Résultats expérimentaux et discussion. *Arch. Sci. Physiol.* 26:387–400.

Kleiber, M. (1947) Body size and metabolic rate. *Physiol. Rev.* 27:511–541.

Lasiewski, R. C., Weathers, W. W., and Bernstein, M. H. (1967) Physiological responses of the giant hummingbird, *Patagona gigas. Comp. Biochem. Physiol.* 23:797–813.

McNab, B. K. (1966) An analysis of the body temperatures of birds. *Condor* 68:47–55.

McNab, B. K. (1970) Body weight and the energetics of temperature regulation. *J. Exp. Biol.* 53:329–348.

McNab, B. K. (1979) The influence of body size on the energetics and distribution of fossorial and burrowing mammals. *Ecology* 60:1010–1021.

Miles, G. H. (1962) Telemetering techniques for periodicity studies. *Ann. N.Y. Acad. Sci.* 98:858–865.

Pohl, H. (1970) Zur Wirkung des Lichtes auf die circadiane Periodik des Stoffwechsels und der Aktivität beim Buchfinken (*Fringilla coelebs* L.). *Z. vergl. Physiol.* 66:141–163.

Schmidt, T. H. (1972). Thermoregulatorische Grössen in Abhängigkeit von Tageszeit und Menstrualzyklus. Medical dissertation, Universität München.

Scholander, P. F., Hock, R., Walters, V., Johnson, F., and Irving, L. (1950) Heat regulation in some arctic and tropical mammals. *Biol. Bull.* 99:237–258.

Spencer, F., Shirer, H. W., and Yochim, J. M. (1976) Core temperature in the female rat: Effect of pinealectomy or altered lighting. *Am. J. Physiol.* 231:355–360.

West, G. C., and Hart, J. S. (1966) Metabolic responses of evening grosbeaks to constant and to fluctuating temperature. *Physiol. Zool.* 39:171–184.

14 Temperature regulation in exercising birds

MARVIN H. BERNSTEIN

Knut Schmidt-Nielsen in his text *Animal physiology* (1979) has pointed out that birds and mammals are so successful at regulating their body temperature that they live through most of their lives with body temperatures fluctuating no more than a few degrees. Birds often fly for hours or even days with rates of heat production that are as much as 10 times greater than resting rates (Tucker, 1968, 1972). To maintain a constant body temperature under these conditions, they must increase rates of heat and energy loss by 10-fold over resting rate. This chapter considers the mechanisms birds have utilized to increase heat loss and maintain a constant body temperature during exercise.

Tolerating high temperatures

Body temperature of birds in the heat and during exercise

Normal body temperatures of nonpasserine birds are 39 to 40 °C, and those of passerines are slightly higher, 40 to 41 °C (Schmidt-Nielsen, 1979). When birds are exposed to hot environments, their body temperature increases by 3 to 4 °C (Table 14.1). During flight body temperatures also rise. Body temperature of the flying starling rises rapidly during the first 2 min of flight and then levels off. The increase in body temperature is relatively independent of air temperature, reaching the same high levels at 22 °C and 0 °C (Torre-Bueno, 1978). The body temperature of the rhea, a large terrestrial bird, also increases dramatically during exercise (Taylor et al., 1971). As pointed out by Schmidt-Nielsen (1972), a high body temperature has two advantages. One is an increase in the thermal gradient for heat loss from the body, or, when ambient exceeds body temperature, a decrease in the rate of heat gain. The second advantage is that for each 1 °C of increase in body temperature, about 3.4 kJ (830 cal) of heat is stored per kilogram of body mass. This heat can in many cases be lost later by conduction and radiation when heat production returns to resting rates. Taylor et al. (1971) showed that in rheas, *Rhea americana*, running for 20 min at 10 km h^{-1}, 75% of the heat produced during the run was stored and dissipated after the animal stopped.

189

190   MARVIN H. BERNSTEIN

Brain temperature

Birds keep their brain cooler than the body core. Even under thermally neutral conditions, brain temperature was found to be about 1 °C below T_b (Kilgore et al., 1973, 1976). Both in birds and in those mammals that undergo brain cooling, the process involves the loss of heat from arterial blood flowing toward the brain to the cooler venous blood returning from the evaporative surfaces of the head and flowing toward the heart. This heat transfer occurs in a countercurrent exchanger in the form of a rete. In mammals the rete is at the base of the brain, and in birds there are two retia, located bilaterally in the temporal region of the skull. A scheme for cranial circulation related to this process in birds is illustrated diagrammatically in Figure 14.1.

There are at least two parallel routes originating from the carotid arteries for the delivery of blood to the avian brain. One route enters the brain case from the posterior, as in mammals, and this blood might undergo some cooling in fowl (Richards, 1967), but not in pigeons (Kilgore et al., 1979). The second route passes through the temporal rete and is capable by itself of providing all the brain's blood requirements (Richards and Sykes, 1967). Nothing is yet known about the differential regulation of blood flow between these two routes.

The importance of cooling venous blood prior to its passage through the rete was explored by causing a pigeon to breathe through a surgical opening in the trachea below the glottis (a tracheostomy), and preventing evaporation from the mouth and throat. This was followed by a reduced brain–body temperature difference, suggesting that the moist bucco-

TABLE 14.1 Steady-state body temperatures in several species of birds at rest, when exposed to air temperatures above core body temperatures at vapor pressures below 10 Torr.

Species	Body mass (g)	Exposure time (min)	Air	Body	Body–air
Zebra finch (*Poephila castanotis*)	11	60	45.3	44.0	−1.3
House finch (*Carpodacus mexicanus*)	21	60	44.2	44.0	−0.2
House sparrow (*Passer domesticus*)	24	60	45.2	44.5	−0.7
Brown towhee (*Pipilo fuscus*)	42	60	44.2	44.1	−0.1
American kestrel (*Falco sparverius*)	—[a]	60	45.0	43.0	−2.0
Mourning dove (*Zenaidura macroura*)	104	120	45.4	43.4	−2.0
Bobwhite quail (*Colinus virginianus*)	166	100	44.8	43.6	−1.2
Gambel's quail (*Lophortyx gambelii*)	166	100	44.8	43.7	−1.1

Temperatures (°C) span Air, Body, Body–air.

[a] No data.
Source: Lasiewski et al., 1966.

pharyngeal surfaces are sites where blood is cooled before flowing through the rete. The temperature difference was not completely eliminated, however, unless heat loss from the eyes, too, was prevented by the use of a water-impermeable blindfold (Bernstein et al., 1979). If, instead, the uncovered eyes were ventilated by a stream of dry air, the temperature difference was increased (B. Pinshow et al., pers. commun.). Perhaps, then, the eye contributes some of the cooled venous blood flowing through the temporal rete. This seems consistent with the rich network of blood vessels in the nonretinal portions of the eye (Figure 14.2).

Brain temperature data obtained during flight are available for only one species, the American kestrel, *Falco sparverius* (Bernstein et al., 1979), and are shown in Figure 14.3. At the beginning of flight, when T_a was 23 °C, T_b rose by 2 °C, but brain temperature increased by only 1 °C. Thus the temperature difference between the body and brain increased. If the eyes play a role in countercurrent cooling of blood, then the increased air flow past the eyes during flight may account in part for the increase in the temperature difference. This idea requires further study, but it is consistent with the observed effects on brain temperature of

FIGURE 14.1 Diagram of cranial circulation relative to heat exchange in birds. Blood flowing toward brain case via the internal carotid arteries can enter via an intracranial or an extracranial branch. The intracranial branch comes into close contact with venous blood, which may in some species receive cool blood from the anterior facial region and the nasal cavity. Cooled venous blood returning from cooling surfaces can also flow through the temporal (ophthalmic) rete, coming into close contact with the extracranial branch of the internal carotid artery. Heat is then transferred from arterial blood to venous blood in the rete, and the arterial blood then enters the brain case from the side, at a lowered temperature. Arterial blood supplied via the rete can fulfill all of the brain's blood requirements. [From Kilgore et al., 1973]

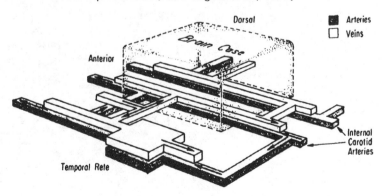

ventilating the cornea in resting birds, as described in the preceding paragraph. In any case, the relatively small increase in brain temperature during flight suggests that much of the stored body heat is confined to noncranial regions.

Dissipating heat

Evaporative cooling

Few data are available on evaporation during flight. The crow, *Corvus ossifragus*, and raven, *C. cryptoleucus*, flying at a T_a of 28 °C, both dissipated 29% of their metabolic heat by respiratory evaporation (Bernstein, 1976; Hudson and Bernstein, 1981). If these birds sustained this evaporation rate while at rest, then they would lose heat approximately 1.7

FIGURE 14.2 Scanning electronic micrograph of blood vessels in the eye of a pigeon. Vessels at left extend circumferentially from the back of eye toward right, halting at the margin of the cornea. Note the honeycombed branching of capillaries, characteristic of the retina. Vessels near cornea with fewer branches may deliver heat for dissipation by evaporation. [Micrograph by Henry P. Adams, Electron Microscope Laboratory, New Mexico State University]

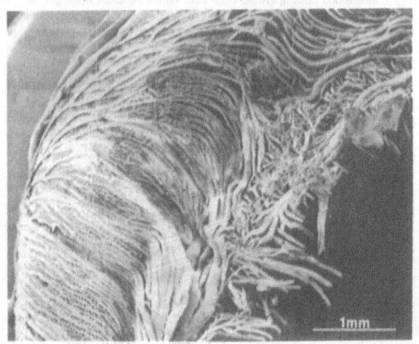

times faster than they produce it metabolically. In flight at T_a above 28 °C, evaporation increased even further, but even at the highest measured rates in budgerigar, *Melopsittacus undulatus* (Tucker, 1968), starling (Torre-Bueno, 1978), and raven (Hudson and Bernstein, 1981), total evaporation dissipated half or less of the heat produced. These results indicate that the majority of heat loss during flight occurs by nonevaporative routes. However, respiratory evaporation, along with that from the eye, may be of special importance in the cooling of the brain.

Evaporation inevitably increases over resting levels in flying birds because of the increased respiratory ventilation associated with elevated gas-exchange requirements. Do birds hyperventilate when they fly, that is, do they increase their ventilation beyond the increase in their CO_2 production or oxygen demands? In flying white-necked ravens, neither oxygen consumption nor CO_2 production changed with body temperature, but ventilation increased by nearly 2 liters BTPS min^{-1} for each 1 °C

FIGURE 14.3 Steady-state colon and brain temperatures in American kestrels, *Falco sparverius*. The kestrels flew at a speed of 10 m s^{-1} and an air temperature of 23 °C. Data preceding the flights (0 flight time) were obtained over the 60 s before takeoff. Data at 5 + min were obtained as mean temperatures over the period between 5 and 15 min after the onset of flight, and represent steady-state values. Postflight data were obtained after reattainment of steady state following each flight. The points represent means for the number of flights indicated. Vertical bars extend 2 standard errors above and below each mean. [From Bernstein et al., 1979]

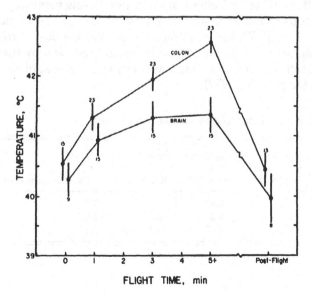

FLIGHT TIME, min

increase in T_b. Thus at the higher temperatures, flying ravens hyperventilated. The unchanged CO_2 output raises the question of whether these birds underwent respiratory alkalosis (Hudson, 1978). Hyperventilation apparently also occurred in flying starlings (Torre-Bueno, 1978) and pigeons (Hart and Roy, 1966). Hyperventilation during flight and hyperventilation in heat-stressed birds at rest are not directly comparable. In the latter, tidal volume decreases (Bernstein and Samaniego, 1981), whereas in the former, it increases (Bernstein, 1976), so it is not possible to predict from resting measurements how the blood gases of flying birds might change during hyperventilation.

In resting birds, cutaneous evaporation accounts for half or more of the total evaporation from the body, even during heat stress. Smith (1969) has reported that in pigeons at T_b of 40 °C, cutaneous exceeded respiratory water loss over fivefold. His results are summarized in Table 14.2. Despite the onset of panting at T_b of 43 °C, cutaneous evaporation tripled over the rate at T_b of 40 °C, and was twice as high as respiratory evaporation. By itself cutaneous water loss accounted for 32% of metabolically produced heat, and this figure increased dramatically with wind convection. Severe heat stress (body temperature 46 °C) was accompanied by a reduction in cutaneous evaporation to half the level at 43 °C, and by a tripling of respiratory evaporation. Perhaps the reduction in cutaneous evaporation at the highest temperature is due to an increase in insulation in an attempt to reduce heat gain.

In the adult painted quail, *Coturnix chinensis*, at a T_a of 42 °C and T_b of 43 °C, cutaneous evaporation equaled respiratory evaporation, even during panting. Both increased about 3.5-fold over levels measured at 25 °C, and cutaneous evaporation accounted for the loss of 32% of metabolic heat (Bernstein, 1976), just as in pigeons. In the ring dove, *Streptopelia risoria*, cutaneous evaporation also increased significantly upon application of heat stress, exceeding respiratory evaporation at T_a of 35 °C two- to threefold (Appleyard, 1979).

TABLE 14.2 Respiratory and cutaneous evaporation in pigeons.

T^b (°C)	Respiratory evaporation		Cutaneous evaporation		
	mg g^{-1} h^{-1}	% of total	mg g^{-1} h^{-1}	% of total	Cutaneous/respiratory
40	0.96	16.2	4.95	83.8	5.2
43	7.45	33.5	14.8	66.5	2.0
46	21.4	71.6	8.47	28.4	0.4

Source: Smith (1969).

Cutaneous evaporation was measured in flying ravens by Hudson (1978), who found that they dissipated 10% of the metabolic heat. At this rate of cutaneous evaporation, a raven would lose 42% of its metabolic heat at rest, which is similar to values actually observed during heat stress in resting birds.

Nonevaporative cooling

As mentioned, more than half of the heat produced by flight in birds is lost by nonevaporative means. In a flying bird, the spread wings help dissipate heat by increasing the surface area for convection and radiation to the cooler environment. To increase the body–air temperature gradient, some birds simply fly at higher altitudes where air temperature is lower (Berger and Hart, 1974; Torre-Bueno, 1976). Many birds extend the feet into the air stream while in flight (Tucker, 1968; Frost and Siegfried, 1975), and this may dissipate a significant fraction of heat production, decreasing the need for evaporative cooling (Baudinette et al., 1976). We still need more data to evaluate the contribution of nonevaporative heat loss from the extremities, including the bill, to the total heat budget of flying birds.

Conclusions

When birds fly, their rate of metabolic energy consumption increases by as much as a factor of 10. About 20% of the energy is utilized for performing external work, and 80% is converted to heat internally. During sustained flights, this heat must be dissipated to prevent body temperature from rising to lethal levels. Some birds reduce heat production during flight by soaring, which requires only a twofold increase in metabolism over resting levels (Baudinette and Schmidt-Nielsen, 1974; Goldspink et al., 1978). About half of the heat is lost by nonevaporative means. High body temperatures observed during flight help increase the rate of nonevaporative heat loss by increasing the gradient between the bird and the surrounding air. The brain appears to be buffered from these increases by a countercurrent heat exchanger, but only one species of bird has been studied. Birds may also increase the gradient for nonevaporative heat loss by flying higher where the air is cooler. Although we lack information about the sites of nonevaporative heat loss during flight, it seems possible that much of the heat may be lost from extremities, (e.g., feet, legs, and beaks). The other half of the heat is lost evaporatively, most of it appearing to be lost from the respiratory tract. Although the skin is an important site for evaporative cooling in birds resting in the heat (accounting for more than half the evaporation), evaporation from the skin does not appear to be as important during flight, accounting for only about 10%.

It is clear that although great progress has been achieved in the past two decades toward understanding mechanisms birds use to maintain a constant body temperature during exercise, much remains to be done in the next 20 years.

This chapter was prepared during a sabbatical leave from New Mexico State University supported by an Arts and Sciences Minigrant, by NSF grant PCM 79-21856, and by a Fulbright Travel Award from the U.S.–Israel Educational Foundation. I thank Dr. Berry Pinshow and the Institute for Desert Research, Ben-Gurion University of the Negev, for their assistance and hospitality.

References

Appleyard, R. F. (1979) Cutaneous and respiratory water losses in the ring dove, *Streptopelia risoria*. M.S. thesis, Washington State University.

Baudinette, R. V., Loveridge, J. P., Wilson, K. J., Mills, C. D., and Schmidt-Nielsen, K. (1976) Heat loss from feet of herring gulls at rest and during flight. *Am. J. Physiol.* 230:920–924.

Baudinette, R. V., and Schmidt-Nielsen, K. (1974) Energy cost of gliding flight in herring gulls. *Nature* 248:83–84.

Berger, M., and Hart, J. S. (1974) Physiology and energetics of flight. In *Avian biology*, vol. 4 (D. S. Farner and J. R. King, eds.), pp. 415–477. New York: Academic Press.

Bernstein, M. H. (1976) Ventilation and respiratory evaporation in the flying crow, *Corvus ossifragus*. *Respir. Physiol.* 26:371–382.

Bernstein, M. H., Curtis, M. B., and Hudson, D. M. (1979) Independence of brain and body temperatures in flying American kestrels, *Falco sparverius*. *Am. J. Physiol.* 237:R58–R62.

Bernstein, M. H., and Samaniego, F. C. (1981) Ventilation and acid–base status during thermal panting in pigeons, *Columba livia*. *Physiol. Zool.* 54:308–315.

Bernstein, M. H., Sandoval, I., Curtis, M. B., and Hudson, D. M. (1979) Brain temperature in pigeons: Effects of anterior respiratory bypass. *J. Comp. Physiol.* 129:115–118.

Frost, P. G. H., and Siegfried, W. R. (1975) Use of legs as dissipators of heat in flying passerines. *Zool. Afr.* 10:101–108.

Goldspink, G., Mills, C., and Schmidt-Nielsen, K. (1978) Electrical activity of the pectoral muscles during gliding and flapping flight in the herring gull (*Larus argentatus*). *Experientia* 34:862–865.

Hart, J. S., and Roy, O. Z. (1966) Respiratory and cardiac responses to flight in pigeons. *Physiol. Zool.* 39:291–306.

Hudson, D. M. (1978) Power input, ventilation, and thermoregulation during steady-state flight in the white-necked raven, Corvus cryptoleucus. Ph.D. dissertation, New Mexico State University.

Hudson, D. M., and Bernstein, M. H. (1981) Temperature regulation and heat balance in flying white-necked ravens, Corvus cryptoleucus. *J. Exp. Biol.* 90:267–282.

Kilgore, D. L., Jr., Bernstein, M. H., and Hudson, D. M. (1976) Brain temperatures in birds. *J. Comp. Physiol.* 110:209–215.

Kilgore, D. L., Jr., Bernstein, M. H., and Schmidt-Nielsen, K. (1973) Brain temperature in a large bird, the rhea. *Am. J. Physiol.* 225:739–742.

Kilgore, D. L., Jr., Boggs, D. F., and Birchard, G. F. (1979) Role of the Rete Mirabile Ophthalmicum in maintaining the body-to-brain temperature difference in pigeons. *J. Comp. Physiol.* 129:119–122.

Lasiewski, R. C., Acosta, A. L., and Bernstein, M. H. (1966) Evaporative water loss in birds – I. Characteristics of the open flow method of determination, and their relation to estimates of thermoregulatory ability. *Comp. Biochem. Physiol.* 19:445–457.

Richards, S. A. (1967) Anatomy of the arteries of the head in the domestic fowl. *J. Zool., Lond.* 152:221–234.

Richards, S. A., and Sykes, A. H. (1967) Responses of the domestic fowl to occlusion of the cervical arteries and veins. *Comp. Biochem. Physiol.* 21:39–50.

Schmidt-Nielsen, K. (1972) Recent advances in the comparative physiology of desert animals. *Symp. Zool. Soc. Lond.* 31:371–382.

Schmidt-Nielsen, K. (1979) *Animal physiology: Adaptations and environment.* Cambridge University Press. 560 pp.

Smith, R. M. (1969) Cardiovascular, respiratory, temperature, and evaporative water loss responses of pigeons to varying degrees of heat stress. Ph.D. dissertation, Indiana University.

Taylor, C. R., Dmi'el, R., Fedak, M., and Schmidt-Nielsen, K. (1971) Energetic cost of running and heat balance in a large bird, the rhea. *Am. J. Physiol.* 221:597–601.

Torre-Bueno, J. R. (1976) Temperature regulation and heat dissipation during flight in birds. *J. Exp. Biol.* 65:471–482.

Torre-Bueno, J. R. (1978) Evaporative cooling and water balance during flight in birds. *J. Exp. Biol.* 75:231–236.

Tucker, V. A. (1968) Respiratory exchange and evaporative water loss in the flying budgerigar. *J. Exp. Biol.* 48:67–87.

Tucker, V. A. (1972) Metabolism during flight in the laughing gull, Larus atricilla. *Am. J. Physiol.* 222:237–245.

15 Behavioral and autonomic thermoregulation in terrestrial ectotherms

EUGENE C. CRAWFORD, JR.

Animals have traditionally been divided into poikilotherms and homeotherms; but it is now clear that many so-called cold-blooded animals can, under appropriate conditions, maintain a relatively constant and often high body temperature. Because they achieve this temperature stability primarily by behavioral exploitation of the external thermal environment, these animals are now called ectotherms.

Among the terrestrial ectotherms, recent studies on temperature control in insects lay the foundation for exciting future investigation for comparative physiologists. However, I shall limit this discussion to vertebrate terrestrial ectotherms, particularly reptiles, which provide the broadest base of information.

Rather than stress unique specific thermal adaptations I shall attempt to address those thermoregulatory features that appear to be common among vertebrates with the hope of providing a broader basis of appreciation and understanding of thermoregulation in general. Such an approach invariably stimulates provocative thoughts concerning the evolution of homeothermy.

Behavioral thermoregulation

The basis of much of the work on behavioral temperature regulation in terrestrial vertebrate ectotherms is founded in the classic report of Cowles and Bogert (1944). They observed that reptiles move toward or away from hot areas in the environment at rather well-defined lower and higher body temperatures, which they called voluntary minimum and maximum temperatures. When animals were within this temperature range, their behavior appeared to be unrelated to body temperature. Voluntary maximum and minimum temperatures thus set the general thermal limits for the normal activity range of a given species. This thermal activity range, rather than the voluntary limits, first attracted the attention of most investigators.

Random measurement of body temperatures of ectotherms in the field during periods of activity typically exhibit a unimodal frequency distribution, as shown in Figure 15.1 (Bogert, 1949; Norris, 1953; Brattstrom,

198

1963, 1965, 1979; Lillywhite, 1970). Although such measurements do provide useful ecological information, their interpretation with regard to thermoregulation should be approached with reservations. It is necessary to observe and record the behavior of the animal as well as environmental and body temperatures. Furthermore, one must be particularly cautious that the observed behavior is, in fact, thermoregulatory.

When animals are placed in a linear temperature gradient chamber, they tend to aggregate, or spend more time within a well-defined thermal region, as shown in Figure 15.2. The thermal zone of congregation is known as the eccritic temperature or, more commonly, the preferred body temperature. For convenience it is often expressed as a single mean body temperature. However, this obscures the central tendency, range of variation, and negative skewness, which are common characteristics of the thermal preference zone. The design of many linear gradient chambers provides greater opportunity for movement along the gradient than movement within a homogeneous thermal zone. This restriction

FIGURE 15.1 Frequency distribution of the body temperatures of two species of lizards randomly measured in the field during the activity period. [From Bogert, 1949]

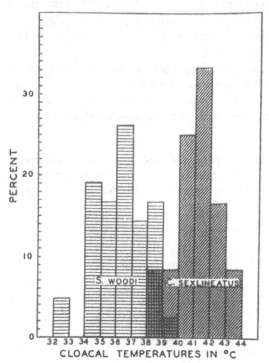

is partially remedied by a concentric gradient chamber in which the animal can move along the gradient from center to periphery or around the chamber in a relatively constant thermal environment. Figure 15.3 shows the distribution of body temperatures of 11 lizards (*Dipsosaurus dorsalis*) observed for 3 days in a concentric gradient (DeWitt, 1967). The preferred body temperature of this species has a range of 32 to 43 °C with a mode of about 38.5 °C.

The similarity of body temperature distribution curves obtained from random field measurements and gradient experiments suggests that both are the result of thermoregulatory behavior. Laboratory thermal gradients are relatively devoid of external stimuli when compared with the richness of the natural habitat. Therefore, after an initial period of escape and exploratory behavior, one might expect that an animal's movements in a thermal gradient are primarily governed by thermoregulatory considerations. Body temperatures of ectotherms in the field probably indicate a range of temperatures acceptable for a variety of behavioral activities rather than thermoregulation alone (Licht et al., 1966). In any event, the preferred body temperature, determined in long-term laboratory gradient experiments, is assumed to reflect behavioral thermoregulation in ectothermic animals (Bligh and Johnson, 1973). The thermal preferendum appears to be species-specific, independent of the thermal history of the animal, and many physiological processes take place at optimal rates within this temperature range (Licht et al., 1966; Licht, 1968; Reynolds and Casterlin, 1979).

In gradient chambers an animal can select from a broad range of ambient temperatures. What happens when the options are limited to only two temperatures, one well below and the other well above the preferred range? In this case an ambient temperature within the thermal

FIGURE 15.2 Frequency distribution of stable flies in an experimental continuous linear thermal gradient chamber. This distribution is characteristic of the "preferred body temperature." [From Fraenkel and Gunn, 1940]

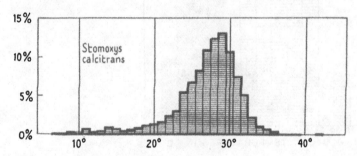

preferendum is not available for selection. Movement within the uniformly hot or cool environment does not result in changes in body temperature. However, an ectotherm can keep its body temperature within the preferred range by migrating between the two extreme temperatures as shown in Figure 15.4 (Heath, 1970). In this case body temperature of the animal continuously increases or decreases, depending upon whether it is in the hot or cool environment.

A description of this behavior provides some insight into the nature of the thermoregulatory system. After an initial period of exploratory behavior, lizards shuttle periodically between hot and cool sides of a shuttle box at rather well-defined body temperatures (Hammel et al., 1967; Heath, 1970; Berk and Heath, 1975; Barber and Crawford, 1979). Upon entering the hot side at a low body temperature, they remain calm and exhibit no escape behavior even though the ambient temperature may be lethal. Body temperature rises until it reaches an upper limit, at which the lizard rapidly leaves the hot side and enters the cool side at a high body temperature. Body temperature then decreases until it reaches a lower limit, and the lizard returns to the hot side. Even though the animal is always in an environment that is either "too hot" or "too cold," no obvious thermoregulatory behavior occurs until body temper-

FIGURE 15.3 Frequency distribution of the body temperatures of 11 lizards (*D. dorsalis*) observed for 3 days in a concentric thermal gradient chamber. [From DeWitt, 1967]

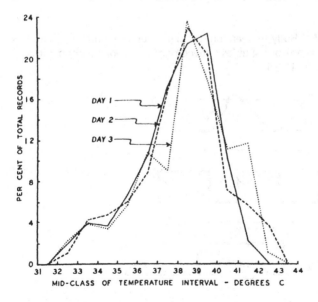

ature reaches upper- or lower-limit temperatures. This behavior is similar to that reported by Cowles and Bogert (1944) with upper- and lower-limit temperatures analogous to their maximum and minimum voluntary temperatures.

These results are consistent with a dual threshold type of thermoregulatory system with upper- and lower-limit temperatures or "set points" at which thermoregulatory responses occur. The limits are separated by a refractory zone. Within this zone, thermoregulatory "drive" is minimal, and the animal is relatively free to pursue other behavioral activity necessary for survival.

Shuttling thermoregulatory behavior results in a rhythmic oscillation of body temperature between limits as predicted by a dual threshold model. Does the same model also predict the unimodal distribution of body temperatures observed in the field and continuous-gradient experiments? Lizards do not always leave hot and cold environments at precisely the same body temperature. Exit temperatures are randomly distributed about a mean value in a more or less normal fashion with the standard deviation of the lower threshold greater than that of the upper threshold (Hammel et al., 1967; Berk and Heath, 1975; Barber and Crawford, 1977), as shown in Figure 15.5. That is, the thresholds are probabilistic rather than discrete. The expectation of an animal's encountering upper and lower threshold temperatures can be derived from the normal probability integral. These expectation curves (Figure 15.6) may be viewed as reflecting thermal barriers expressing the probability that an animal moving in a temperature gradient will reverse its direction of

FIGURE 15.4 Body temperature oscillation between high (T_h) and low (T_l) limits as a lizard shuttles between hot and cool environments. [From Heath, 1970]

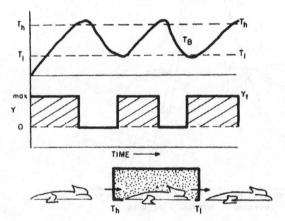

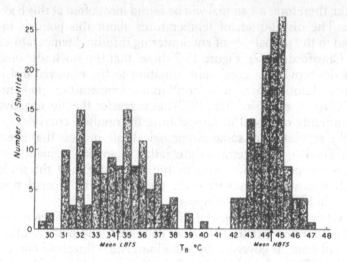

FIGURE 15.5 The distribution of body temperatures of *Dipsosaurus dorsalis* when it leaves the cool side (LBTS) and hot side (HBTS) of a two-temperature shuttle chamber. [From Berk and Heath, 1975]

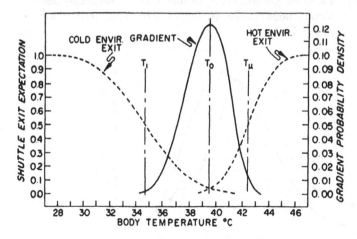

FIGURE 15.6 Graph in which the solid line (gradient) represents the predicted distribution of body temperatures of a lizard in a continuous-gradient chamber, whose thermoregulatory behavior is governed by a dual threshold control system. The dashed lines represent the expectations of encountering lower and upper threshold temperatures and reversing the direction of motion. T_l and T_u are the mean lower and upper threshold temperatures and T_o the mode of the derived preferred body temperature. [From Barber and Crawford, 1979]

motion. The probability of thermoregulatory motion is least at the point of intersection; therefore, an animal will be found most often at this body temperature. The distribution of temperatures about this point is inversely related to the probability of encountering threshold temperatures (Barber and Crawford, 1977). Figure 15.7 shows that the stochastic dual threshold model provides a good approximation to the observed distribution of body temperatures in a continuous temperature gradient. DeWitt (1967; Dewitt and Friedman, 1979) has suggested that the negative skewness commonly observed in temperature distribution curves originates from the regulation of some physiological rate process that is exponentially related to body temperature rather than the regulation of body temperature per se. However, in the stochastic dual threshold model, negative skewness arises from the greater standard deviation of the lower threshold relative to the upper.

Both theoretical and experimental analyses provide convincing evidence that the thermoregulatory behavior of lizards, and perhaps other vertebrate ectotherms, is governed by a stochastic dual threshold control system (Heath, 1970; Berk and Heath, 1975; Barber and Crawford, 1979; Garrick, 1979). Given appropriate environmental conditions, these ectotherms are capable of maintaining a relatively constant body temperature. In nature these conditions are usually met during daylight hours; so behavioral thermoregulation is primarily limited to daytime activity. If these animals were experimentally provided environmental conditions

FIGURE 15.7 Comparison of the predicted preferred body temperature of *Diposaurus dorsalis* assuming stochastic dual threshold regulation (solid line) with the preferred body temperature experimentally determined in a continuous gradient chamber (dashed line). [From Barber and Crawford, 1977]

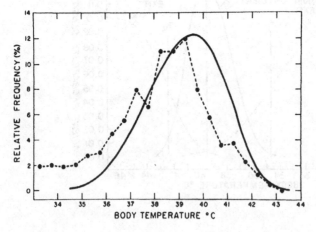

so that they could behaviorally thermoregulate continuously, would they do so? The answer is apparently no. Lizards kept in a thermal gradient on a light–dark cycle chose cooler temperatures during the dark period (Regal 1967; Engbretson and Hutchison, 1976). Moreover, the rhythm persists in constant darkness with a period of about 24 h (Cowgell and Underwood, 1979). The shuttling behavior of the desert iguana in two-temperature selection experiments is biphasic. They shuttle during morning hours and in the late afternoon. At night and for several hours around noon they remain inactive on the cool side of the shuttle. If moved to the hot side during an inactive period, the lizard returns to the cool compartment when its body temperature reaches threshold and remains there until its normal shuttling period begins (Barber and Crawford, 1979). This rhythmic biphasic thermoregulatory behavior persists for up to 5 days in constant light (Barber, unpubl. data). These preliminary results suggest that thermoregulatory behavior is linked to a circadian clock. Further critical analysis of the circadian aspects of behavioral thermoregulation in ectotherms will likely contribute to a better understanding of the relationships between this and other circadian activities and their ecological significance.

Autonomic thermoregulation

Evidence for autonomic thermoregulation in terrestrial vertebrate ectotherms rests mainly on the observations that many reptiles warm up faster than they cool off (Bartholomew and Tucker, 1964; Bartholomew and Lasiewski, 1965; Spray and May, 1972) and increase evaporative water loss at high body temperatures (Templeton, 1970; Crawford and Kampe, 1971; Weathers and White, 1971; Spotila et al., 1977; Smith, 1979).

Differential heating and cooling rates have been explained on the basis of alterations in cardiac output and blood flow to the skin. An increase in cardiac output and cutaneous vasodilation during warming increases the rate of convective heat transport from surface to core. Vasoconstriction and a decrease in cardiac output reduce heat transport from core to surface during cooling. These have been widely interpreted as thermoregulatory responses, under autonomic control, which extend the period of time that body temperature remains within the preferred range.

Heating and cooling curves are typically obtained by subjecting animals to step changes in ambient temperature and recording changes in core temperature with little or no reference to skin temperature. If an animal has significant mass, a thermal gradient will exist between skin and core such that skin temperature will always be greater than core during heating and less than core during cooling. Changes in cutaneous blood flow could be due to a local effect of temperature on the peripheral vasculature and lead to thermal hysteresis of core temperature. Localized heating results in cutaneous blood flow changes which are independent of core temperature and heart rate. Moreover, these local responses are not abolished by sympathetic blocking agents (Weathers and Morgareidge, 1971; Baker et al., 1972). Differences in heart rate during heating and cooling may well be due to the effect of temperature on the heart itself (Spray and Belkin, 1972) and reflex responses to shifts in blood pressure associated with changes in peripheral resistance (Weathers and White, 1971; Crawshaw, 1980). However, heating and cooling the hypothalamus cause substantial changes in blood pressure (Heath et al., 1968), and section of dorsal roots abolishes the difference between heating and cooling rates of turtles (Spray and May, 1972). These results suggest that peripheral and central neuronal components may be involved in the general cardiovascular responses to heating and cooling. Whether the response is mediated by the autonomic nervous system or is the result of a local temperature effect unrelated to temperature regulation remains an open question.

Although both aquatic and terrestrial ectotherms may utilize cardiovascular responses for thermoregulation, increased evaporative cooling is an option limited to terrestrial animals. A variety of mechanisms are employed: mucus secretion by amphibians (Lillywhite, 1970), salivation and cloacal discharge by turtles (Morgareidge and Hammel, 1975), gaping

FIGURE 15.8 Respiratory evaporation of *Sauromalus obesus* at different body temperatures. The increase at 43.5 °C results from panting.

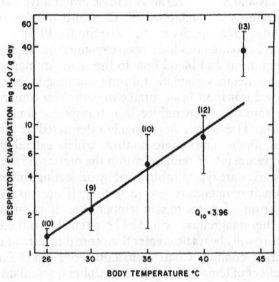

in crocodiles (Spotila et al., 1977; Smith, 1979), and panting in lizards (Templeton, 1970; Crawford and Kampe, 1971). Of these, panting has been most extensively investigated.

Evaporation of water from the respiratory tract of the chuckwalla (*Sauromalus obesus*) is shown in Figure 15.8. The increase at 44 °C is accomplished by an increase in pulmonary ventilation resulting from an increase in breathing frequency and a decrease in tidal volume. This respiratory pattern is typical of panting in birds and mammals. Total evaporative cooling is sufficient to maintain body temperature several degrees below ambient for several hours. Panting has a greater cooling effect on the brain and leads to head–body temperature differences (DeWitt, 1967; Crawford, 1972; Crawford et al., 1977). These steady-state head–body temperature gradients, during panting, could be the result of selective physiological regulation of brain temperature, as is the case in certain mammals. However, critical evidence in support of this hypothesis for ectotherms is lacking at the present time.

Ventilation in ectotherms is governed by requirements for oxygen consumption and acid–base balance (Howell et al., 1970; Chapter 6). The ventilation necessary to support significant evaporative cooling is greater than that required to meet these metabolic demands. In the chuckwalla, a significant portion of the excess ventilation apparently passes over the gas-exchange surfaces of the lung, resulting in respiratory hypocapnia and alkalosis. However, the acid–base disturbance is transitory, persisting only during panting which probably is of short duration under natural environmental conditions. These results suggest an interesting feature of regulatory systems, that is, inherent priorities for regulation. In this case the demand for evaporative cooling during heat stress apparently overrides the ventilatory drive usually associated with acid–base regulation (Crawford and Gatz, 1974). This competitive interplay between regulatory systems seems a fruitful area for further investigation.

Neuronal aspects of thermoregulation

The complex behavioral and physiological thermoregulatory responses of terrestrial ectotherms imply the existence of the cybernetic neuronal components that control those biological responses necessary for regulating body temperature. Such systems consist of sensors providing input information about body temperature, comparators that compare actual body temperature with a desirable body temperature (set point), and some central integrative mechanism that provides for an appropriate effector output. The system is closed through negative feedback loops. Although our understanding of the neurophysiological basis of thermoregulation is incomplete, some progress has been made.

The preoptic region of the hypothalamus is generally accepted as the area of the brain containing the vertebrate thermostat (Heller et al., 1978). Much of our understanding of thermoregulatory systems has been derived from selective heating and cooling of this region of the brain. Ted Hammel and his colleagues (1967) have been successful in modifying behavioral thermoregulation in the blue tongue lizard (*Tiliqua scincoides*) by altering brain temperature. The core temperature at which lizards leave a hot box is significantly lower when the brain is heated and higher when it is cooled. Panting can be initiated or inhibited by heating or cooling the brain of chuckwallas (Crawford and Barber, 1974). These results suggest that hypothalamic temperature plays an integral role in eliciting these thermoregulatory responses. However, neither shuttling nor panting responses can be activated by hypothalamic heating until other body temperatures increase to some appropriate value. Furthermore, when spontaneous panting is inhibited by cooling of the brain, slight increases in skin temperature reinitiate panting. It would appear, therefore, that peripheral temperatures serve as important additional thermal inputs.

What body temperature is controlled? We may gain some insight into this problem by analyzing transient rates of temperature change when a terrestrial ectotherm is subjected to step changes in ambient heat load (Barber and Crawford, 1979). Because of thermal lag, core (T_C), brain (T_B), and skin (T_S) temperatures change at different rates, such that, $dT_C/dt < dT_B/dt < dT_S/dt$ (Figure 15.9). If a thermoregulatory response

FIGURE 15.9 Schematic diagram of the rate of change of skin, brain, and core temperatures when a lizard is subjected to a step change in ambient heat load.

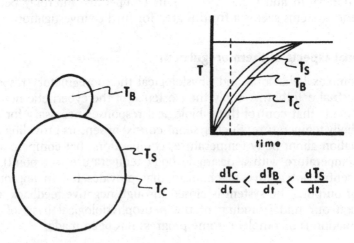

is initiated by skin temperature, then the T_S at which the response just begins will be unrelated to the square wave heat load. That is, the slope of T_S/T_A will be zero, and T_B/T_A and T_C/T_A will have negative slopes, depending upon the degree of thermal coupling between head and core compartments (Figure 15.10). Utilizing this technique, Barber and Crawford (1979) found that the desert iguana (*Dipsosaurus dorsalis*) leaves a hot environment at essentially the same skin temperature regardless of hot environment temperature (Figure 15.11). Similarly, the skin temperature at the onset of panting is independent of ambient heat load (Crawford, unpubl. results). These results suggest that skin or perhaps other peripheral temperatures play a predominate role in activating both thermoregulatory responses. However, both peripheral thresholds can be modified by changes in hypothalamic temperature, implying that thermoregulation in terrestrial ectotherms is a complex function of peripheral and brain temperatures.

The thermoregulatory responses of vertebrate ectotherms are apparently mediated by temperature-sensitive neuronal components of the nervous system similar to those of endotherms. Utilizing thermodes for intracranial heating and cooling, Cabanac et al. (1967) recorded the electrical activity of individual neurons in the preoptic region of the hypo-

FIGURE 15.10 Schematic representation of the relative rates of change of skin and core temperatures when a lizard is subjected to increasing step heat loads. T_p represents the thermoregulatory response. The figures to the right are predicted patterns of core and skin temperatures when the response occurs as a function of ambient heat load. (See text for further explanation.) [Adapted from Barber and Crawford, 1979]

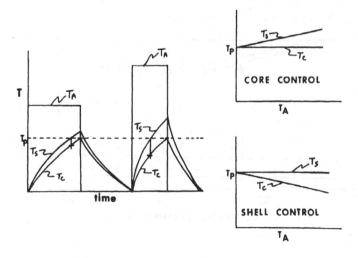

thalamus of *T. scincoides*. Although most neurons appear insensitive to temperature, some increase firing rate when heated, whereas others increase firing rate when cooled. The thermal sensitivity of warm neurons is greater than that of cold neurons (Figure 15.12). The intersection of the activity curves could represent the set point in a single-set-point control system. But Berk and Heath (1975) have suggested that an interaction between temperature-insensitive neurons (dashed line in Figure 15.12) and warm and cold neurons could give rise to upper and lower set points in a dual-threshold system. Furthermore, the low thermal sensitivity of cold neurons could account for the greater standard deviation of the lower behavioral threshold.

The apparent importance of peripheral temperature in initiating thermoregulatory responses suggests that temperature receptors occur in the skin of ectotherms. Furthermore, the dual-threshold hypothesis for behavioral thermoregulation is consistent with two populations of thermal sensors, cold receptors corresponding to the lower threshold and warm receptors corresponding to the upper threshold. Indeed, the thermal response of cutaneous cold and warm receptors in primates (Iggo, 1969) is remarkably similar in form to the stochastic distribution of lower and

FIGURE 15.11 Graph showing that the skin temperature (UTPSU) at which *Dipsosaurus dorsalis* exits a hot environment is independent of ambient temperature, whereas core temperature (UTPC) has a negative slope. This pattern is consistent with a shell-control hypothesis.

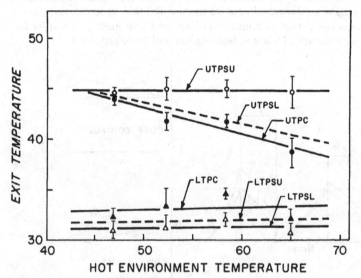

upper behavioral thresholds of *D. dorsalis*. Whether such receptors, in fact, exist in the skin of reptiles has not been verified. However, temperature receptors have been found in the skin of frogs (Spray, 1974). They respond only to cooling and are classified as cold receptors because their response is in the same temperature range as that of mammalian cold receptors. Curiously no warm receptors have been identified. Based on the existence of cutaneous thermoreceptors in frogs and mammals, it is surprising that the search for similar receptors in the skin of reptiles has been unsuccessful. Perhaps they reside in some other peripheral site such as the air passages, tongue, or spinal cord. The indirect evidence for their existence should be sufficient stimulus to continue the search.

The behavioral responses that give rise to the preferred body temperature appear common to all vertebrate species, and all seem to be mediated by neuronal components of similar organization. The panting response of reptiles is only activated at near-lethal temperatures and only then when behavioral options are unavailable. Our understanding of the neuronal basis of their autonomic response is just beginning.

A more complete understanding of the neurophysiological basis of thermoregulation, particularly the interaction between peripheral and

FIGURE 15.12 Unit recordings from temperature-sensitive neurons in the preoptic region of the brain of *Tiliqua scincoides*. The dashed line schematically depicts temperature-insensitive neurons. [Redrawn from Cabanac et al., 1967, and Berk and Heath, 1975]

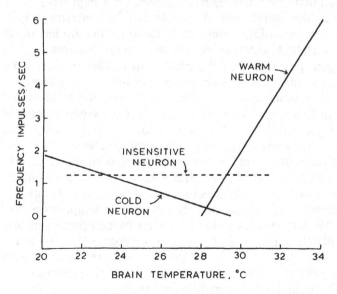

central neurons, remains a major challenge of the future, and significant contributions are likely to come from comparative studies. Several models, based upon accumulated results from behavioral and physiological studies, have been developed as aids in understanding thermoregulation in vertebrate ectotherms (Berk and Heath, 1975; Heller et al., 1978; Barber and Crawford, 1979; Crawshaw, 1980). Many features of these models are strikingly similar to models proposed for endothermic temperature regulation. Moreover, the discovery of pyrogen-induced behavioral fever in lizards (Vaughn et al., 1974), fish (Reynolds et al., 1976), and newborn rabbits (Satinoff et al., 1976) suggests that thermoregulatory systems of ectothermic and endothermic vertebrates share common features of earlier phylogenetic origin (Kluger, 1979). Although it is questionable whether these observations provide any insight into the evolution of homeothermy, they do provoke speculation.

Recall that endothermy requires a high capacity for aerobic metabolism for both heat production and sustained activity and some mechanism for reducing heat loss from the body surface – insulation in the form of fur, feathers, subcutaneous fat, or a large body size which reduces the relative surface from which body heat is lost. The critical factor in the course of endothermic evolution could well have been the development of a high aerobic capacity and consequently an effective oxygen-transport and energy-processing system. The efficacy of oxygen transport and utilization can be increased by: (1) a lung with a diffusing capacity capable of delivering O_2 to the pulmonary blood at the required rate, (2) a heart that separates oxygenated from deoxygenated blood, (3) a high oxygen-carrying capacity of the blood, and (4) a mitochondrial enzyme activity sufficient for energy transformation. All of these factors are less developed in extant vertebrate ectotherms compared to endotherms.

It is tempting to postulate that the early stem reptiles possessed the cybernetic components for thermoregulation but little or no autonomic capacity. The thecodonts and therapsids, which gave rise to birds and mammals, gradually acquired an increasingly effective oxygen-processing system, which increased the aerobic capacity for sustained activity. The increase in metabolic capacity permitted endothermy, which, with the preexisting control system and later-acquired insulation, culminated in endothermic homeothermy. Bennett and Ruben (1979) have skillfully argued that if this were the case, selection would favor, at least initially, an increased aerobic capacity for activity rather than temperature regulation per se. Modern reptiles without effective oxygen processing and insulation are, like their ancestral stem reptiles, incapable of endothermic homeothermy. The output of their vestigial temperature control system is limited principally to behavioral regulation, which is conservative of energy expenditure, and a rather ineffective panting.

References

Baker, L. A., Weathers, W. W., and White, F. N. (1972) Temperature-induced peripheral blood flow changes in lizards. *J. Comp. Physiol.* 80:313–323.

Barber, B. J. and Crawford, E. C., Jr. (1977) A stochastic dual-limit hypothesis for behavioral thermoregulation in lizards. *Physiol. Zool.* 50:53–60.

Barber, B. J., and Crawford, E. C., Jr. (1979) Dual threshold control of peripheral temperature in the lizard *Dipsosaurus dorsalis*. *Physiol. Zool.* 52:250–263.

Bartholomew, G. A., and Lasiewski, R. C. (1965) Heating and cooling rates, heart rate and simulated diving in the Galapagos marine iguana. *Comp. Biochem. Physiol.* 16:573–582.

Bartholomew, G. A., and Tucker, V. A. (1964) Size, body temperature, thermal conductance, oxygen consumption, and heart rate in Australian varanid lizards. *Physiol. Zool.* 37:341–354.

Bennett, A. F., and Ruben, J. A. (1979) Endothermy and activity in vertebrates. *Science* 206:649–654.

Berk, M. L., and Heath, J. E. (1975) An analysis of behavioral thermoregulation in the lizard, *Dipsosaurus dorsalis*. *J. Therm. Biol.* 1:15–22.

Bligh, J., and Johnson, K. G. (1973) Glossary of terms for thermal physiology. *J. Appl. Physiol.* 35:941–961.

Bogert, C. M. (1949) Thermoregulation in reptiles: A factor in evolution. *Evolution* 3:195–211.

Brattstrom, B. H. (1963) A preliminary review of the thermal requirements of amphibians. *Ecology* 44:238–255.

Brattstrom, B. H. (1965) Body temperatures of reptiles. *Am. Midl. Nat.* 73:376–422.

Brattstrom, B. H. (1979) Amphibian temperature regulation studies in the field and laboratory. *Am. Zool.* 19:345–366.

Cabanac, M., Hammel, H. T., and Hardy, J. D. (1967) *Tiliqua scincoides*: Temperature sensitive units in lizard brain. *Science* 158:1050–1051.

Cowgell, J., and Underwood, H. (1979) Behavioral thermoregulation in lizards: A circadian rhythm. *J. Exp. Zool.* 210:189–194.

Cowles, R. B., and Bogert, C. M. (1944) A preliminary study of the thermal requirements of desert reptiles. *Bull. Am. Mus. Nat. Hist.* 83:261–296.

Crawford, E. C., Jr. (1972) Brain and body temperatures in a panting lizard. *Science* 177:431–433.

Crawford, E. C., Jr. and Kampe, G. (1971) Physiological responses of the lizard *Sauromalus obesus* to changes in ambient temperature. *Am. J. Physiol.* 220:1256–1260.

Crawford, E. C., Jr. and Barber, B. J. (1974) Effects of core, skin and brain temperature on panting in the lizard *Sauromalus obesus*. *Am. J. Physiol.* 226:569–573.

Crawford, E. C., Jr. and Gatz, R. N. (1974) Respiratory alkalosis in a panting lizard (*Sauromalus obesus*). *Experientia* 30:638–640.

Crawford, E. C., Jr., Palomeque, J., and Barber, B. J. (1977) A physiological basis for head–body temperature differences in a panting lizard. *Comp. Biochem. Physiol. 56A*:161–163.

Crawshaw, L. I. (1980) Temperature regulation in vertebrates. *Annu. Rev. Physiol. 42*:473–491.

DeWitt, C. B. (1967) Precision of thermoregulation and its relation to environmental factors in the desert iguana, *Dipsosaurus dorsalis. Physiol. Zool. 40*:49–66.

DeWitt, C. B., and Friedman, R. M. (1979) Significance of skewness in ectotherm thermoregulation. *Am. Zool. 19*:195–209.

Engbretson, G. A., and Hutchison, V. H. (1976) Parietalectomy and thermal selection in the lizard *Sceloporus magister. J. Exp. Zool. 198*:29–38.

Fraenkel, G. S., and Gunn, D. L. (1940) *The orientation of animals: Kineses, taxes and compass reactions*. New York: Oxford University Press. 352 pp.

Garrick, L. D. (1979) Lizard thermoregulation: Operant responses for heat at different thermal intensities. *Copeia* no. 2:258–266.

Hammel, H. T., Caldwell, F. T., and Abrams, R. M. (1967) Regulation of body temperature in the blue-tongued lizard. *Science 156*:1260–1262.

Heath, J. E. (1970) Behavioral regulation of body temperature in poikilotherms. *Physiologist 13*:399–410.

Heath, J. E., Gasdorf, E., and Northcutt, R. G. (1968) The effect of thermal stimulation of anterior hypothalamus on blood pressure in the turtle. *Comp. Biochem. Physiol. 26*:509–518.

Heller, H. C., Crawshaw, L. I., and Hammel, H. T. (1978) The thermostat of vertebrate animals. *Sci. Am. 239*:102–113.

Howell, B. J., Baumgardner, F. W., Bondi, K., and Rahn, H. (1970) Acid–base balance in cold-blooded vertebrates as a function of body temperature. *Am. J. Physiol. 218*:600–606.

Iggo, A. (1969) Cutaneous thermoreceptors in primates and sub-primates. *J. Physiol. 200*:403–430.

Kluger, M. J. (1979) Fever in ectotherms: Evolutionary implications. *Am. Zool. 19*:295–304.

Licht, P. (1968) Response of the thermal preferendum and heat resistance to thermal acclimation under different photoperiods in the lizard *Anolis carolinensis. Am. Midl. Nat. 79*:149–158.

Licht, P., Dawson, W. R., Shoemaker, V. H., and Main, A. R. (1966) Observations on thermal relations of Western Australian lizards. *Copeia* no. 1:97–110.

Lillywhite, H. B. (1970) Behavior thermoregulation in the bullfrog, *Rana catesbeiana. Copeia 1970*:158–168.

Morgareidge, K. R., and Hammel, H. T. (1975) Evaporative water loss in box turtles: Effects of rostral brainstem and other temperatures. *Science 187*:366–368.

Norris, K. S. (1953) The ecology of the desert iguana, *Dipsosaurus dorsalis. Ecology 34*:265–287.

Regal, P. J. (1967) Voluntary hypothermia in reptiles. *Science 155*:1551–1553.

Reynolds, W. W., and Casterlin, M. E. (1979) Behavioral thermoregulation and the "final preferendum" paradigm. *Am. Zool. 19*:211–224.

Reynolds, W. W., Casterlin, M. E., and Covert, J. B. (1976). Behavioral fever in teleost fishes. *Nature* 259:41–42.

Satinoff, E., McEwen, G. N., Jr., and Williams, B. A. (1976) Behavioral fever in newborn rabbits. *Science* 193:1139–1140.

Smith, E. N. (1979) Behavioral and physiological thermoregulation of crocodilians. *Am. Zool.* 19:239–247.

Spotila, J. R., Terpin, K. M., and Dodson, P. (1977) Mouth gaping as an effective thermoregulatory device in alligators. *Nature* 265:235–236.

Spray, D. C. (1974) Characteristics, specificity, and efferent control of frog cutaneous cold receptors. *J. Physiol.* 237:15–38.

Spray, D. C., and Belkin, D. B. (1972) Heart rate–cloacal temperature hysteresis in the iguana is a result of thermal lag. *Nature* 239:337–338.

Spray, D. C., and May, M. L. (1972) Heating and cooling rates in four species of turtles. *Comp. Biochem. Physiol.* 41A:507–522.

Templeton, J. R. (1970) Reptiles. In *Comparative physiology of thermoregulation*, vol. 1 (G. Causey Whittow, ed.), pp. 205–209. New York: Academic Press.

Vaughn, L. K., Bernheim, H. A, and Kluger, M. J. (1974) Fever in the lizard *Dipsosaurus dorsalis*. *Nature* 252:473–474.

Weathers, W. W., and Morgareidge, K. R. (1971) Cutaneous vascular responses to temperature change in the spiny-tailed iguana, *Ctenosauro hemilopha*. *Copeia* no. 3:548–551.

Weathers, W. W., and White, F. N. (1971) Physiological thermoregulation in turtles. *Am. J. Physiol.* 221:704–710.

16

Warm fish

FRANCIS G. CAREY

Some of the actively swimming fish of the open ocean are warm. Fishermen have known for centuries that tuna are warm to the touch, yet until recently physiologists gave little credance to this "lay knowledge." This chapter explores the phenomenon of warm fish. First, it considers the question of how warm, then the mechanisms fish use to keep different parts of the body warm, and finally reasons fish might have for keeping warm.

How warm are "warm fish"?

Temperatures in the warmest muscle of large bluefin tuna, *Thunnus thynnus*, are commonly 10 to 15 °C above water temperature, and we have recorded differences of as great as 21 °C. Other tunas are also warmer than their environment: 2 °C for blackfin, *Thunnus atlanticus*, 3 °C for yellowfin, *T. albacares*, 6 °C for skipjack, *Katsuwonus pelamis*, 8 °C for bigeye, *T. obesus*, and 12 °C for albacore, *T. alalunga* (Barrett and Hester, 1964; Carey et al., 1971; Stevens and Fry, 1971; Graham, 1973; Sharp and Vlymen, 1978). Lamnid sharks are also warm, and the mako, *Isurus oxyrinchus*, averages 5 °C, the white shark, *Carcharodon*, 5 °C, and the porbeagles, *Lamna nasus* and *Lamna ditropis*, 10 °C warmer than the water (Carey and Teal, 1969a; Carey, unpubl. data). The above temperatures have been measured on freshly caught fish, on fish held in captivity, and by telemetry from free-ranging fish in the open ocean. As might be expected, the temperatures observed vary with the situation.

Freshly caught fish may be warmer than "normal" for a number of reasons. In a fish struggling to avoid capture, anaerobic metabolism of glycogen to lactic acid could raise its body temperature 2 to 3 °C (Burton and Krebs, 1953; Barrett and Connor, 1964; Guppy et al., 1979), and increased aerobic metabolism will add to this increase. Also, many of the warm fish are tropical species where heating from the sun and from a hot deck after capture can rapidly increase the temperature of the smaller specimens. Additionally, good fishing areas are often characterized by sharp thermoclines and inversion layers, and there is uncertainty about what water temperature the fish has been exposed to. These factors add

216

variability to temperatures taken from freshly caught fish, yet large size and higher temperatures of larger fish make the errors small, and temperatures measured on freshly caught specimens of large fish are probably within a few degrees of normal temperatures. It seems likely that the temperatures measured on freshly caught fish give the high end of the range.

Captive fish may have temperatures lower than normal. Captive skipjack, yellowfin, and kawakawa (*Euthynnus affinis*) kept in an aquarium facility at NMFS Kewalo Basin Laboratory, Hawaii are usually only several degrees warmer than the water (Nakamura, 1972). Their relatively low temperatures may be related to their small size, for they usually weigh only 1 to 2 kg. It is also possible that captivity has a depressing effect on the temperature. It seems likely that the temperatures measured on captive fish give the low end of the range.

Temperatures measured by telemetry from free-swimming fish give the "normal temperatures" under a variety of natural situations, but the measurements are difficult to make. The fish are delicate and must be handled rapidly, and there is some uncertainty as to the exact location of the probe. Activity-related variations in maximum body temperature of tunas and some sharks may be more than 5 °C in some of the smaller species. Although more telemetry experiments from free-swimming specimens are needed, it seems likely that there is no rigidly fixed or "set" temperature for the warm fish, and that body temperature of individuals at a given water temperature may vary by several degrees.

How do fish keep warm?

Why is it unusual for a fish to be warmer than the water?

Most fish are cold because they breathe water. They must extract oxygen from water, a medium that has one-fortieth the oxygen content and 3000 times the heat capacity of air. The potential for heat loss in extracting the same amount of oxygen is 10^5 times greater for water than for air. This large factor overwhelms other routes of heat loss.

Over the long term, the heat produced by the fish comes from aerobic metabolism. Each time a milliliter of blood passes through the gills, it comes into thermal equilibrium with the water as it extracts oxygen because heat diffuses more than 10 times as rapidly as oxygen. Thus, the oxygenated arterial blood entering the tissue will be close to water temperature (Stevens and Sutterlin, 1976). Metabolic heat is continually carried away to the gills, and warming is limited to the amount of heat released during a single pass of the blood.

An oxygen-rich blood, such as that of bluefin tuna, has an oxygen capacity of 18 vol % and will release about 0.1 ml O_2 ml^{-1} blood as it

TABLE 16.1 Heat exchanges in some pelagic fish.

Scientific name	Common name	Heat exchangers[a]		
		Muscle	Viscera	Eye, brain
Xiphiidae				
Xiphias gladius	Swordfish	0	0	+
Istiophoridae				
Istiophorus platypterus	Sailfish	0	0	+
Tetrapterus albida	White marlin	0	0	+
T. audax	Striped marlin	0	0	+
T. agustirostris	Spearfish	0	0	+
Makaira nigricans	Blue marlin	0	0	+
M. indica	Black marlin	??	0	?
Scombridae[b,c]				
Auxis rochei	Bullet tuna	+	0	?
A. thazard	Frigate tuna	+	0	?
Euthynnus affinis	Kawakawa	+	0	?
E. allatteratus	Little tunny	+	0	?
E. lineatus	Black skipjack	+	0	+
Katsuwonus pelamis	Skipjack tuna	+	0	+
Thunnus tonggol	Longtail tuna	+	0	?
T. atlanticus	Blackfin tuna	+	0	?
T. albacares	Yellowfin tuna	+	0	+
T. alalunga	Albacore	+	+	+
T. obesus	Bigeye tuna	+	+	+
T. thynnus	Bluefin tuna	+	+	+
T. maccoyii	Southern blue-fin tuna	+	+	?
Gasterochisma melampus		??	??	+
Miscellaneous teleosts				
Lepidocybium flavobrunneum	Escolar	0	0	0
Sphyraena barracuda	Barracuda	0	0	0
Coryphaena hippurus	Dolphin fish	0	0	0
Lampris regius	Opah	0	0	?
Luvaris imperialis	Luvar	??	??	??
Elasmobranchs[d]				
Lamnidae				
Isurus oxyrinchus	Mako	+	+	+
I. paucus	Long-finned mako	+	+	+
Lamna nasus	Porbeagle	+	+	+
L. ditropis	Salmon shark	+	+	+
Carcharodon carcharias	White shark	+	+	+
Alopiidae				
Alopias superciliosus	Big-eyed thresher	weakly developed	0	+
Alopias vulpinus	Thresher	?	0	+

TABLE 16.1 (cont.)

Scientific name	Common name	Heat exchangers[a]		
		Muscle	Viscera	Eye, brain
Cetorhinidae				
Cetorhinus maximus	Basking shark	0	0	0
Carcharhinidae				
Carcharhinus obscurus	Dusky shark	0	0	0
Prionace glauca	Blue shark	0	0	0
Sphyrnidae				
Sphyrna mokarran	Great hammerhead shark	0	0	0??
Mobuloidae				
Manta birostris	Manta ray	0	0	0

[a] The presence of a rete mirabile, which may serve as a heat exchanger, is indicated by +. Its absence or probable absence is indicated by 0. Species that I have not examined or seen described in the literature, but which probably have a rete are indicated by ?. Wild guesses are indicated by ??.
[b] Some 31 species of Scombridae, mostly tropical and warm temperate fish, do not have retia. These include various "mackerels," the wahoo Acanthocybium solandri, and the "bonitos."
[c] Among the tunas, the tropical species Auxis through Thunnus albacares have both lateral retia originating from the cutaneous vessels and central retia in the hemal canal. For Auxis, Euthynnus, and Katsuwonus, the central rete is the main heat exchanger. In the genus Thunnus, the lateral retia are well developed and are the only retia serving the muscle in the cold-water or deep-swimming species T. alalunga, T. obesus, T. thynnus, and T. maccoyii. In the Scombridae, visceral retia occur only in the cold-water species of Thunnus. Many fish species have a rete formed by a variety of configurations of the carotid artery in the blood supply to the brain and eye.
[d] Retia are absent in most elasmobranchs.

passes through the tissues. This oxygen will generate 0.5 cal of metabolic heat, and the tissue will warm about 0.5 °C. In a steady state, the venous blood leaving the tissues will be 0.5 °C warmer than the entering arterial blood.

Heat exchangers for muscles

Tunas and other warm fish have developed heat exchangers that transfer heat from the warm venous blood to the cold arterial blood. Heat is not carried away to the gills but accumulates, raising the tissue temperatures. The heat exchangers are formed by the retia mirabilia, masses of vascular tissue arranged for close contact in the countercurrent flow of venous and arterial blood. They occur in a variety of configurations in many species of fish (Table 16.1), but all offer a large interface between arteries and veins for the exchange of heat.

In tunas of the genus *Thunnus*, the major heat exchanger serving the muscle arises from two paired sets of cutaneous vessels. Each set of cutaneous vessels gives rise to many small, parallel arteries and veins that form a slab of vascular tissue on the dorsal and ventral surface of the dark muscle (Kishinouye, 1923; Godsil and Byers, 1944; Carey and Teal, 1966).

Whereas the most conspicuous rete serves the dark muscle, there is another system serving the light muscle in *Thunnus* (Figures 16.1 and 16.2). A series of segmental vessels arise from the cutaneous vessels and run dorsally and ventrally between the skin and the surface of the muscle. These give rise to bands of small arteries and veins that run into the light muscle in a radial direction (Carey and Teal, 1966). The vascular bands are two-dimensional arrays of parallel vessels arranged side by side, arteries alternating with veins. Within the plane of the vascular band the arteries and veins are in good thermal contact. Thermal contact in the dimension normal to the vascular bands is reduced by the interposed muscle fibers, but because the bands are numerous and the distance between them is small, the entire mass of blood vessels and muscle fibers acts as a heat exchanger in the blood supply to the light muscle. Thermal gradients in the tissue indicate that it works well despite being less elaborate than the major heat exchanger to the dark muscle. Although only one layer of vessels thick, the vascular bands are several centimeters long and apparently provide adequate heat exchange at the lower rates of blood flow in the light muscle.

A centrally located rete is the major heat exchanger in the muscle of skipjacks (*Katsuwonus, Euthynnus*, etc.), and a smaller central rete is also present in yellowfin (Stevens et al., 1974; Graham and Diener, 1978). In these fish the hemal canal of the vertebrae is greatly enlarged in the region over the body cavity. The dorsal aorta and postcardial vein which run along the bottom of the hemal canal give rise to many small vessels that fill the lumen with a mass of vertically oriented, parallel, arterioles and venules. These small vessels coalesce into a system of distribution vessels that exit near the top of the hemal canal and run out along the vertebral spines to the muscle. The arterioles and venules in the central

FIGURE 16.1 Vascular bands in the light muscle of *Thunnus*. Vascular bands originate from segmental arteries and veins that run between the surface of the muscle and the skin. The band of alternately arranged blood vessels penetrates radially into the muscle with many branchings and recombinations. The segmental vein is about 1.5 mm in diameter, and the portion of vascular band shown is about 10 mm long.

rete of skipjack are small, 36 and 84 μm in diameter, considerably smaller than those in the lateral retia of the large tunas. The vessels are longer in the large tunas, however, and the total area of interface between arteries and veins per unit of warm muscle is similar (Stevens, pers. commun.).

The lamnid sharks all have laterally located retia. In the mako, *Isurus oxyrinchus*, the rete is a thick layer of vascular tissue with the vessels running in along the dorsal surface of the dark muscle, an arrangement similar to the dorsal half of the bluefin tuna system. The other lamnids, *Carcharodon* and the porbeagles, *Lamna nasus* and *Lamna ditropis*, have a more diffuse rete. The dark muscle in these fish is located close to the vertebral column. A multitude of small arteries and veins arise from the

FIGURE 16.2 Vascular bands in the light muscle of *Thunnus*. (a) Diagram of a vascular band in cross section. There is good thermal contact between the arteries (a) and veins (v). The vascular bands run through the muscle fibers in thin layers of fatty connective tissue. The band illustrated is about 1 mm wide. The relative dimensions in these specimens, which were injected with colored gelatin, fixed, and cleared, may be altered from those in life. (b) Diagram of muscle from a bigeye tuna, *T. obesus*, which was injected with red gelatin in the arteries and black in the veins, fixed, and cleared. This tissue has been cut in parasaggital section in a plane where the main lateral heat exchanger, which shows as the dark horizontal line near the bottom of the specimen, has been reduced to a layer a few vessels thick. The vascular bands consist of 4 to 12 arteries and veins in parallel. They are abundant and should provide a significant heat-exchange effect in the light muscle.

(a)   (b)

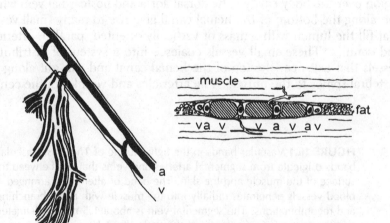

dorsal cutaneous vessels and pass through the intervening light muscle as an array of vascular bands (Burne, 1923) rather than as a solid slab of vascular tissue. The arrangement is similar to the vascular bands in *Thunnus* where light muscle fibers are interposed between the strands of blood vessels.

Heat exchangers for brains and eyes

The brains and eyes of albacore, bigeye, bluefin, and skipjack (Stevens and Fry, 1971; Linthicum and Carey, 1972), and probably of other tunas, are warm. The brain temperatures are generally somewhat lower than the warmest muscle and in bluefin seem to be fairly well regulated over a wide range of ambient temperatures. The eye is somewhat cooler than the brain. The mako and the porbeagle and almost certainly the other lamnid sharks have warm brains and eyes.

Heat exchangers for visceral organs

The visceral organs may also be warm. This is true for bluefin, bigeye, and albacore tunas and for the lamnid sharks. Visceral temperatures are erratic and show wide variation in freshly caught fish. In bluefin tuna there is a clear cycle of temperature changes associated with feeding (Stevens et al., 1978). For an unfed bluefin in 15 °C water, the stomach temperature will be 5 or 6 °C warmer than the water. A few hours after ingestion of a large meal, the temperature starts to rise, and it may reach 15 °C above water temperature in the space of 12 h. The stomach gradually cools as it empties, and 36 h after feeding it is back to 5 or 6 °C above water temperature. It seems likely that the bigeye, albacore tunas, and lamnid sharks will also show changes in visceral temperature associated with feeding.

All of the warm fish are characterized by multiple parallel blood vessels. Where a normal fish may have single, separate veins and arteries, the warm fish will have several arteries and veins running together in close contact.

Experimental evidence for importance of heat exchangers

Some recent experiments illustrate the role of the circulation in transferring heat between a normal fish and its environment and the degree of uncoupling from environmental temperatures made possible by the presence of heat exchangers. Body temperatures and water temperatures were telemetered from several species of large sharks as they swam free in the ocean. We were fortunate in having some of these animals make vertical movements that repeatedly took them through the thermocline and allowed us to observe the effect of changing water temperature on body temperature. The experiments included two blue sharks, *Prionace glauca,* poikilothermic fish that swam up and down between the surface and a depth of 250 m. These excursions were made with a regular 2- to 3-h period and involved a water temperature change of 5 °C or more (Figure 16.3). Muscle temperature (measured at a point 10 cm in from the skin in the region below the dorsal fin) followed water temperature closely. When the sharks remained at constant depth for a period of an hour or so, body temperature came to within the 0.2 °C experimental error of water temperature. The rate of temperature change in the muscle was more than 20 times greater than would be possible with heat transfer by conduction only. A model that assumed that most of the heat was transferred by the circulation was used to calculate blood flow from the observed gradients and rates of muscle temperature change (Barcroft and Millen, 1939). An average blood flow of 69 ml kg^{-1} min^{-1} when the shark was swimming up and 28 ml kg^{-1} min^{-1} as it was gliding down would account for the observed changes in muscle temperature. These blood-flow values are reasonable for active fish muscle (Satchell, 1971) and would account for all of the observed temperature change at the point of measurement. Convective heat transfer by the blood is clearly

FIGURE 16.3 Temperature telemetry experiment with a blue shark, *Prionace glauca.* Muscle temperature: heavy continuous line; water temperature: light interrupted line. Muscle temperature measured at a point 10 cm deep from the surface changes rapidly with water temperature. The blue shark is a poikilotherm and comes to water temperature when it remains at constant depth for an hour or so as at the beginning of this record. Each graph represents 24 h, plotted from midnight to midnight.

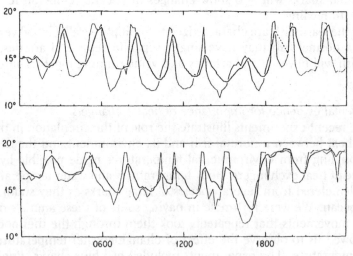

the dominant mode of heat transfer in this fish and ties the animal closely to the water temperature.

The rapid transport of heat by the circulation in the blue shark can be contrasted with results from two warm species. One of these was a 200-kg female mako shark that we followed for 4 days in an area of clear, warm water north of the Bahamas. She repeatedly passed through the thermocline in regular 2- to 3-h excursions between the surface and 500 m (Figure 16.4). The temperature in the stomach of this shark showed almost no response to fluctuations in water temperature with a period

FIGURE 16.4 Stomach temperature for a mako shark, *Isurus oxyrinchus.* Upper, smooth line: stomach temperature; lower varying line: water temperature. Tissue temperatures remained constant as this shark swam up and down through the thermocline. The suprahepatic rete mirabile in the visceral circulation greatly retards heat transfer between the stomach and the environment. Heat transfer by conduction and through small inefficiencies in heat exchange causes the stomach temperature to vary slowly with long-term changes in average water temperature.

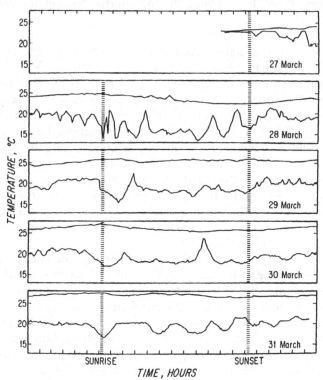

of 2 to 3 h, although it did follow slow changes in average water temperature with a period on the order of a day. The large suprahepatic rete in the viscera (Burne, 1923) had eliminated most of the heat transfer through the circulation, and conductive heat transfer was too slow to produce any temperature change on a time scale of 2 to 3 h.

In another experiment muscle temperature was telemetered from a 900-kg white shark, *Carcharodon carcharias* (Carey, unpubl. experiments). The thermistor harpoon was placed 30 cm deep in the muscle in the region below the dorsal fin. Although this fish did not swim up and down in a regular fashion like the blue and the mako, it did pass through some changes in water temperature. *Carcharodon* has a well-developed lateral heat exchanger, and, as expected, muscle temperature did not respond to rapid changes in water temperature. Heat transfer by the circulation is proportional to blood flow and independent of the size of the fish. In large, active fish where conduction of heat is a slow process, convective heat transfer by the circulation is the dominant process in temperature change. When heat exchangers are present, convective heat transfer will be retarded, and the fish will respond very slowly to environmental temperature changes. It will have a large thermal inertia (Neill et al., 1976).

How efficient are heat exchangers?

In order to achieve the higher temperatures observed in some of the warm fish, the heat exchangers must work at a high efficiency. Maximum muscle temperatures in bluefin tuna may be more than 15 °C warmer than the water. The source of heat is aerobic metabolism with oxygen supplied by blood with a capacity of 0.18 ml O_2 ml^{-1} blood. If about half of this oxygen is utilized in the tissues, 0.5 cal of metabolic heat will be produced per milliliter of blood flow. The venous blood leaving the warmest tissue, however, will carry an excess of 15 cal ml^{-1} with respect to the entering arterial blood (Figure 16.5). To maintain a constant temperature, heat loss must equal heat production, and only 0.5 cal ml^{-1} blood can be lost in the gill. This requires that 14.5 of the 15 cal ml^{-1} venous blood be transferred to the arterial stream in the exchanger. The exchanger must work at 96.7% efficiency.

This is the case if the only route of heat loss is through the circulation. However, as body temperature rises, conduction of heat to the surface becomes an important route of heat loss. Brill et al. (1978) estimated that 30 to 70% of metabolic heat loss in 2-kg skipjack occurs through the body surface. Surface heat loss in large tuna should be proportionately lower, but still may be an important fraction of total heat production because of the higher temperatures and steep thermal gradients. If we assume that half of the 0.5 cal of metabolic heat generated per milliliter of blood flowing through the tissue in a warm fish is lost through the body surface, only 0.25 cal ml $^{-1}$ can be lost through the circulation if the temperature is to remain constant. This requires that the exchanger transfer 14.75/15.00 or 98.3% of the heat.

Still higher efficiencies may actually be required. In swimming, about 20% of the energy produced by metabolism is transferred to the environment (Webb, 1971). This means that less heat is available for warming the body, and the efficiency of the heat exchanger must be still further increased.

A small change in efficiency should produce a large change in temperature. With heat production as in the above example, a decrease in exchange efficiency from 98.3% to 95% would result in a decrease in the temperature differential from 15 °C to 5 °C. Temperature control by shunting blood around the rete or otherwise altering the efficiency of the exchanger should be practical, with only a small rerouting of blood flow required to produce a large effect on temperature.

High efficiencies can create new problems

A consequence of the high efficiencies required of the heat exchanger is that some oxygen may begin to pass from the arterial to the venous stream. Because heat diffuses more rapidly than oxygen, the exchanger dimensions seem appropriate for the effective transfer of heat with a

FIGURE 16.5 Diagram of heat production and loss in a bluefin tuna. Temperature (in °C) and heat are referenced to water. Heat production by the tissue is dependent on oxygen from the gill, and is expressed as calories per unit of blood flow. The oxygen extracted from the blood allows the generation of 0.5 cal ml $^{-1}$ blood. Of this it is assumed that half will be lost through the body surface, half in respiration. The venous blood leaving the tissue carries away 15 cal ml $^{-1}$ blood, but only 0.25 cal ml $^{-1}$ of this is available from the arterial blood for replacing respiratory losses. To maintain constant temperature the heat-exchange rete must transfer 14.75 cal ml $^{-1}$ or 98.3% of the heat from the venous to the arterial stream.

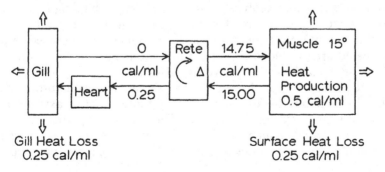

minimal loss of oxygen. However, as the efficiencies required for heat exchange approach 99%, an increasing loss of oxygen from the arterial to the venous stream can be expected.

The hemoglobins of bluefin tuna and porbeagle shark have properties that suggest that loss of oxygen from the arterial stream in the heat exchanger is a problem. Both of these fish have high body temperatures requiring very efficient transfer of heat in the heat exchanger. As arterial blood with a normal hemoglobin traversed the rete, it would bind oxygen less firmly as it warmed, and the rise in P_{O_2} would increase the diffusion of oxygen from the arterial to the venous stream. In the porbeagle, oxygen binding is unaffected by temperature (Andersen et al., 1973), and this increase in P_{O_2} would not take place. Bluefin tuna is even more unusual in having a reversed temperature effect on oxygen binding from that seen in other vertebrates (Carey and Gibson, 1977). In the bluefin the mechanism of temperature stability involves an effect on the allosteric parameter L (Monod et al., 1965), such that the hemoglobin, which is predominantly in the low-affinity T state when cold, shifts to the high-affinity R state when warm. This temperature effect on the hemoglobin configuration more than counteracts the normal positive effect of temperature on the dissociation constant (Gibson, 1959), the result being a hemoglobin in which the effect of temperature on oxygen binding is the reverse of normal. Oxygen is bound less tightly as temperature decreases. It is most interesting that the porbeagle has achieved a temperature stability in quite a different way. In this shark the allosteric change in configuration is not strongly influenced by temperature (Dickinson and Gibson, unpubl. data), but the rate of the association reaction is unusually temperature-sensitive (Andersen et al., 1973). This balances the normally large effect of temperature on the dissociation constant and results in an equilibrium curve that is temperature-stable. Thus the tuna and the shark that maintain the greatest temperature elevation above the environment of any species in their groups have both achieved a remarkable temperature stability in the oxygenation of their hemoglobin, but have evolved quite different molecular mechanisms for doing so. This implies that temperature invariance of oxygen binding is not some second-order effect, but is of primary importance to these fish. It is probably important for the reason we have suggested: the need to pass oxygenated arterial blood through a heat exchanger of high efficiency without appreciable loss to the venous stream.

Calculations of oxygen loss in the rete indicate that with normal mammalian hemoglobin the arterial blood might lose 9% of its oxygen to the venous stream in traversing the rete. The special properties of bluefin hemoglobin would reduce this loss to about 3%. At first sight this improvement is not very impressive, possibly because of some inappropriate assumptions in the calculation. Its importance is amplified by the assumed coefficient of oxygen utilization of 0.5, which would require an 18% increase in cardiac output to supply the tissues with oxygen using a mammalian hemoglobin, but only a 6% increase with bluefin hemoglobin. To achieve the high temperatures observed in bluefin and porbeagle, it is important that oxygen be supplied with a minimum flow of cooling blood through the tissues. By reducing the cardiac output required to deliver oxygen to the tissue, the reverse temperature effect on oxygen binding helps accomplish this.

Because oxygen must be delivered with a minimum flow of blood, to achieve elevated tissue temperatures the warm fish would be expected to have high blood oxygen-carrying capacities. This in fact is the case: Blood from the cold blue shark has an oxygen capacity of 4 vol %, that from the porbeagle 14 vol %. Blood from salmon may carry 9 vol %, blood from bluefin tuna 18.5 vol % (Holeton and Randall, 1967; Carey, unpubl. data). We would also expect warm fish to have unusually large coefficients of oxygen utilization. Realistic measurements of arterial and venous oxygen content in the blood of such active and delicate animals will be difficult to obtain, but from the above considerations one would expect to find that a large fraction of the oxygen was extracted from the blood and utilized by the tissues.

Why should a fish be warm?

What advantage might a fish obtain by being warm? The following suggestions have been made, but none of them seems overwhelmingly compelling.

Thermoregulation

Being warmer than its environment gives an animal the option of regulating its temperature. This suggestion is weakened by lack of evidence that most of these fish thermoregulate. The only species that clearly seems to control its temperature is bluefin tuna (Carey and Teal, 1969b).

Increased power from muscle

A frog muscle completes its contraction–relaxation cycle three times faster when warmed up by 10 °C, whereas the contractive force remains the same (Hartree and Hill, 1921). This suggests that a threefold increase in power might be obtained by warming the muscle 10 °C. Power is important for speed, and tunas are fast-swimming fish. However, fish such as the wahoo, *Acanthocybium solandri*, can swim as fast without being warm (Walters and Fiersteine, 1964), and tuna muscle does not seem to show any peculiar increase in contractive properties on warming (Brill and Dizon, 1979). Being warm is not the only way to achieve high burst-swimming speeds.

Thermal inertia

The heat exchangers may allow the fish to make excursions into the cold water beneath the thermocline without its body temperature dropping rapidly (Neill et al., 1976). This was clearly demonstrated in the mako shark experiment (Figure 16.4). The heat exchangers may thus function to enlarge the thermal environment available to the fish. The adaptive value of slowing the changes in body temperature might explain the presence of heat exchangers in fish such as the yellowfin tuna which normally does not appear to be very warm. The advantages of increased thermal inertia may have been important in the evolution of the countercurrent heat exchange system. A rudimentary system that was not efficient enough to produce physiologically significant 5 to 10 °C temperature elevations could still be effective in increasing thermal inertia.

Recovery from burst activity

Most fish recover slowly from severe exercise. Typically 12 to 24 h may elapse before lactate in fish muscle and blood returns to pre-exercise levels (Stevens and Black, 1966). In yellowfin and skipjack tuna, lactate levels return to normal in about 2 h, much more quickly than in normal fish. The fish warm up during and after severe exercise, and the increased temperature may speed recovery (Stevens and Neill, 1978). This could be a decided advantage to an active predator pursuing schools of fast-swimming prey.

Increased rate of aerobic metabolism

Tunas can swim for hours at speeds in excess of 8 knots. They are true marathon performers, able to sustain high levels of muscle activity for prolonged periods of time. Stevens and Carey (1981) suggest that their elevated body temperatures allow tuna to sustain a high level of muscle activity by aiding the delivery of oxygen from the capillaries to the mitochondria. Oxygen supplied by simple diffusion is not increased by warming. The increasing rate of diffusion is balanced by a decreasing solubility, resulting in a negligible increase in oxygen flux. Tuna muscle is notably rich in myoglobin, however; the dark muscle is almost black, and even the light muscle is red. The myoglobin significantly increases the solubility of oxygen, and as much as half of the oxygen supply to the muscle may be carried to the mitochondria bound to myoglobin. The rate of such facilitated diffusion increases with temperature, and a 10 °C rise in temperature might increase the flux of oxygen to the mitochondria by 40% in a muscle where the myoglobin concentrations are as high as found in tuna dark muscle. This is a significant increase and may help the tuna achieve its maximum performance.

It seems likely that the selective advantages of being warm encompass a wide variety of effects, some of which perhaps have not yet occurred to us.

References

Andersen, M. E., Olson, J. S., Gibson, Q. H., and Carey, F. G. (1973) Studies on ligand binding to hemoglobins from teleosts and elasmobranchs. *J. Biol. Chem.* 248:331–341.

Barcroft, H., and Millen, J. L. E. (1939) The blood flow through muscle during sustained contraction. *J. Physiol.* 97:17–31.

Barrett, I., and Connor, A. R. (1964) Muscle glycogen and blood lactate in yellowfin tuna, *Thunnus albacares*, and skipjack, *Katsuwonus pelamis*, following capture and tagging. *Bull. Inter-Am. Trop. Tuna Comm.* 9(4):218–268.

Barrett, I., and Hester, F. J. (1964) Body temperature of yellowfin and skipjack tunas in relation to sea surface temperature. *Nature* 307:96–97.

Brill, R., and Dizon, A. (1979) Effect of temperature on isotonic twitch of white muscle and predicted maximum swimming speeds of skipjack tuna. *Environ. Biol. Fish* 4:199–205.

Brill, R. W., Guernsey, D. L., and Stevens, E. D. (1978) Body surface and gill heat loss rates in restrained skipjack tuna. In *The physiological ecology of tunas* (G. D. Sharp and A. E. Dizon, eds.), pp. 261–276. New York: Academic Press.

Burne, R. H. (1923) Some peculiarities of the blood vascular system of the porbeagle shark (*Lamna cornubica*). *Phil. Trans. R. Soc. Lond.* 212B:1923–1924.

Burton, K., and Krebs, H. A. (1953) The free energy changes associated with the individual steps of the tricarboxylic acid cycle, glycolysis and alcoholic fermentation and with the hydrolysis of the pyrophosphate groups of adenosine triphosphate. *Biochem. J.* 54:94–107.

Carey, F. G., and Gibson, Q. H. (1977) Reverse temperature dependence of tuna hemoglobin oxygenation. *Biochem. Biophys. Res. Commun.* 78:1376–1382.

Carey, F. G., and Teal, J. M. (1966) Heat conservation in tuna fish muscle. *Proc. Natl. Acad. Sci. U.S.A.* 56:1464–1469.

Carey, F. G., and Teal, J. M. (1969a) Mako and porbeagle: Warm bodied sharks. *Comp. Biochem. Physiol.* 28:199–204.

Carey, F. G., and Teal, J. M. (1969b) Regulation of body temperature by the bluefin tuna. *Comp. Biochem. Physiol.* 28:205–213.

Carey, F. G., Teal, J. M., Kanwisher, J. W., Lawson, K. D., and Beckett, J. S. (1971) Warm-bodied fish. *Am. Zool.* 11:137–145.

Gibson, Q. H. (1959) The kinetics of reactions between hemoglobin and gases. *Prog. Biophys. Chem.* 9:1–53.

Godsil, H. C., and Byers, R. D. (1944) A systematic study of the Pacific tunas. *Calif. Dept. Fish Game Fish. Bull.* 60:1–31.

Graham, J. B. (1973) Heat exchange in the black skipjack and the blood–gas relationship of warm-bodied fishes. *Proc. Natl. Acad. Sci. U.S.A.* 70:1964–1967.

Graham, J. B., and Diener, D. R. (1978) Comparative morphology of the central heat exchangers in the skipjacks *Katsuwonus* and *Euthynnus*. In *The physiological ecology of tunas* (G. D. Sharp and A. E. Dizon, eds.), pp. 113–133. New York: Academic Press.

Guppy, M., Hurlbert, W. C., and Hochachka, P. W. (1979) Metabolic sources of heat and power in tuna muscles. II. Enzyme and metabolite profiles. *J. Exp. Biol.* 82:303–320.

Hartree, W., and Hill, A. V. (1921) The nature of the isometric twitch. *J. Physiol.* 55:389–411.

Holeton, G. F., and Randall, D. J. (1967) The effect of hypoxia upon the partial pressure of gases in the blood and water afferent and efferent to the gills of rainbow trout. *J. Exp. Biol.* 46:317–327.

Kishinouye, K. (1923) Contributions to the study of the so-called scombrid fishes. *J. Coll. Agric. Imp. Univ. Tokyo* 8:394–475.

Linthicum, D. S., and Carey, F. G. (1972) Regulation of brain and eye temperatures by the bluefin tuna. *Comp. Biochem. Physiol.* 43A:425–433.

Monod, J., Wyman, J., and Changeux, J. P. (1965) On the nature of allosteric transitions: A plausible model. *J. Mol. Biol.* 12:88–118.

Nakamura, E. L. (1972) Development and uses of facilities for studying tuna behavior. In *Behavior of marine animals*, vol. 2 (H. E. Winn and B. L. Olla, eds.), pp. 245–277. New York: Plenum Press.

Neill, W. H., Chang, R. K. C., and Dizon, A. G. (1976) Magnitude and ecological implications of thermal inertia in skipjack tuna *Katsuwonus pelamis* (Linnaeus). *Environ. Biol. Fish* 1:61–80.

Satchell, G. H. (1971) *Circulation in fishes.* Cambridge University Press. 131 pp.

Sharp, G. D., and Vlymen, W. J., III (1978) The relation between heat generation, conservation and the swimming energetics of tunas. In *The physiological ecology of tunas* (G. D. Sharp and A. E. Dizon, eds.), pp. 213–259. New York: Academic Press.

Stevens, E. D., and Carey, F. G. (1981) One "why" of the warmth of warm-bodied fishes. *Am. J. Physiol.* 240:R151–R155.

Stevens, E. D., and Black, E. C. (1966) The effect of intermittent exercise on carbohydrate metabolism in rainbow trout. *J. Fish. Res. Bd. Can.* 23:471–485.

Stevens, E. D., Carey, F. G., and Kanwisher, J. W. (1978) Changes in the visceral temperatures of bluefin tuna. *Collective volume of scientific papers, Intl. Comm. Conserv. Atlantic Tuna* 7:383–388.

Stevens, E. D., and Fry, F. E. J. (1971) Brain and muscle temperatures in ocean caught and captive skipjack tuna. *Comp. Biochem. Physiol.* 38A:203–211.

Stevens, E. D., Lam, H. M., and Kendall, J. (1974) Vascular anatomy of the countercurrent heat exchanger of skipjack tuna. *J. Exp. Biol.* 61:145–153.

Stevens, E. D., and Neill, W. H. (1978) Body temperature relations of tunas, especially skipjack. In *Fish physiology*, vol. 7 (W. S. Hoar and D. J. Randall, eds.), pp. 316–359. New York: Academic Press.

Stevens, E. D., and Sutterlin, A. M. (1976) Heat transfer between fish and ambient water. *J. Exp. Biol.* 65:131–145.

Walters, V., and Fiersteine, H. L. (1964) Measurement of the swimming speeds of yellowfin tuna and wahoo. *Nature* 202:208–209.

Webb, P. W. (1971) The swimming energetics of trout. II. Oxygen consumption and swimming efficiency. *J. Exp. Biol.* 55:521–540.

PART FOUR
Water

Overview

Part Four considers how animals osmoregulate.

In Chapter 17 Peter Bentley takes the unusual approach of treating the skin of vertebrates as an "osmoregulatory organ." Looked at in this way, skin exhibits rich anatomical and physiological diversity in which differences reflect the needs of the organisms.

In Chapter 18 Malcolm Peaker reviews the progress that has been made in understanding how salt glands work since Knut Schmidt-Nielsen discovered their function nearly 25 years ago. The cellular mechanisms for producing highly concentrated salt solutions by these glands is still not well understood, but Peaker is optimistic, predicting that new tools may soon provide answers.

In Chapter 19 Reinier Beeuwkes discusses the countercurrent multiplier mechanism used by mammalian kidneys to form a concentrated urine. New experiments on isolated tubule segments, together with mathematical modeling, have led to a new proposal for explaining how the mechanism works: a "passive" countercurrent hypothesis in which urea plays a major role.

In Chapter 20 Simon Maddrell discusses the osmotic mechanisms and adaptations of insects. Insects are perhaps the "premier" osmoregulators among animals. Studies of the osmoregulatory roles of their integument, excretory system, and hindgut have uncovered fascinating mechanisms. Maddrell points out that the role of the respiratory system in limiting water loss during oxygen uptake is not understood and should be a fruitful area for future research.

17 Roles of vertebrate skin in osmoregulation

P. J. BENTLEY

The skin of vertebrates provides an admirable exoskeleton, which is consistent with the maintenance of their shape while allowing flexibility for growth and locomotion. It may also take on many hues and acquire a variety of appendages, such as scales, feathers, and hair, which may aid survival. It is, however, probably less appreciated that the skin also has several important physiological functions in addition to these essential morphological roles. In tetrapod vertebrates the skin is synonymous with the integument, though in fish and larval amphibians the gills also make a major contribution to the external surface of the animal.

The skin contributes to the normal physiological functioning of the animal by limiting exchanges with its surroundings, in either direction, of water, salts, nutrients, and heat. In some instances the skin can be quite selective and facilitate the uptake or loss of specific substances so that it may play a dynamic role in homeostasis. Such roles may be dictated phyletically or reflect the nature of the environment where the animal normally lives. Thus the skin of amphibians contributes to their respiration and that of many mammals to their thermoregulation. Species that live in dry desert regions may have skin that strongly restricts evaporation, and aquatic animals an integument that is almost impermeable to ions and water.

The present chapter will be concerned with the various roles of the skin in helping to maintain the body's water and salt content (i.e., osmoregulation). Vertebrates live in a great variety of ecological situations where the environmental stresses on their abilities to maintain their water and salt content differ. Such environments include the sea, fresh-water rivers and lakes, tropical rain forests, and hot dry deserts, and these conditions are often associated with special adaptations in the properties of the skin.

Exchanges of water and salts across the skin depend on the external conditions and will be influenced by the beast's size, surface area, and metabolic rate. In terrestrial vertebrates the skin is usually a major, if not the principal, avenue of water loss, whereas in many species that live in fresh water, especially the amphibians, it can be a major route for the gain of water and loss of salts. Some mammals, especially those living

237

in hot tropical conditions, may also lose considerable quantities of salts as a result of sweating through their skin. Many species have, however, adopted morphological, physiological, and behavioral strategies that limit or compensate for such cutaneous exchanges.

Morphology of the skin

The morphology of the skin of vertebrates reflects its general protective role (see Scheuplin and Blank, 1971; Flaxman, 1972), and the skin of all vertebrates displays certain basic structural similarities. However, there are also many differences, both between the major phyletic groups, such as the fish and tetrapods, and also between species, such as among amphibians, that are more closely related.

Vertebrate skin contains two main morphological regions that have different embryonic origins: the epidermis, which is ectodermal, and the dermis, or corium, which is mesodermal. The epidermis consists of stratified layers of epithelial cells, an arrangement that is unique to the vertebrates. The dermis provides support and sustenance to the epidermis, and it is the site of blood vessels, nerves, and pigment cells. In fish, scales and denticles arise in this layer of the skin. Many multicellular glands and appendages, such as feathers and hair, appear to have a joint dermal–epidermal origin. The dermis contains collagen fibers in a mucopolysaccharide matrix, and it can be quite thick, so that it can provide a strong and substantial protective barrier. This function, however, appears to be mainly of a mechanical nature because when the overlying epidermis is removed, transcutaneous molecular exchanges increase considerably. Thus, the principal permeability barrier in the skin appears to reside in the epidermis.

The epidermal cells usually originate in the basal regions of this layer of tissue, the stratum germinativum, and then migrate toward the outer surface. It appears that in some bony fish the formation of epidermal cells may not be confined to such a basal layer. In tetrapod vertebrates the outward migration of the epidermal cells is associated with a process of their differentiation that involves the formation of a characteristic protein called keratin. This protein is insoluble and is laid down as fibers within the epidermal cells, which may eventually become packed with it (keratinization or cornification). These cells die, but continue to form a contiguous outer layer, the stratum corneum, in which the permeability barrier of the skin appears to reside. The intercellular spaces between these cornified epithelial cells contain lipids. Cornified cells may be continually or periodically discarded and replaced, the latter process occurring during molting.

The skin of amphibians is only lightly keratinized as compared with other tetrapod vertebrates. Keratin is the protein from which hair, feathers, and reptile scales are made.

Fish skin does not contain keratin, and cornification of the epidermis does not occur. These vertebrates appear to have adopted an alternative cutaneous protective strategy, which is consistent with their manner of life; their epidermis contains numerous mucus-secreting cells. The amphibian skin also secretes mucus, but it is formed by multicellular glands. The role of mucus in cutaneous function is controversial. It appears to have a lubricating function and can exclude microorganisms, but its role as a permeability barrier is doubtful (see Marshall, 1978). Thus, water and ions can diffuse through mucus quite freely. Appendages such as secretory glands, feathers, hair, and even pigment cells, may contribute to the effectiveness of the skin as a permeability barrier.

Pathways for movement across skin

Molecules and ions may cross the skin by two general pathways, which can be designated as percutaneous and secretory, respectively.

The percutaneous route

This route involves movement either through the cells or between them (paracellular pathway). Ions and polar molecules such as water generally cross the skin by the percutaneous route less readily than lipid-soluble molecules. However, there are considerable interspecies differences in the skin, one of the best-known examples being the skin of amphibians, which usually is quite permeable to water and ions when compared to the integument of other vertebrates (see Tables 17.1–17.3). Percutaneous movements of molecules usually occur by diffusion, but in some fish and amphibians the epithelial cells may contribute active ion "pumps."

The precise barrier for transcutaneous molecular exchanges has not been defined, but in tetrapod vertebrates it appears to be present in the stratum corneum (see Scheuplin and Blank, 1971). Diffusion through the densely packed keratin appears to be slow, and the intercellular lipids may have an important role in restricting movement of hydrophilic molecules. A surface layer of lipids, such as may be contributed by mammalian sebaceous glands or amphibian lipid glands, may also contribute toward limiting the permeability of the skin. Some molecules may be able to bypass the main epithelial barrier by diffusing through hair follicles and feather tracts.

Secretion

Secretion of water and solutes may play an important role in the movements of ions or molecules across the skin. This process may occur as a result of the activities of single epithelial cells, such as mucous cells and chloride cells in fish, or multicellular secretory glands, such as the mucus glands of amphibians and the sweat glands of mammals, which secrete fluids that keep the skin moist or act as an aid to the dissipation of heat.

Role of the skin in osmoregulation

In order to appreciate the role of the skin in the osmoregulation of vertebrates it is necessary to recall the nature of the physicochemical gradients that may exist between different species and their normal external environments. Two general types of exchange occur across the skin: those between two liquid phases, and those between a gas and a

TABLE 17.1 Passive unidirectional fluxes of sodium, in vitro, across the skin of various species of vertebrates (n Equiv cm^{-2} h^{-1}).

Mammals[a]	
Rabbit	370
Pig	110
Man (forearm)	5
Reptiles[b]	
Garter snake (*Thamnophis radix*, terrestrial)	9
Water snake (*Natrix sipedon*, fresh water)	25
Yellow-bellied sea snake (*Pelamis platurus*)	0
Amphibians	
Frog (*Rana pipiens*)[c]	130
African clawed toad (*Xenopus laevis*, fresh water)[d]	90
Fishes	
Spiny dogfish (*Squalus acanthias*)[e]	80
Dogfish (*Scyliorhinus acanthias*)[f]	370
Trout (*Salmo irideus*)[g]	150 (Cl)
Lamprey (*Lampetra fluviatilis*)[h]	70 (Cl)

[a] Tregear (1966).
[b] Dunson (1978).
[c] Bentley and Yorio (1976).
[d] Yorio and Bentley (1978).
[e] Horowicz and Burger (1968).
[f] Payan and Maetz (1970).
[g] Krogh (1937).
[h] Bentley (1962).

liquid phase. Species from aquatic habitats experience the former and terrestrial animals the latter.

Liquid–liquid interface

Exchange at a liquid–liquid interface involves the diffusion of molecules in water and, if they are separated by a semipermeable membrane, movements of water in response to differences in osmotic pressure. I shall only briefly summarize the occurrence of the osmotic gradients that may occur across the integument of different vertebrates.

The concentrations of body fluids of vertebrates can be broadly placed in two groups: those with an osmolality of about 1 and those where it is 0.3 to 0.4. Sea water has an osmolality of about 1, and the myxinoid cyclostome fish (hagfish), the cartilaginous fish, the coelacanth (*Latimeria*), and the crab-eating frog (*Rana cancrivora*) all have body fluids that are nearly equal or slightly hyperosmotic to this external solution. Thus they have avoided the potential problem of dehydration due to an osmotic loss of water. However, apart from the hagfish, they all have a considerably lower salt concentration in their body fluids than that in the sea water in which they live, so that they tend to gain salts by diffusion.

TABLE 17.2 Permeability coefficients of tritiated water in vertebrate skins bathed on both sides by aqueous solutions $(K_{trans} \times 10^{-7} \text{ cm s}^{-1})$

Mammal skin	
Man (stratum corneum)[a]	2.8
Reptile skin	
Water snake (*Natrix sipedon*)[b]	6
Marine snake (*Cerberus rhynchops*)[b]	2
Amphibian skin	
Frog (*Rana pipiens*)[c]	5200
Mudpuppy (*Necturus maculosus*)[d]	310
Fish skin	
Dogfish (*Scyliorhinus canicula*)[e]	1000
Cellular membranes[a]	
Red blood cell	130 000
Arbacia eggs	5000
Artificial membrane[a]	
Saran plastic	0.2

[a] Scheuplin (1965).   [b] Dunson (1978).
[c] Bentley and Yorio (1976).   [d] Bentley and Yorio (1977).
[e] Payan and Maetz (1970).

Marine bony fish (except the coelacanth), petromyzontid cyclostomes (lampreys), and marine reptiles and mammals all have body fluids with an osmolality of 0.3 to 0.4. They are thus hypoosmotic to sea water and will tend to lose water by osmosis, as well as gain salts by diffusion. In contrast, vertebrates that live in fresh water are hyperosmotic to their surroundings and will tend to gain water by osmosis but lose salts by diffusion.

Although such gradients in concentration of water and solutes are a potential problem, they clearly do not limit the life of such animals in their natural media, owing to two main types of physiological mechanisms: (1) the conservative role of the skin in reducing movements of water and solutes along their concentration gradients; and (2) an ability of marine species, when necessary, to excrete salts while retaining water.

TABLE 17.3 Evaporative water loss from the skin *in vivo* of various tetrapod vertebrates (μg cm^{-2} h^{-1} mm Hg saturation deficit). (See text for details.)

Mammals[a]	
Man	48
White rat	46
Camel	27
Birds	
Painted quail (*Excalfactoria chinensis*)[b]	20
Zebra finch (*Taeniopygia castanotis*)[c]	
Water ad lib	18
No drinking water	6
Reptiles[a]	
Caiman (*Caiman sclerops*)	65
Water snake (*Natrix sipedon*)	41
Iguana (*Iguana iguana*)	10
Chuckwalla (*Sauromalus obesus*)	3
Amphibians[d]	
Leopard frog (*Rana pipiens*)	9000
Mexican tree frog (*Agalychnis dacnicolor*)	300[e]
Planar water surface	(1500)

[a] Schmidt-Nielsen (1969).
[b] Calculated from Bernstein (1971).
[c] Calculated from Lee and Schmidt-Nielsen (1971).
[d] Bentley and Yorio (1979).
[e] Much lower evaporation than this has been observed in *Chiromantis* and *Phyllomedusa* (see text) but insufficient information for calculation in this way.

In bony fish this excretion is due to the ability of chloride cells in the integument, especially the gills, to actively secrete chloride against its electrochemical gradient, back into the surrounding sea water. Some cartilaginous fish and reptiles possess glands ("salt glands") that can secrete solutions with a higher sodium and chloride concentration than that in the sea water.

Liquid–gas interface

A liquid–gas interface exists in terrestrial vertebrates which have a basically different type of osmoregulatory problem from the fish – they tend to lose water by evaporation from their body surface. Such water loss may occur unavoidably, or, in homeothermic mammals, the process may even be promoted in order to facilitate loss of heat. Water loss by evaporation involves movements of water between the liquids in the body and the gases of the atmosphere. Knut Schmidt-Nielsen in 1969 referred to this as the "neglected interface" because so little is known about factors that determine such water losses in animals.

The process of the diffusion of water, or evaporation, across a liquid–gas interface had been described in physical terms by Davies and Rideal (1963). The resistance to molecular diffusion can be divided into three parts, contributed (1) by the unstirred layer of the liquid, R_L; (2) by the boundary, or interface, between the two phases, R_I; and (3) by the gas phase at the outer surface, R_G. The total resistance $R = R_L + R_I + R_G$. For evaporation of water R_L is zero, and R_I is usually quite small for pure water. However, the addition of monolayers at the interface R_I, such as by higher alcohols or long-chain fatty acids, can increase this resistance by as much as 10^4 times.

Impurities in the liquid will also result in an increase of R_I. The R_G is usually the predominant component of the resistance to evaporation, but it can be reduced by intensive stirring, such as with a jet of dry air, or by reducing the atmospheric pressure. Evaporation from a liquid–gas interface will also be influenced by its geometry. If the water is in droplets, the rate of evaporation per unit surface area will be greater than from a flat surface. This effect reflects the geometrical effect of the curved surface in decreasing R_G. Thus, evaporation may be greater if the water is in droplets as compared with a continuous film.

Precise information is not available about the nature of the cutaneous liquid–gas interface and the effects of various external factors, such as temperature, barometric pressure, and air movement. An approximate estimate of the driving force for evaporation can be provided by the saturation deficit, which is the difference between the vapor pressure for water over the water phase at the skin surface (at the temperature of the observation) and the vapor pressure of water in the external air. On this

TABLE 17.4 Possible strategies for changing the permeability of skin to water and solutes

Strategy	Example
To limit exchanges	
Increased thickness	Layers of epithelial cells increased, cocoons in estivating lungfish and amphibians
Increased density of insoluble protein (increased activation energy for diffusion)	Keratinization of epidermis in tetrapods, stratum corneum
Utilization of hydrophobic properties of lipids and waxes	Intercellular lipids, surface lipids, and waxes (e.g., amphibian skin, bird feathers, mammal skin?)
Establishment of unstirred layers of fluid and air (increased resistance to diffusion)	Fur, plumage, scales (?)
Pigmentation (light color reduces heating from radiation)	Animal coloration
Reduction of surface area and/or shape alteration	Postural changes
Ion "pumps"	Na "pump" in epithelial cells of amphibians (opposes leakage)
Transcutaneous electrical potential difference	Corium electropositive in amphibians (limits anion efflux)
Behavior	Avoidance of environments that enhance evaporation and osmosis (choice of temperature, humidity, and salinity)
To enhance exchanges	
Secretory cells and glands	Mucous cells in fish and amphibians, Cl cells in fish skin, sweat glands in mammals (neural and humoral control)
Controlled permeability of epithelial skin	Na, Cl, and water uptake in amphibians (may involve hormonal control)
Mechanical collection of fluids	Sculptured surface (helps uniformly wet skin in amphibians and drinking in lizard, *Moloch*), feathers (can be used to collect water for young in birds)
Increased blood flow	Increased evaporation and enhanced osmotic water uptake in amphibians
Regional differences in cutaneous permeability	Water uptake in amphibian skin, different rates of evaporation from parts of skin in man

basis one can also make comparisons of the evaporation from the skins of different species. The results tabulated in Table 17.3 show considerable differences, which have a phyletic pattern but are also related to the types of environments where each species normally lives.

Some of the strategies that vertebrates may utilize to restrict or enhance exchanges of water and solutes across their skins are summarized in Table 17.4.

Phyletic review of the role of skin in osmoregulation

Fish

Fish are aquatic, and their integument must moderate molecular exchanges at a liquid–liquid interface. Their skin consists of several layers of epidermal cells in a viscous cement. Mucous cells are characteristically plentiful in the epidermis, which also often contains cells having a network of filaments, and certain specialized types of cells including "club" cells and chloride cells (see Henrikson and Matoltsy, 1968a,b,c).

Fish skin is usually considered to be almost impermeable to water and salts (see Parry, 1966), but there are only a few direct measurements of this property available. It is notable that damage, such as can even occur from handling the fish, can result in marked increases in permeability. It thus seems that the skin of fish can be a somewhat tenuous and delicate permeability barrier. The limited information available suggests that some exchanges of water and salts can occur (Tables 17.1–17.3). The skin of cyclostomes (Bentley, 1962; Munz and McFarland, 1964) and cartilaginous fish (Horowicz and Burger, 1968; Payan and Maetz, 1970) may be quite permeable to water, and diffusional ion exchanges occur. Krogh (1937) found that trout (*Salmo irideus*, Teleostei) skin was slightly permeable to heavy water and chloride. In the eel (*Anguilla anguilla*) indirect measurements with tritiated water and ^{36}Cl indicate that exchanges of these molecules across the skin are negligible (Motais et al., 1969; Kirsch, 1972). It was suggested that the low permeability of the skin may reflect its thickness (eel skin is eight times as thick as frog skin, the epidermis containing 8–12 layers of cells) and poor vascularization.

The epidermis of a number of fish, including the killifish (*Fundulus heteroclitus*; Karnaky et al., 1977), the guppy (*Poecilia reticulata*; Schwerdtfeger, 1979), and the goby (*Gillichthys mirabilis*; Marshall and Bern, 1980), contains mitochondria-rich cells that can secrete chloride (chloride cells). The hormones epinephrine and prolactin have been variously shown to either increase or decrease Cl secretion from such cells, and at present the picture is not a consistent one. The mucus and slime secretion of the Pacific hagfish (*Epatretus stouti*) contains ions at higher concentrations than are present in the extracellular fluids; the ratio slime/serum is 30 for K^+, 2.3 for Ca^{2+}, and 3 for Mg^{2+}. It has thus been suggested (Munz and McFarland, 1964) that the mucous cells may provide a mechanism for ion excretion in these cyclostome fish.

When the *tetrapod vertebrates* emerged from an aquatic life and inhabited a terrestrial environment, they encountered a new type of problem, the evaporation of water from their skin. There are, however, a number of species of tetrapods that live in aquatic environments, especially among the Amphibia but also among reptiles and mammals. Thus the problem associated with molecular exchanges at a cutaneous liquid–liquid interface still persists among the tetrapods.

Amphibians

The skin of amphibians is usually kept moist, it is only lightly keratinized, and it is well supplied with cellular mucus glands. In almost all amphibians it is quite permeable to water and salts (Tables 17.1–17.3). In fresh water the exchanges of these molecules can be quite substantial, and in air evaporation can occur rapidly.

Diffusional loss of salts from the body fluids to fresh water may be reduced by the activity of a Na-pump in the epidermal cells; this process may even produce a net uptake of ions. A transcutaneous electrical difference (PD, corium side positive) is also created by the Na-pump, which acts to limit anion losses. This ion pump mechanism can be under the physiological control of the adrenocortical hormone aldosterone. Not all amphibians possess such a cutaneous ion pump; it is, for instance, lacking in the mudpuppy, *Necturus maculosus*, which nevertheless osmoregulates effectively without it (Bentley and Yorio, 1977).

Water can cross the liquid–liquid interface quite rapidly, and it is interesting that amphibians, which normally do not drink, utilize this process for their rehydration. Such osmotic water uptake across the skin can be increased by the neurohypophysial hormone vasotocin, which is the natural analogue of mammalian antidiuretic hormone. The pelvic regions of the skin of some amphibians may be specialized for this osmotic activity and be extremely permeable to water (McClanahan and Baldwin, 1969; Christensen, 1974; Yorio and Bentley, 1977). Such adaptations may involve an increase in local vascularization and a sculpturing of the surface of the epidermal cells to form small channels and grooves. The crab-eating frog, *Rana cancrivora*, can live in sea water (Gordon et al., 1961) and so may be expected to have a number of unique cutaneous problems as compared with other amphibians. The physiology of its skin, however, has not been adequately investigated.

In 1970 J. P. Loveridge made a startling observation: He described an African frog, *Chiromantis xerampelina*, that lost water by evaporation very slowly, at about the same rate as a lizard. This frog was still, however, able to gain water rapidly by osmosis, through the ventral skin. Subsequently a South American tree frog, *Phyllomedusa sauvagei*, possessing skin with similar properties, was also described (Shoemaker et al., 1972). The skin of this hylid frog and several other related species was shown to contain glands that can secrete lipids that are methodically spread over the skin surface by the fore and hind limbs (Blaylock, et al., 1976). These lipids consist mainly of wax esters (McClanahan et al., 1978), which create a monolayer that can restrict evaporation. The African species of *Chiromantis*, which belongs to a different family (Rhacophoridae), lacks such lipid glands (Drewes et al., 1977); so the basis for its low rate of evaporation appears to be different from that of the tree frogs.

Most amphibians maintain a moist skin and must replace cutaneous water that is lost as a result of evaporation. This water appears to be largely obtained by percutaneous absorption, for which an adequate blood supply may be important (Lillywhite, 1975). A periodic secretion by mucus glands may occur, the fluid dispersing as a film over the body surface (Lillywhite and Licht, 1975; McClanahan et al., 1978). The skin of some amphibians also has a sculptured pattern of grooves, channels, and furrows on the surface of the epidermis (Christensen, 1974; Lillywhite and Licht, 1974; Brown and Brown, 1980) through which fluid may be collected from a pool of water or even a damp surface that is apposed to the ventral side of the animal. This water may move, utilizing surface tension forces, to the dorsal surface where it is available for evaporation, body water thus being conserved. It is also possible that such sculpturing may provide unstirred layers of fluid that could impede molecular exchanges.

A number of anuran amphibians that estivate during periods of drought form protective cocoons from the outer, shed, parts of the stratum corneum (Mayhew, 1965; Lee and Mercer, 1967; Loveridge and Craye, 1979). These structures can reduce evaporation from the animal by as much as 90%. The aquatic urodele amphibian *Siren intermedia* forms a different type of cocoon when it estivates. A parchmentlike structure is formed as a result of secretion from mucus-type skin glands (Reno et al., 1972). This cocoon appears to resemble that formed by the African lungfish *Protopterus*.

Reptiles

Until it was questioned by Knut Schmidt-Nielsen in 1964, it was almost a zoological dictum that the skin of reptiles was "impermeable." It is certainly much less permeable than the integument of amphibians, possibly reflecting its thickness, its heavy keratinization (Maderson et al., 1978), and the presence of lipids between the epidermal cells (Roberts and Lillywhite, 1980). The role of scales in restricting permeability is uncertain; aberrant scaleless individual water snakes (*Natrix sipedon*) have, for instance, been shown to lose water by evaporation at the same rate as fully scaled individuals (Bennett and Licht, 1975). However, such observations may not be generally applicable to all reptiles.

There is considerable variability in the permeability of the skin of reptiles (Tables 17.1–17.3); evaporation can in fact be quite high, especially in normally amphibious species such as the caiman, water snakes, and turtles, but is very low in desert dwellers such as the chuckwalla and gopher tortoise. The skin of marine reptiles has been shown to have a very low permeability to water and salts (Tables 17.1, 17.2; Dunson, 1978).

Other specializations may occur in reptiles' skin that could influence their water metabolism. Australian desert lizards of the genus *Amphibolurus* "blanch" at high body temperatures (Bradshaw and Main, 1968). This response reflects changes in the distribution of pigment in their dermal melanophores that thus could, by reducing the absorption of radiation, indirectly influence evaporation. The skin of another Australian desert reptile, *Moloch horridus*, possesses on its epidermal surface small channels along which water can flow. This water, which can thus be collected from damp surfaces, is subsequently drawn into the mouth and swallowed (Bentley and Blumer, 1962).

Birds

Birds maintain a body temperature of about 40 °C, which would be expected to enhance evaporation from the integument. They lack cutaneous secretory glands, but evaporation occurs from the skin, presumably via the percutaneous route and the feather tracts. In the painted quail (*Excalfactoria chinensis*) evaporation from the skin accounts for about 60% of the total evaporative water loss (Bernstein, 1971). About 45% of such water loss in the zebra finch (*Taeniopygia castanotis*) occurs in this way when the birds have drinking water available (Lee and Schmidt-Nielsen, 1971). However, when drinking water is withdrawn, cutaneous water loss decreases to very low levels. The mechanism of this interesting effect is unknown.

It has been suggested (Wunder, 1979) that birds with a dark plumage lose water by evaporation more rapidly in sunlight than birds of a lighter hue. Thus a pale plumage could be an advantage to desert birds that do not have free access to water. This interesting possibility needs further study.

The possible role of the plumage in restricting cutaneous evaporation by helping to maintain unstirred layers of air apparently has not been examined. Such an effect could be especially important during flying.

Mammals

The skin of mammals possesses two phyletically characteristic appendages: hair and secretory glands that produce milk. It may also possess a number of other secretory glands including the sebaceous glands and apocrine (or epitrichial) sweat glands, which are associated with the hair follicles, and eccrine (or atrichial) sweat glands. Such appendages can influence the animals' osmoregulation in various ways. The fur provides an insulating layer of relatively unstirred air, which can retard exchanges of water and heat. Its color can also influence the absorption of radiant heat, and hence affect evaporation (see, for instance, D'miel et al., 1979). The sweat glands can secrete sweat that, when evaporated, can help dissipate excess body heat. This latter process mainly occurs in man and some apes, large ungulates (for instance horses, cattle, and camels) and some marsupials. This mechanism for dissipating heat is especially prominent in large mammals that come from hot tropical regions. It has been suggested that the lipids that are secreted by the sebaceous glands may provide waxes, fatty acids, and alcohols that, when spread over water at the skin surface, could form monolayers that might retard evaporation (Nicholaides, 1974). This suggestion does not appear to have been adequately explored, though in man (see Scheuplin and Blank, 1971) careful removal of the sebum from the skin surface does not appear to influence water loss.

The water that is evaporated from the skin surface in mammals is partly derived from that which diffuses across the stratum corneum and through the hair follicles. This percutaneous water movement ("insensible perspiration") is unavoidable and reflects the general permeability properties of the skin. It may, however, vary in magnitude in different regions of the body. The secretion of sweat is regulated and may be under neural control. Circulating epinephrine, from the adrenal medulla, offers an alternate control mechanism in bovid mammals, and this hormone may also facilitate sweating, even when neural pathways are also present, especially during exercise (see Robertshaw, 1975). Evaporation of water from the skin is often decreased in animals that are dehydrated, but the mechanism for this effect is unknown.

Sweat is a salt solution containing substantial amounts of Na, K, and Cl. Losses of salts can be considerable in man in hot environments. There is, however, a process of adaptation, which takes place over a period of several days when the sodium content of human sweat declines. This conservation is mediated by the hormone aldosterone (see Bentley and Scott, 1978), which promotes the reabsorption of salt across the ducts of the eccrine sweat glands. Whether such a response exists in the apocrine sweat glands, which are the predominant type of sweat glands in mammals, is unknown. Aldosterone is a hormone that is concerned with the conservation of salt, but its more usual effector site is the kidney. However, its effect on skin is not unique to mammals, as it also acts on amphibian skin where it increases transport of sodium across the epithelial cells.

Conclusion

The skin of vertebrates is a morphologically unique integument compared with that of other animals, but it also exhibits considerable anatomical and physiological diversity. The differences, such as the keratinization of the skin of tetrapod vertebrates and the mucous secretion that is prevalent in fish skin, often follow a broad phyletic pattern. Such morphological differences are consistent with the animals' physiological requirements, such as life in water or on dry land. The skin may wear several physiological "hats," probably the most important of which is to help conserve the volume and composition of the body fluids.

Acknowledgment

Most of the author's research that was quoted was supported by the National Science Foundation.

References

Bennett, A. F., and Licht, P. (1975) Evaporative water loss in scaleless snakes. *Comp. Biochem. Physiol.* 52A:213–215.

Bentley, P. J. (1962) Permeability of the skin of the cyclostome *Lampetra fluviatilis* to water and electrolytes. *Comp. Biochem. Physiol.* 6:95–97.

Bentley, P. J., and Blumer, W. F. C. (1962) Uptake of water by the lizard, *Moloch horridus*. *Nature, Lond.* 194:699–700.

Bentley, P. J., and Scott, W. N. (1978) The actions of aldosterone. In *General, comparative and clinical endocrinology of the adrenal cortex*, vol. 2 (I. Chester Jones and I. W. Henderson, eds.), pp. 498–564, New York: Academic Press.

Bentley, P. J., and Yorio, T. (1977) The permeability of the skin of a neotenous urodele amphibian, the mudpuppy *Necturus maculosus*. *J. Physiol. Lond.* 265:537–547.

Bentley, P. J., and Yorio, T. (1977) The permeability of the skin of a neotenous urodele amphibian, the mudpuppy *Necturus maculosus*. *J. Physiol. Lond.* 265537–547.

Bentley, P. J., and Yorio, T. (1979) Evaporative water loss in anuran Amphibia: A comparative study. *Comp. Biochem. Physiol.* 62A:1005–1009.

Bernstein, M. H. (1971) Cutaneous and respiratory evaporation in the painted quail, *Excalfactoria chinensis*, during ontogeny of thermoregulation. *Comp. Biochem. Physiol.* 38A:611–617.

Blaylock, L. A., Ruibal, R., and Platt-Aloia, K. (1976) Skin structure and wiping behaviour of Phyllomedusine frogs. *Copeia*, 293–295.

Bradshaw, S. D., and Main, A. R. (1968) Behavioral attitudes and regulation of temperature in *Amphibolurus* lizards. *J. Zool. Lond.* 154:193–221.

Brown, S. C., and Brown, P. S. (1980). Water balance in the California newt *Taricha torosa*. *Am. J. Physiol.* 238:R113–R118.

Christensen, C. U. (1974) Adaptations in the water economy of some anuran Amphibia. *Comp. Biochem. Physiol.* 47A:1035–1049.

Davies, J. T., and Rideal, E. K. (1963) *Interfacial phenomena*, pp. 301–308, New York: Academic Press.

D'miel, R., Robertshaw, D., and Choshniak, I. (1979) Sweat gland secretion in the black Bedouin goat. *Physiol. Zool.* 52:558–564.

Drewes, R. C., Hillman, S. S., Putnam, R. W., and Sokol, O. M. (1977) Water, nitrogen and ion balance in the African tree frog *Chiromantis petersi* Boulenger (Anura: Rhacophoridae) with comments on the structure of the integument. *J. Comp. Physiol.* 116:257–267.

Dunson, W. A. (1978) Role of the skin in sodium and water exchange of aquatic snakes placed in seawater. *Am. J. Physiol.* 235:R151–R159.

Flaxman, B. A. (1972) Cell differentiation and its control in vertebrate epidermis. *Am. Zool.* 12:13–25.

Gordon, M. S., Schmidt-Nielsen, K., and Kelly, H. M. (1961) Osmotic regulation in the crab-eating frog (*Rana cancrivora*). *J. Exp. Biol.* 38:659–678.

Henrikson, R. C., and Matoltsy, A. G. (1968a) The fine structure of teleost epidermis. I. Introduction and filament containing cell. *J. Ultrastruct. Res.* 21:194–212.

Henrikson, R. C., and Matoltsy, A. G. (1968b) The fine structure of teleost epidermis. II. Mucous cells. *J. Ultrastruct. Res.* 21:213–221.

Henrikson, R. C., and Matoltsy, A. G. (1968c) The fine structure of teleost epidermis. III. Club cells and other cell types. *J. Ultrastruct. Res.* 21:222–232.

Horowicz, P., and Burger, J. W. (1968) Unidirectional fluxes of sodium ions in the spiny dogfish *Squalus acanthias*. *Am. J. Physiol.* 214:635–642.

Karnaky, K. J., Degnan, K. J., and Zadunaisky, J. A. (1977) Chloride transport across isolated opercular epithelium of killifish: A membrane rich in chloride cells. *Science* 195:203–205.

Kirsch, R. (1972) The kinetics of peripheral exchanges of water and electrolytes in the silver eel (*Anguilla anguilla*, L.) in fresh water and sea water. *J. Exp. Biol.* 57:489–512.

Krogh, A. (1937) Osmotic regulation in freshwater fishes by active absorption of chloride ions. *Z. vergl. Physiol.* 24:656–666.

Lee, A. K., and Mercer, E. H. (1967) Cocoon surrounding desert dwelling frogs. *Science* 157:87–88.

Lee, P., and Schmidt-Nielsen, K. (1971) Respiratory and cutaneous evaporation in the zebra finch: Effect on water balance. *Am. J. Physiol.* 220:1598–1605.

Lillywhite, H. B. (1975) Physiological correlates of basking in amphibians. *Comp. Biochem. Physiol.* 52A:323–330.

Lillywhite, H. B., and Licht, P. (1974) Movement of water over toad skin: Functional role of epidermal sculpturing. *Copeia,* 165–171.

Lillywhite, H. B., and Licht, P. (1975) A comparative study of integumentary secretions in amphibians. *Comp. Biochem. Physiol.* 51A:937–941.

Loveridge, J. P. (1970) Observations on nitrogenous excretion and water relations of *Chiromantis xerampelina. Arnoldia* 5:1–6.

Loveridge, J. P., and Craye, G. (1979) Cocoon formation in two species of southern African frogs. *S. Afr. J. Sci.* 75:18–20.

McClanahan, L., and Baldwin, R. (1969) Rate of water uptake through the integument of the desert toad, *Bufo punctatus. Comp. Biochem. Physiol.* 28:381–389.

McClanahan, L., Stinner, J. N., and Shoemaker, V. H. (1978) Skin lipids, water loss and energy metabolism in a South American tree frog (*Phyllomedusa sauvagei*). *Physiol. Zool.* 51:179–187.

Maderson, P. F. A., Zucker, A. H., and Roth, S. I. (1978) Epidermal regeneration and percutaneous water loss following cellophane striping of reptile epidermis. *J. Exp. Zool.* 204:11–32.

Marshall, W. S., (1978) On the involvement of mucous secretion in teleost osmoregulation. *Can. J. Zool.* 56:1088–1091.

Marshall, W. S., and Bern, H. A. (1980) Ion transport across the isolated skin of the teleost, *Gillichthys mirabilis.* In *Epithelial transport in lower vertebrates* (B. Lahlou, ed.), pp. 337–350, Cambridge University Press.

Mayhew, W. W. (1965) Adaptations of the amphibian *Scaphiopus couchi* to desert conditions. *Am. Midl. Nat.* 74:95–109.

Motais, R., Isaia, J., Rankin, J. C., and Maetz, J. (1969) Adaptive changes of the water permeability of the teleostean gill epithelium in relation to external salinity. *J. Exp. Biol.* 51:529–546.

Munz, F. W., and McFarland, W. N. (1964) Regulatory function of a primitive vertebrate kidney. *Comp. Biochem. Physiol.* 13:381–400.

Nicolaides, N. (1974) Skin lipids: Their biochemical uniqueness. *Science* 186:19–26.

Parry, G. (1966) Osmotic adaptation in fishes. *Biol. Rev.* 41:392–444.

Payan, P., and Maetz, J. (1970) Balance hydrique et minérale chez les élasmobranchs: Arguments en faveur d'un contrôle endocrinien. *Bull. Inform. Scientifiques et Techniques du Commissariat à l'Energie Atomique (France).* 146:77–96.

Reno, H. W., Gehlbach, F. R., and Turner, R. A. (1972) Skin and aestivational cocoon of the aquatic amphibian, *Siren intermedia,* Le Conte. *Copeia,* 625–631.

Roberts, J. B., and Lillywhite, H. B. (1980) Lipid barrier to water exchange in reptile epidermis. *Science* 207:1077–1079.

Robertshaw, D. (1975) Catecholamines and control of sweat glands. In *Handbook of physiology,* section 7, Endocrinology, vol. 6, *Adrenal gland* (H.

Blascho, G. Sayers, and A. D. Smith, eds.), pp. 591–603. Washington, D.C.: American Physiological Society.

Scheuplin, R. J. (1965) Mechanism of percutaneous absorption. 1. Routes of penetration and the influence of solubility. *J. Invest. Dermatol.* 45:334–346.

Scheuplin, R. J., and Blank, I. H. (1971) Permeability of the skin. *Physiol. Rev.* 51:702–747.

Schmidt-Nielsen, K. (1964) *Desert animals. Physiological problems of heat and water.* Oxford: University Press.

Schmidt-Nielsen, K. (1969) The neglected interface: The biology of water as a liquid–gas system. *Quart. Rev. Biophys.* 2:283–304.

Schwerdtfeger, W. K. (1979) Morphometrical studies of the ultrastructure of the epidermis of the guppy, *Poecilia reticulata* Peters, following adaptation to seawater and treatment with prolactin. *Gen. Comp. Endocrinol.* 38:476–483.

Shoemaker, V. H., Balding, D., Ruibal, R., and McClanahan, L. (1972) Uricotelism and low evaporative water loss in a South American frog. *Science* 175:1018–1020.

Tregear, R. T. (1966) The permeability of mammalian skin to ions. *J. Invest. Dermatol.* 46:16–23.

Wunder, B. A. (1979) Evaporative water loss from birds: Effects of artificial radiation. *Comp. Biochem. Physiol.* 63A:493–494.

Yorio, T., and Bentley, P. J. (1977) Asymmetrical permeability of the integument of tree frogs (Hylidae). *J. Exp. Biol.* 67:197–204.

Yorio, T., and Bentley, P. J. (1978) The permeability of the skin of the aquatic anuran *Xenopus laevis* (Pipidae). *J. Exp. Biol.* 72:285–289.

18 Salt glands: A perspective and prospective view

M. PEAKER

An animal can excrete solutes with very little use of water by excreting the solutes as a very concentrated solution. Mammals do this by forming a highly concentrated urine, but because the bird kidney is incapable of forming an equally concentrated urine, the problem is solved somewhat differently in birds. Birds excrete the major nitrogenous excretory products as precipitated uric acid, and marine birds excrete excess soluble ions in a highly concentrated solution from a nasal gland, the so-called salt gland.

This function of the salt gland was discovered by Knut Schmidt-Nielsen and his colleagues (Schmidt-Nielsen et al., 1957). With those and other collaborators, Schmidt-Nielsen laid a firm foundation for the future study of these fascinating and important organs.

The importance of salt glands lies in their ability to secrete highly hypertonic solutions, to secrete within minutes after a bird receives a salt load, and to adapt to loads by hypertrophy of the gland. Functionally analogous glands appear in different groups of reptiles, but are not always anatomically homologous to the avian glands. The further study of salt glands has drawn together workers from many different disciplines, from fields as far apart as membrane biochemistry and applied ecology.

After many years of active research on the salt gland, I wish to give an account of certain aspects in which research is active and in which progress is being made, or likely will be made. I have made no attempt to review the early literature because the late Jim Linzell and I did so in our monograph published in 1975 (Peaker and Linzell, 1975).

Mechanism of secretion

What is the cellular mechanism by which the salt glands produce a highly concentrated sodium chloride solution? Presumably a single layer of cells in the secreting tubules serves to move sodium and chloride ions from the extracellular fluid at the basal end of the cell to the lumen at the apical end (see Figure 18.1). Are the ions moved transcellularly, through the cells, or is the pathway through the intercellular junctions?

In the early 1970s it seemed inconceivable that the junctions between

254

the adjoining cells in the salt gland could be other than physiologically tight because they supposedly separate two markedly different fluids. Therefore, hypotheses of transport were based entirely on transcellular movements of the secreted ions. However, it has since been suggested that the junctions (*zonulae occludentes*) are in fact "leaky," and other hypotheses for the ionic movements have been proposed.

The scheme shown in Figure 18.1 was based on a number of findings, notable among them that the intracellular concentrations of Na^+ and Cl^- are much lower than in the secretion and do not change during secretory activity (Peaker, 1971a; Peaker and Stockley, 1973; B. Schmidt-Nielsen, 1976); therefore, the main concentration gradient for Na^+ and Cl^- would be across the luminal (apical) cell membrane.

From the electrophysiological studies of Thesleff and Schmidt-Nielsen (1962) it appeared that with the lumen of the duct becoming positive with respect to blood during stimulation of the secretory nerve, the electrochemical gradient for Na^+ was established across the luminal membrane; Cl^- would then follow Na^+ pumping passively. Therefore, according to this scheme, the properties of the luminal membrane are clear: an electrogenic, Na^+-extruding pump with Cl^- following passively; low passive permeability to ions; low osmotic permeability to water.

In order for secretion to occur, the rate of Na^+ and Cl^- movements

FIGURE 18.1 A possible mechanism to account for ion transport in the salt gland. Solid star, electrogenic pump; open star, nonelectrogenic pump. [From Peaker, 1971b]

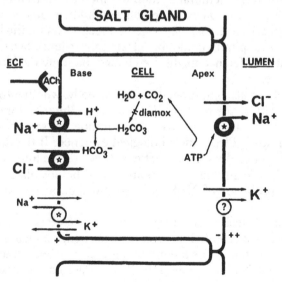

into the cell must increase, and I proposed that this occurred by coupled exchanges of Na^+ for H^+ and Cl^- for HCO_3^- across the basolateral membrane (Peaker, 1971b). The arguments for this scheme as opposed to one involving an increase in passive permeability can be found in Peaker (1971b) and Peaker and Linzell (1975). However, the essential feature is that there is a direct link between events at the luminal membrane and those at the basolateral membrane through respiration and the production of CO_2; thus, by this coupling, efflux matches influx, and intracellular concentrations of Na^+ and Cl^- remain unchanged.

Kirschner (1977) has pointed out that passive movements of Na^+ and Cl^- across the basolateral membrane might still occur, if the findings of Thesleff and Schmidt-Nielsen (1962) on intracellular potential differences prove to be incorrect. These workers found no difference in intracellular potential difference during secretion (induced by stimulation of the secretory nerve) compared with the resting gland. However, the cells were clearly damaged during impalement because the potential difference fell rapidly from 40–80 mV to 20–30 mV. Kirschner (1977) argued that more work is needed because acetylcholine could act to increase the passive permeability of the basolateral membrane to Na^+ and Cl^- – this would be a return to my earlier scheme (Peaker, 1971a) – and depolarization would occur. Although I would be the first to argue that more work should be done on this aspect, would there not have been some indication of depolarization, had it occurred, in the Thesleff and Schmidt-Nielsen experiments?

A corollary of Kirschner's (1977) arguments is that if there is depolarization of the basolateral membrane, then the increase in transepithelial potential difference at the onset of secretion could supervene without the involvement of an additional electrogenic mechanism on the luminal membrane (i.e., the pump could be electrically silent) because the cell–lumen potential difference might not change, but only the blood–cell potential difference.

However, Kirschner also questioned, quite correctly, whether the duct-positive potential difference really reflects events in the secretory tubule and whether the ducts are as inactive in elaborating the secretion as we have supposed or argued from morphological evidence. If the duct-positive potential difference does not reflect events across the luminal membrane, then an electrically silent Cl^--pump could be postulated, which would overcome problems of Na/K ATPase distribution (see following discussion).

Even if we accept Kirschner's comments on the secretory mechanism, we are still left with the essential features of the scheme shown in Figure 18.1: a concentrating step across the luminal membrane and some means of getting additional Na^+ and Cl^- into the cell across the basolateral membrane. These are the features of a transcellular route which are not very different from any scheme based on the classic Koefoed-Johnsen and Ussing hypothesis. But we must now examine later evidence and hypotheses to see whether they offer a more complete explanation of morphological and physiological findings.

Two recent hypotheses attempt to reconcile the morphological appearance of the junctions that connect neighboring cells and the location of the Na/K ATPase in the cell membrane. All reliable methods indicate that the Na/K ATPase is present on the basolateral membrane but not on the luminal membrane (Ernst, 1972; Ernst and Mills, 1977; Hossler et al., 1978). If Na/K ATPase is the sole enzyme system responsible for Na$^+$ translocation, then the transcellular scheme for secretion cannot obtain unless, of course, the pump on the luminal membrane is not a Na$^+$-pump. Moreover, the structure of the junctions has been studied by freeze-fracture methods and their structure interpreted as consisting of one to two contiguous strands without discontinuities (Ellis et al., 1977; Riddle and Ernst, 1979). By analogy with other tissues these authors suggested that these junctions should be classified as "leaky."

Ellis et al. (1977) proposed the scheme shown in Figure 18.2. The terminal cells of the tubule (peripheral cells in the terminology of Peaker and Linzell, 1975) secrete an isotonic solution that flows past the principal

FIGURE 18.2 Diagram of a single secretory tubule in a salt gland. The number of cell junctions as well as the complexity of the intercellular channels is greatly reduced, and the diameter of the lumen of the tubule is much enlarged. The terminal segment (a) of the tubule is composed of small peripheral cells; the remainder of the tubule (b) is lined by principal cells. The central excretory canal (c) is continuous with an excretory duct (not shown). The intercellular channels (IC) are closed at their luminal ends by leaky cell junctions. The density of the stipple in the lumen and in the intercellular channels indicates the concentration of Na$^+$. [From Ellis et al., 1977]

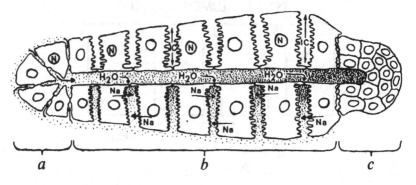

cells. Na^+ (and Cl^-) is pumped into the intercellular spaces, and H_2O is drawn osmotically from the primary secretion into these spaces through the leaky junctions. Therefore, the secretion is a hypertonic solution equal in tonicity to the maximum concentration achieved in the "standing gradient" of the intercellular clefts, and the whole mechanism works by water reabsorption.

Ernst and Mills (1977) and Riddle and Ernst (1979) proposed that Na^+ is pumped into the intercellular clefts to form a solution of high concentration and passes through the leaky junctions to form the secretion (Figure 18.3A); Cl^- then passes passively across the apical membrane. I see no reason to provide a separate route for Cl^- because, if the scheme operates in this way, an anion (presumably Cl^-) must be present with Na^+ in the clefts, and I have drawn this slightly modified scheme in Figure 18.3B. Therefore in both schemes Na^+ and Cl^- enter the principal cell, Na^+ is pumped into the intercellular clefts by the basolaterally located Na/K ATPase, and the leaky junctions are involved in forming the secretion.

The scheme of Ellis et al. (1977) can be criticized on several grounds. First, do the relatively small numbers of peripheral cells form an isotonic secretion? To do so they would have to perform prodigious feats because they would need to produce up to 12 times the final rate of secretion.

FIGURE 18.3 (A) A diagram of the scheme suggested by Ernst and Mills (1977) and Riddle and Ernst (1977) to account for salt-gland secretion. (B) The same but modified to show Cl^- moving with Na^+.

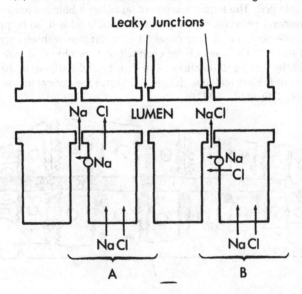

These cells are relatively undifferentiated and appear to be the site of cell division in the tubule (see Peaker and Linzell, 1975; Knight and Peaker, 1979). To propose a major transport role for them is to create as many problems as are solved, as I did when I suggested a somewhat similar scheme (Peaker, 1971b). Second, if the junctions really are leaky, then Na$^+$ would pass into the lumen, and osmotic flow through the junctions would be negligible.

While we await modeling to see whether either of these mechanisms is compatible with reasonable values of permeability, pumping rates, and intercellular cleft sizes, we might as well examine the evidence for the leakiness of the junctions.

Both groups of authors failed to mention that disaccharides (sucrose) and polysaccharides (inulin) fail to enter the secretion in significant amounts (Schmidt-Nielsen, 1960; Peaker and Hanwell, 1974). Surely if the junctions are as leaky as they claim and are important routes of movement, then sucrose should enter the secretion. Other explanations for the failure of sucrose to enter the secretion are unconvincing.

However, this still leaves us with the question of the location of the Na/K ATPase as well as another puzzling feature of the avian salt gland. Although the intercellular folds are elaborate, the basal membrane is folded in such a manner that the folds do not appear to be part of the intercellular space. Why is there such a high concentration of Na/K ATPase in this basal infolding if it really is involved in the main secretory process? The questions are clear enough, but how do we obtain the answers? Indeed, the difficulties in providing an overall hypothesis have increased since 1974. How can we explain, for example, the salt gland secretion that appears in cold-stressed birds with an osmolality lower than that of blood plasma (Ensor et al., 1976)? Clearly, our present methods are inadequate, and perhaps organ culture – growing salt gland cells as confluent monolayers – will provide many of these answers. We must also hope that electron probe microanalysis will deliver what it has promised to the transport field. I look forward to seeing the results. Whatever emerges, there is no doubt that the salt gland will continue to provide a challenge to the transport physiologist.

Control of secretion

In the early 1970s our work on the control of secretion culminated in the scheme shown in Figure 18.4 for the initiation and maintenance of secretion. According to this scheme, the secretory reflex works as follows: Osmoreceptors situated in or near the heart activate secretion via afferent fibers in the vagus and efferent fibers in the "secretory nerve" to the gland (Hanwell et al., 1972). Although many questions remained unanswered (they still do), more recent work has been concerned with the modulation of secretory activity by changes in blood or extracellular fluid volume.

By 1974 we thought we had shown that an increase in blood volume alone was an insufficient stimulus to secretion, and that the solutions used by some workers to raise plasma volume were sufficiently hypertonic to stimulate the osmoreceptors (Peaker and Linzell, 1975). Although there has been another claim that blood volume expansion will induce secretion in the absence of an osmotic stimulus (Zucker et al., 1977), I was again able to show that this claim could be attributed to the solutions used for volume expansion being hypertonic (Peaker, 1978a). Therefore, except near the osmotic threshold for secretion, I do not believe that volume expansion alone will initiate secretion. We also showed that a rise in blood volume is not necesssary for secretion to occur (Hanwell et al., 1971).

However, convincing evidence has now been obtained that changes in blood volume can affect secretory rate. Deutsch et al. (1979) found that in salt-adapted domestic ducks, salt gland activity, plasma osmolality, and Na^+ concentration were poorly correlated during balanced states of salt input and output, implying the involvement of another factor. Secretion was inhibited by withdrawal of blood and restored when the blood was returned. Kaul and Hammel (1979) altered dehydrated ducks by infusing NaCl solutions hyperosmotic to the salt gland secretion. By determining the plasma osmolality at which salt gland secretion started during infusion of a weaker NaCl solution before and after this acute dehydration, they suggested that the osmotic threshold for secretion increased as extracellular fluid volume decreased. In geese, I have also

FIGURE 18.4 The proposed reflex arc for stimulation of the salt gland in the goose. [From Hanwell et al., 1972]

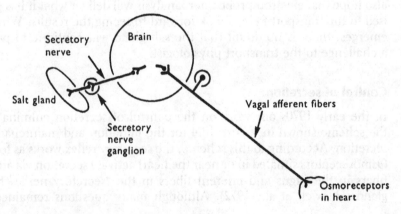

found that blood withdrawal inhibits secretion. When homologous blood was given intravenously to secreting geese, the effects were variable: In those secreting at a relatively low rate secretion was enhanced; in those secreting at a relatively high rate there was little or no apparent effect (unpubl.).

Kaul and Hammel (1979) stated that "the patterns of secretion observed might be explained by postulating that the salt glands are driven by a proportional controller sensitive to ECF tonicity and that the reference tonicity is inversely proportional to ECF volume." But it is at this stage that I must sound a note of caution. Are volume signals being integrated with osmotic signals to alter the impulses through the secretory nerve, or is another system involved? In this respect we cannot ignore the adrenergic innervation of the salt gland because decreases in blood volume could lead to sympathetic discharge and attenuation of secretion (see Peaker and Linzell, 1975). Indeed there is experimental evidence on this point. Douglas and Neely (1969) found that secretion in response to a salt load was impaired in dehydrated herring gulls; treatment with an α-adrenergic blocking agent largely reversed this inhibition. Clearly there is scope for further work because it seems pointless to pursue integrative systems within the CNS if another relatively independent pathway is involved in inhibiting secretion when blood volume is altered.

But Kaul and Hammel (1979) take their hypothesis one stage further by postulating that the receptors in or near the heart (Figure 18.4) are volume receptors and not osmoreceptors, and that the osmoreceptors are cranial. They unfortunately neglect to mention that we (Hanwell et al., 1972) failed to find the osmoreceptors for secretion in the head of birds by the classical Verrey (1974) technique of infusing hypertonic NaCl solutions into the carotid artery. It was this failure to find such receptors in the head that led us to look elsewhere!

Although there is evidence that a number of hormones can affect salt gland secretion, we still have no real demonstration that they modulate secretion under physiological conditions. For adrenocorticosteroids we know the paradoxical situation in which receptors for corticosteroids have been found in the glands (Sandor and Fazekas, 1974; Allen et al., 1975; Sandor et al., 1977), but, contrary to earlier evidence, the gland will function virtually normally in adrenalectomized birds. In early studies (Phillips et al., 1961) secretion was virtually abolished by adrenalectomy and restored by cortisol administration. In later studies using an improved technique, a transient secretion was obtained (Thomas and Phillips, 1975), but very recently Butler (1980) has shown that in ducks given 14 days to recover after adrenalectomy and kept in "good condition," salt gland secretion was virtually normal, whereas corticosterone could not be detected in the plasma.

Other aspects

During the next few years I believe we shall see exciting developments made at the interface between the external control of secretion (nerves, hormones) and internal events in the cells of the salt gland (secretion, hypertrophy, and hyperplasia). The gland is particularly suited to this sort of work because its secretion is simple, and events within the gland during adaptation of birds to salt water are rapid and dramatic. Thus there is evidence that external calcium is necessary for, and cyclic-GMP probably involved in, excitation–secretion coupling (Peaker, 1978b; Stewart et al., 1979); that hypertrophy of the cells during salt-adaptation is dependent on stimulation by the secretory reflex (Hanwell and Peaker, 1975); and that a burst of cell division follows hypertrophy (Knight and Peaker, 1979). All of these aspects require further study to determine the controlling processes that are involved within the gland.

But the study of salt glands does not involve just the glands. The pattern of salt and water balance control is a fertile field because aspects such as control of ingestion, absorption from the gut, and exchanges between body compartments must all be integrated to enable birds to survive in and exploit environments as taxing as the sea, estuaries, inland salt lakes, or deserts.

Reptilian salt glands

In many reptiles, the salt glands are not anatomically homologous with the nasal salt glands of birds (see Schmidt-Nielsen and Fänge, 1958; Peaker and Linzell, 1975; Dunson, 1976), and indeed "new" salt glands are being discovered in snakes (see, e.g., Dunson and Dunson, 1979).

A fascinating field awaits exploitation in reptiles. Their glands serve the same function as in birds, but as the glands are of different anatomical origin, how is their function controlled? The sort of question that springs to mind is: Is the secretory reflex similar in all reptiles; is it similar to that in birds; where are the receptors; are vagal afferent impulses as important in sea-snakes, say, as they are in birds for the initiation and maintenance of secretion? A highly interesting exercise in true comparative physiology should soon develop.

Credo

I have outlined some areas in which interesting, sometimes exciting and important advances should emerge in the not-too-distant future. Physiology of the salt glands is but part of the overall field of general and comparative physiology. But comparative studies are not only good in an abstract sense for the advancement of physiological and zoological science. I am convinced that they are good for an individual's own work. Thus while working on the mechanism of salt gland secretion I was also studying the mechanism of milk secretion – at that time an unexplored area of mammalian physiology. The hypotheses that emerged on the mammary gland (see Linzell and Peaker, 1971; Peaker, 1977, 1978c), I feel, owe as much to the very different mechanism we were studying in the salt gland as to direct studies on the mammary gland.

Finally, I believe research should be enjoyed. I have enjoyed salt glands as a player; I shall continue to do so as a spectator.

References

Allen, J. A., Abel, J. H., and Takemoto, D. J. (1975) Uptake and binding of labelled corticosterone by the salt gland of the duck (*Anas platyrhynchos*). *Gen. Comp. Endocrinol.* 26:217–225.

Butler, D. G. (1980) Functional nasal salt glands in adrenalectomized domestic ducks (*Anas platyrhynchos*). *Gen. Comp. Endocrinol.* 40:15–26.

Deutsch, H., Hammel, H. T., Simon, E., and Simon-Oppermann, C. (1979) Osmolality and volume factors in salt-gland control of Pekin ducks after adaptation to chronic salt loading. *J. Comp. Physiol.* 129:301–308.

Douglas, D. S., and Neely, S. M. (1969) The effect of dehydration on salt gland performance. *Am. Zool.* 9:1095.

Dunson, W. A. (1976) Salt glands in reptiles. In *Biology of the reptilia*, vol. 5, *Physiology A* (C. Gans and W. R. Dawson, eds.), pp. 413–445. London: Academic Press.

Dunson, W. A., and Dunson, M. K. (1979) A possible new salt gland in a marine homalopsid snake (*Cerbenus rhynchops*). *Copeia*, 661–672.

Ellis, R. A., Goertmiller, C. C., and Stetson, D. L. (1977) Significance of extensive 'leaky' cell junctions in the avian salt gland. *Nature* 268:555–556.

Ensor, D. M., Phillips, J. G., and O'Halloran, M. J. (1976) The effects of extreme cold stress on nasal gland secretion in the domestic duck, *Anas platyrhynchos*. *Gen. Comp. Endocrinol.* 30:223–227.

Ernst, S. A. (1972) Transport adenosinetriphosphatase cytochemistry. 2. Cytochemical localization of ouabain-sensitive, potassium-dependent phosphatase activity in the secretory epithelium of the avian salt gland. *J. Histochem. Cytochem.* 20:23–38.

Ernst, S. A., and Mills, J. W. (1977) Basolateral plasma membrane localization of ouabain-sensitive sodium transport sites in the secretory epithelium of the avian salt gland. *J. Cell Biol.* 75:74–94.

Hanwell, A., Linzell, J. L., and Peaker, M. (1971) Cardiovascular responses to salt-loading in conscious domestic geese. *J. Physiol.* 213:389–398.

Hanwell, A., Linzell, J. L., and Peaker, M. (1972) Nature and location of the receptors for salt-gland secretion in the goose. *J. Physiol.* 226:453–472.

Hanwell, A., and Peaker, M. (1975) The control of adaptative hypertrophy in the salt glands of geese and ducks. *J. Physiol.* 248:193–205.

Hossler, F. E., Sarras, M. P., and Barrnett, R. J. (1978) Ouabain binding during plasma membrane biogenesis in duck salt gland. *J. Cell Sci.* 31:179–197.

Kaul, R., and Hammel, H. T. (1979) Dehydration elevates osmotic threshold for salt gland secretion in the duck. *Am. J. Physiol.* 237:R355–R359.

Kirschner, L. B. (1977) The sodium chloride excreting cells in marine vertebrates. In *Transport of ions and water in animals* (B. L. Gupta, R. B. Moreton, J. L. Oschmann, and B. J. Wall, eds.), pp. 427–452. London: Academic Press.

Knight, C. H., and Peaker, M. (1979) Adaptive hyperplasia and compensatory growth in the salt glands of ducks and geese. *J. Physiol.* 294:145–151.

Linzell, J. L., and Peaker, M. (1971) Mechanism of milk secretion. *Physiol. Rev.* 51:564–597.

Peaker, M. (1971a) Intracellular concentration of sodium, potassium and chloride in the salt gland of the domestic goose and their relation to the secretory mechanism. *J. Physiol.* 213:399–410.

Peaker, M. (1971b) Avian salt glands. *Phil. Trans. R. Soc. B* 262:289–300.

Peaker, M. (1977) The aqueous phase of milk: Ion and water transport. In *Comparative aspects of lactation* (M. Peaker, ed.), pp. 113–134. Symposia of the Zoological Society of London 41. London: Academic Press.

Peaker, M. (1978a) Do osmoreceptors or blood volume receptors initiate salt-gland secretion in birds? *J. Physiol.* 276:66–67P.

Peaker, M. (1978b) Salt glands in marine birds: What triggers secretion and what makes them grow? In *Comparative physiology – water, ions and mechanics* (K. Schmidt-Nielsen, L. Bolis, and S. H. P. Maddrell, eds.), pp. 207–212. Cambridge University Press.

Peaker, M. (1978c) Ion and water transport in the mammary gland. In *Lactation*, vol. 4 (B. L. Lárson, ed.), pp. 437–462. New York: Academic Press.

Peaker, M., and Hanwell, A. (1974) Transepithelial ^{14}C sucrose movements in the goose salt gland. Relation to the secretory mechanism. *Pflügers Arch.* 352:363–366.

Peaker, M., and Linzell, J. L. (1975) *Salt glands in birds and reptiles.* Cambridge University Press.

Peaker, M., and Stockley, S. J. (1973) Lithium secretion by the salt gland of the goose. *Nature* 243:297–298.

Phillips, J. G., Holmes, W. N., and Butler, D. G. (1961) The effect of total and subtotal adrenalectomy on the renal and extra-renal response of the domestic duck (*Anas platyrhynchos*) to saline loading. *Endocrinology* 69:958–969.

Riddle, C. V., and Ernst, S. A. (1979) Structural simplicity of the *zonula occludens* in the electrolyte secreting epithelium of the avian salt gland. *J. Membr. Biol.* 45:21–35.

Sandor, T., and Fazekas, A. G. (1974) Corticosteroid-binding macromolecules in the nasal gland of the domestic duck. *Gen. Comp. Endocrinol.* 22:348–349.

Sandor, T., Mehdi, A. Z., and Fazekas, A. G. (1977) Corticosteroid-binding macromolecules in the salt-activated nasal gland of the domestic duck (*Anas platyrhynchos*). *Gen. Comp. Endocrinol.* 32:348–359.

Schmidt-Nielsen, B. (1976) Intracellular concentrations of the salt gland of the herring gull *Larus argentatus*. *Am. J. Physiol. 230*:514–521.

Schmidt-Nielsen, K. (1960) Salt-secreting gland of marine birds. *Circulation 21*:965–967.

Schmidt-Nielsen, K., and Fänge, R. (1958) Salt glands in marine reptiles. *Nature 182*:783–785.

Schmidt-Nielsen, K., Jörgensen, G. B., and Osaki, H. (1957) Secretion of hypertonic solutions in marine birds. *Fed. Proc. 16*:113–114.

Stewart, D. J., Sax, J., Funk, R., and Sen, A. R. (1979) Possible role of cyclic GMP in stimulus-secretion coupling by salt gland of the duck. *Am. J. Physiol. 237*:C200–C204.

Thesleff, S., and Schmidt-Nielsen, K. (1962) An electrophysiological study of the salt gland of the herring gull. *Am. J. Physiol. 202*:597–600.

Thomas, D. H., and Phillips, J. G. (1975) Studies in avian adrenal steroid function. IV. Adrenalectomy and the response of domestic ducks (*Anas platyrhynchos* L.) to hypertonic NaCl loading. *Gen. Comp. Endocrinol. 26*:427–439.

Verrey, E. B. (1947) The anti-diuretic hormone and the factors which determine its release. *Proc. R. Soc. B 135*:25–106.

Zucker, I. H., Gilmore, C., Dietz, J., and Gilmore, J. F. (1977) Effect of volume expansion and veratrine on salt gland secretion in the goose. *Am. J. Physiol. 232*:R185–R189.

19 Renal countercurrent mechanisms, or how to get something for (almost) nothing

REINIER BEEUWKES III

In general, animals that can survive in very arid habitats have the ability to form urine of very high solute concentration. That is, they are able to maintain solute balance and excrete metabolic wastes with very little water loss. On the other hand, animals in water-rich habitats have not found water conservation an important selective force, and generally form relatively dilute urine. Table 19.1 compares the maximum urine concentration of several species. Beavers, which usually have free access to water, can make urine have a total solute concentration of 600 mOsm kg^{-1} water – about twice the plasma concentration. Humans concentrate to about 1400 mOsm – almost five times the plasma concentration. Desert rodents can concentrate their urine to spectacular degrees. The hopping mouse nearly reaches 10 000 mOsm. To make such a concentrated solution in the laboratory, you would have to·add about 600 g of common salt or urea to 1000 g (1 liter) of water. The details of the mechanism by which concentrated urine is formed remain unknown. However, the broad outline of the mechanism – the so-called countercurrent multiplier – has been known for many years. Recent experimental studies have markedly enhanced our understanding of the structure of the renal tubules involved and the solute and water flows that probably occur between them during the urine-concentrating process. To appreciate these new understandings, however, we must begin at the beginning.

Basic mechanisms

Urine formation begins with the generation of a protein-free plasma ultrafiltrate in the glomeruli. The rate of formation of this filtrate is large, typically 100 times the rate of urine flow. This ultrafiltrate flows into the renal tubules where most of the salt and water is reabsorbed (Figure 19.1). Proximal convoluted tubules reabsorb about two-thirds of the filtered salt and water, together with essentially all filtered sugar and amino acids. The last part of each proximal tubule is straight and extends toward the renal medulla. It thus constitutes the first part of the loop of Henle. Because of proximal reabsorption, the volume of fluid passing through the thin descending and ascending segments of loops is substantially less

266

than the amount filtered. Samples drawn from tubules at the tip of the loop are found to have high solute concentrations compared to the original ultrafiltrate. (Remember that, except for protein, the ultrafiltrate has essentially the same solute concentration as in plasma.) In the early distal convoluted tubules at the kidney surface, the tubular fluid is less concentrated than plasma, indicating that a large amount of solute has been removed during the passage of the fluid up the ascending limb. Some water reabsorption has also occurred, because the fluid flow rate in the distal tubule is only about 10% of the rate of filtration. Along the distal convoluted tubules further salt and water reabsorption occurs before the fluid passes into collecting ducts.

The water permeability of collecting ducts is controlled by antidiuretic hormone (ADH). In the presence of this hormone, fluid within the duct can come to osmotic equilibrium with the interstitial fluid surrounding the duct. In the kidney cortex, where the interstitial osmolality is close to that of systemic plasma, this equilibration could lead only to the production of urine isosmotic with plasma. However, as the fluid passes through collecting ducts in the medulla, the flow of water down its osmotic gradient from duct lumen to the adjacent medullary interstitium concentrates the solutes remaining within the duct to a very high degree. This osmotic removal of water is understood to be the basic mechanism by which the final collecting duct fluid (i.e., urine) becomes concentrated. Because osmotic water flow can only occur from less concentrated to more concentrated solutions, this mechanism for urine concentration requires that the papillary interstitium have a total solute osmotic concentration that is at least as high as that of the urine being formed, although the solute species in urine and in papillary interstitium need not be the same. Direct chemical analysis of slices cut from the renal medulla shows that this region does indeed contain a high concentration

TABLE 19.1 Urine-concentrating ability.

Animal	Maximum concentration (mOsm)	Ratio to plasma
Beaver	520	2
Pig	1100	4
Human	1400	5
White rat	3000	10
Kangaroo rat	5500	14
Hopping mouse	9400	25

Modified from Schmidt-Nielsen, 1979.

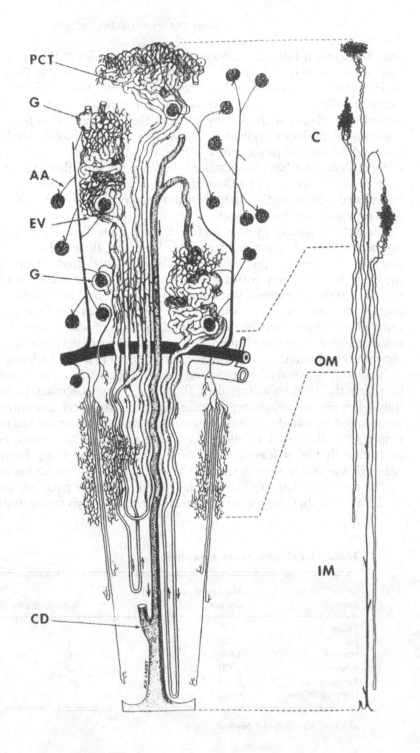

of solutes (Figure 19.2), and that a solute gradient exists with highest concentrations near the papillary tip where the final urine is formed (Ullrich et al., 1961). Knowing that the medullary interstitium has high solute concentration, and that antidiuretic hormone allows collecting duct fluid to equilibrate with this highly concentrated region, it is easy to understand how collecting duct fluid becomes concentrated, and how the level of concentration is regulated by the hormone. However, real intellectual problems remain. How does the medullary interstitium accumulate osmotically active solutes? How is this high solute concentration maintained despite the continuous addition of water from collecting ducts?

There is only one mechanism by which the medulla can accumulate solutes: The rate of solute addition to the medulla (i.e., flow in) must exceed the rate of solute removal (i.e., flow out). The only way the interstitial solute concentration can be maintained despite the continuing entry of water from collecting ducts is for the imbalance of solute flows (solute "source") to match the rate of water entry (water "source"). The only way a gradient of solute can be formed is for the flow of solutes toward the tip to exceed, temporarily, the flow of solutes away from the

FIGURE 19.1 Diagram of renal vascular and tubular organization. Only three nephrons are shown (a human kidney contains about one million nephrons, a rat kidney about 50 000), and the vascular structures have been extensively simplified and the vertical scale compressed. (To the right, the same nephrons are shown in their true proportions.) Major zones include: cortex (C), outer medulla (OM), and inner medulla (IM). Afferent arterioles (AA), glomeruli (G), and efferent vessels (EV) are shown together with parts of the peritubular capillary network. Outer medullary vascular bundles arise from the efferent vessels of inner cortical glomeruli. From the glomeruli, fluid flows first into proximal convoluted tubules (PCT) and enters the descending parts of Henle loops. In the outer medulla the thin descending parts of Henle loops begin. Thin ascending limbs begin at the loop turning points in the inner medulla. Thick ascending limbs begin at the border between outer and inner medulla and continue upward into the cortex. In the outer medulla, descending thin limbs of short loops are close to the vascular bundles, whereas thin limbs of long loops are generally located together with collecting ducts and thick ascending limbs in the region between vascular bundles. Distal convoluted tubules (dark hatching) lie among proximal convolutions. Connecting segments join distal convolutions to collecting ducts (CD), which descend through the cortex and medullary regions to open at the papillary tip. [Modified from Beeuwkes and Bonventre, 1975]

tip. These requirements follow directly from the physical principle of mass balance. Some simple examples may help make this clear.

Mass balance and countercurrent multiplication

Consider a closed chamber with two pipes connecting it to the outside world (Figure 19.3*a*). If a flow of salt solution is started into one of the pipes, then after the chamber has filled up, the flow in the pipe leaving the chamber will become equal to the flow entering. In this steady state,

FIGURE 19.2 Osmotic and solute gradients in kidney as measured in tissue slices. Left: Osmolality is found to rise from isotonic (zero) to maximum (100) as the papilla tip is approached. The increase begins in the outer zone of the medulla (O.Z.). Right: Concentrations of sodium (open circles), chloride (closed circles), and urea (crosses) in slices taken from different zones. At the papilla tip, the total osmotic contribution of sodium chloride equals that of urea. These three solutes account for nearly all of the osmotic activity within the papillary tissue. [From Ullrich et al., 1961]

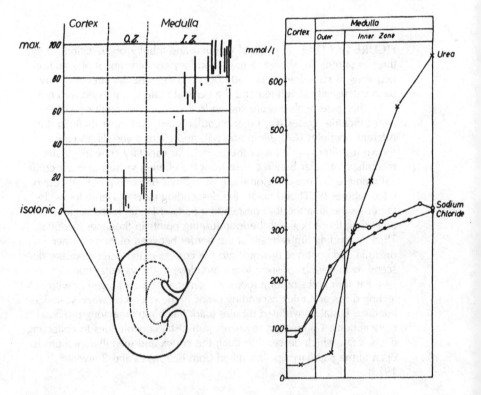

the concentration of salt at any point within the chamber is exactly equal to that of the incoming solution, and does not change with time. The total amount of salt within the chamber is similarly constant. The amount of salt within the chamber can only be increased if the rate of salt entry is made greater than the rate at which salt leaves. For example, if the outflow port were closed with a semipermeable membrane that passed water but not salt, then the outflowing fluid would be more dilute than the incoming fluid (Figure 19.3b). This difference in salt concentration, multiplied by the flow rate, would be the rate of accumulation of salt within the chamber. Because the volume of the chamber is fixed, and the amount of salt in it is increasing, the concentration of salt within the chamber must increase. As long as the imbalance of salt flows continues, the salt concentration within the chamber must rise.

To form a gradient within the chamber, it is necessary that the flow *toward* one end exceed the flow *away* from that end. Imagine a chamber with a barrier that extends from the top almost to the bottom (Figure 19.4a). If we start a flow of solution in one side, without any restrictions to flow of water or salt, the flow extends around the end of the barrier

FIGURE 19.3 Mass balance as applied to solute accumulation. (a) If the flow of water and solute into and out of a compartment are equal, then the concentration of solute within that compartment will be constant and equal to that of the inflowing and outflowing solution. (b) If the flow of solute out of the compartment is made less than the amount flowing in (if, e.g., the outlet port is closed by a semipermeable membrane), then the amount of solute within the compartment will increase.

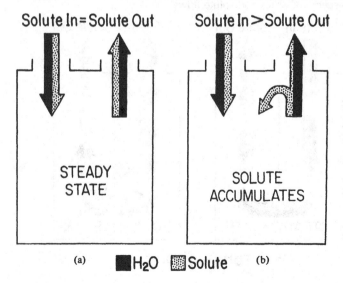

Solute In = Solute Out   Solute In > Solute Out

STEADY STATE   SOLUTE ACCUMULATES

(a)   ■H₂O   ▦Solute   (b)

and continues to the outflow pipe. The composition of the fluid is the same everywhere, but now the compartment contains two oppositely directed flows – one toward the bottom and one away from the bottom. The chamber has now become a countercurrent system. To accumulate salt within this system, it is necessary to make the outgoing solution more dilute than the solution coming in. To make the salt most concentrated near the bottom (i.e., form a gradient), it is necessary to make the *downward* flow of salt exceed the *upward* flow. This could be accomplished by moving salt from the ascending to the descending flow across the barrier between them. Suppose, for example, that there were a transport mechanism located in the barrier. Each sodium or chloride ion moved from the ascending flow to the descending flow would dilute the ascending flow and increase the salt concentration within the descending flow. (Such a transport system, if it existed, would have to be active.) The fluid leaving the chamber would be more dilute than that entering, and the salt flow toward the bottom would be greater than the flow away from the bottom (Figure 19.4*b*). Even though the rate of salt transfer from ascending to

FIGURE 19.4 Mass balance as applied to gradient formation. (a) If a compartment is divided by a barrier, then a countercurrent system is established. If the rate of solute inflow and outflow are equal in such a system, it remains in steady state. (b) if solute outflow is made less than solute inflow, solute will accumulate within the system. If the ascending flow is less than the descending flow, then this solute will accumulate near the turning point, and a gradient will be formed. Once established, such a gradient can be maintained by an equality of descending and ascending water and solute flows.

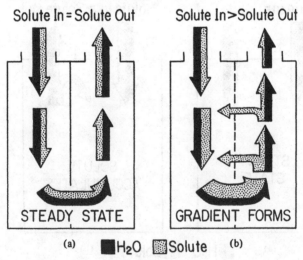

descending flows might be small, with time the amount of salt and the salt concentration at the bottom would still increase (hence the title of this chapter!). This type of system is called a countercurrent multiplier because even a vanishingly small single effect that reduces the salt concentration in the ascending side or increases the descending concentration may be multiplied to form a fluid of very high concentration (Kuhn and Ryffel, 1942; Hargitay and Kuhn, 1951; Kuhn and Ramel, 1959). It is important to note that actual addition of salt to the descending side is not necessary. For example, movement of salt from the ascending side to an adjacent interstitial region (which might be passive) could also make the salt upflow less than that flowing down and lead to gradient formation. On the other hand, transfer of salt from the descending flow to the interstitium or to the ascending flow would prevent the accumulation of salt and formation of the gradient because then the descending salt flow would be less than the ascending flow. Movement of water from the descending flow to the interstitium, perhaps as a result of high interstitial solute concentration, would not hinder the multiplier directly because it would not change the amount of salt flowing toward the tip.

For a given loop length and flow rate, the limit to the concentration that can be attained at the tip of a countercurrent multiplier is set by the rate of solute removal from the ascending limb and the loads on the system (Garner et al., 1978). In the real world, however, we know that diffusion along the length of the system must become increasingly important as the concentration builds up. Similarly, we would expect that if the concentrations became high enough, salt might leak out of the compartment, or water might enter. Indeed, if this mechanism is to do any useful work in urine concentration, we must allow water to enter the system down its own concentration (i.e., osmotic) gradient. And, if this mechanism is to take place in animals, we must try to establish a correspondence between this very hypothetical countercurrent system and the real structures of the kidney.

In the kidney, the structures that correspond to the classical countercurrent multiplier are obviously the loops of Henle (Figure 19.1). Direct measurements made on the kidney surface show that fluid at the start of the descending limb is isosmotic with plasma, whereas the fluid at the top of the ascending limb is, indeed, more dilute with plasma. In the medulla, interstitium includes Henle loops together with collecting ducts. Thus, a concentration gradient formed by loss of salt from ascending limbs into the interstitium could readily provide the osmotic drive for water removal from the ducts. Because the water dilutes the salt in the interstitial compartment, this water addition limits the maximum concentration achievable. Also, because movement of water out of collecting ducts or the descending sides of loops reduces flow in these structures, mass balance requires that there be an equal fluid outflow from the interstitial compartment, or else it will swell up without limit. In the kidney this outflow occurs by way of blood vessel loops that lie beside the tubular structures. Because these vessels can passively gain and lose solute as they descend and ascend, they act as countercurrent isolators that can carry off water and salt without washing out the gradient (Berliner et al., 1958). The same principle is used in many animals for the preservation of body heat. (For a full discussion of countercurrent isolation, see p. 255 of *Animal physiology*, 2nd ed., by Knut Schmidt-Nielsen.) In the kidney in steady state, salt from the ascending limb, together with water removed from both descending limbs and collecting ducts, leaves the medulla by way of vascular loops, and mass balance is maintained.

Problems with the classical multiplier

A classical countercurrent multiplier, based on active transport of salt from the ascending limb for urine concentration is illustrated in Figure 19.5. This mechanism could account for the establishment of the medullary gradient, the concentration of solutes at the tip of the loop, and their subsequent dilution as found in the early distal tubule. Sodium transport from the lumen of the ascending limb to the adjacent interstitium would provide the energy necessary for the performance of osmotic work. In the 1960s, when this proposal became widely accepted, the available experimental evidence clearly favored it. Active pump mechanisms for ions, sodium in particular, had been described, and the electron microscope had not yet shown the structural correlates of transport. It was known, however, that the medullary descending and ascending parts of the Henle loop were composed of tubles with extremely thin walls. The cells in these walls also had very few mitochondria. Electron microscopy soon showed that cells with high sodium transport rates (such as the proximal and distal convoluted tubules) were thick, and that their numerous mitochondria were associated with deep basal infoldings. The absence of these features in the thin *descending* limbs of Henle loops was not a cause for concern because this segment was presumed only to equilibrate with the adjacent interstitium by water loss. On the other hand, this countercurrent multiplier depended upon the presence of active sodium transport from the inner medullary *ascending* thin limb, and there the structure was clearly inconsistent. In the outer medulla, the ascending limb cells are thick and have many mitochondria, so that active sodium reabsorption from this segment (between outer medulla and the cortical surface) could easily result in dilution of the tubular fluid. But transport in this thick segment could not be associated with the establishment of a steep solute gradient in the inner medulla. To be sure, the outer medulla had other interesting structural features including the presence of distinctive bundles of blood vessels and a characteristic overlap between thick ascending limbs and thin descending limbs. But the only established correlation between overall structure and function was the finding of higher urine-concentrating ability in animals with relatively long papillas, structures that contain only thin limbs together with collecting ducts and blood vessels (Schmidt-Nielsen, 1979). Finally, the role of urea seemed not satisfactorily accounted for. Animals fed high protein diets, yielding large amounts of urea as a nitrogenous product, can concentrate their urine better than those in which urea is scarce. The gain in concentrating ability associated with urea is greater than that

FIGURE 19.5 Schematic representation of the now obsolete classical countercurrent multiplier. In this hypothesis active salt reabsorption from the entire ascending limb (including thin segment) made ascending salt flow less than descending salt flow, and caused salt accumulation. Numbers indicate total osmolality at each location.

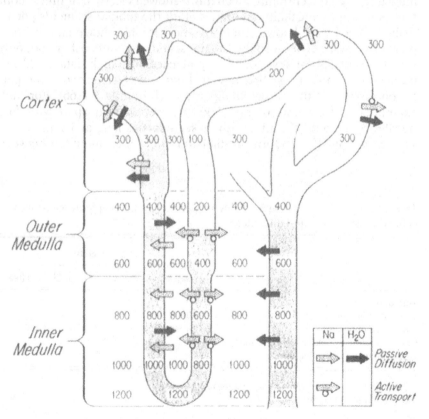

which would be expected if urea were merely present as an additional urinary solute (Gamble et al., 1934; Levinsky and Berliner, 1959). Urea thus seems to have a more fundamental role in urine concentration than suggested by the classic countercurrent mechanism.

New data from isolated tubule experiments

These paradoxes have led to renewed experimental efforts. Because formation of the solute gradient required that there be a salt source in the inner medulla and that it be localized to the ascending limbs of Henle loops, the countercurrent hypothesis was tested by micropuncture collection of fluid samples from different points along ascending thin limbs. These difficult experiments showed that the fluid in the ascending limb was indeed dilute compared to that in the descending limb, and that the dilution was due to the loss of sodium chloride (Jamison et al., 1967). However, this finding did not establish that the mechanism by which sodium left the tubule was an active transport process. Also, computer models of the concentrating mechanism showed clearly that unless both transport and permeability properties of *all* the tubular segments potentially involved were known, there was no basis for choice among competing hypotheses. Because the tubular segments of interest are not only very tiny but also almost completely inaccessible, such data could not be obtained until an exciting new technique was developed – the perfusion of isolated tubule segments in vitro (Burg et al., 1966). For such experiments, fragments of rabbit kidney were teased apart with fine needles so as to obtain intact tubule segments as long as 1 mm. These segments, only 15 to 30 μm in diameter, were then mounted between

TABLE 19.2 Transport and permeability properties of rabbit nephron segments as defined in isolated perfused tubules.

	Active salt transport	Permeability		
		H_2O	NaCl	Urea
Thin descending limb	0	+ + + +	0	+
Thin ascending limb	0	0	+ + +	+
Thick ascending limb	+ + + +	0	0	0
Cortical and outer medullary collecting duct	+	+ + +[a]	0	0
Inner medullary collecting duct	+	+ + +[a]	0	+ + +

[a] Collecting duct water permeability assumes ADH present.
Modified from Kokko and Rector, 1972.

micropipets so that the tubular lumen could be perfused with fluid of known composition and containing appropriate radioactive tracers. In such preparations, the movement of solutes across the tubular wall could be directly measured as a function of imposed driving forces.

Some results from these difficult and exciting experiments are shown in Table 19.2. Neither the descending nor ascending *thin* limbs of Henle loops show an active sodium transport mechanism – a finding consistent with the relative lack of surface infoldings and mitochondria in the cells of these segments. However, the thin limbs are quite different in their passive permeability properties. The thin descending limb is much more permeable to water than to salt, whereas the thin ascending limb is much more permeable to salt than it is to water. The *thick* ascending limbs are relatively impermeable to both water and salt, but contain an active transport system for sodium chloride capable of reabsorbing salt from the tubular lumen against a substantial gradient. The collecting ducts have different properties in their cortical and outer medullary part from those of the inner medullary region. Although water permeability in both regions is controlled by antidiuretic hormone, urea can leave collecting ducts only in the inner medullary region. The finding that active salt transport is essentially restricted to the thick ascending limbs appears inconsistent with the countercurrent mechanism, for we know that gradient formation requires a mechanism for removal of salt from the thin ascending side of the loop. How can active salt reabsorption from the thick ascending limbs of the outer medulla satisfy this requirement?

A new mechanism: the two-solute model

It is, in fact, possible to couple active transport in the thick ascending limb to passive salt reabsorption in the thin ascending limb. The coupling mechanism, based upon recent theoretical evaluation, depends upon both the physical arrangement of the tubules and their permeability properties (Stephenson, 1978). The physical structure includes not only the Henle loop passing through the medulla, but also a second loop formed by the thick ascending limb, distal tubule, and collecting duct within the cortex. In this second loop, tubular fluid which has been modified by thick ascending limb active transport processes is returned to the medulla by way of the collecting duct. Because the thick ascending limb and distal convoluted tubule reabsorb salt, tubular fluid returning to the medulla must be relatively low in salt but may be high in urea. Passive water removal from the cortical and medullary collecting duct would increase the concentration of all solutes remaining within the duct lumen, thus raising duct urea to a very high level. Water loss from the thin descending limb would similarly increase the concentration of solute there, but in the thin descending limb the solute is primarily salt. Thus, when fluid in the thin descending limb turns the corner into the ascending side, it is very high in salt. Although the total osmolality in both thin ascending limb and inner medullary collecting duct may be the same, chemical energy is available because the collecting duct contains a relatively pure urea solution, and the ascending limb contains a relatively pure salt solution. The diffusion of each solute down its own concentration gradient into the interstitium will take the system to a lower energy state. If appropriate permeability barriers existed, part of this energy could perform osmotic work by driving net solute reabsorption from the thin ascending limb. Such a passive solute source would meet the requirements for gradient formation.

Because the concepts involved in the performance of osmotic work through the dilution of relatively pure solutions are not obvious, a hypothetical example may be helpful (Stephenson, 1972). Consider a compartment surrounded by a membrane that is permeable to salt but relatively impermeable to water and urea. Now imagine another compartment nearby that is quite impermeable to salt but freely permeable to water and urea. These compartments correspond to the thin ascending limb (salt-permeable) and the inner medullary collecting duct (water- and urea-permeable). The space between them corresponds to the interstitial compartment (Figure 19.6a). Now place a pure sodium chloride solution at an osmotic concentration of 300 mOsm in the ascending limb compartment. Place a solution of pure urea at the same osmotic concentration in the collecting duct and interstitial compartments (Figure 19.6b). Because there is no salt in the interstitial space, and the ascending limb is permeable to salt, salt will move down its concentration gradient from the ascending limb into the adjacent interstitium. Such passive loss of salt from the ascending limb will result in dilution of this compartment. In the interstitial space, there is now some salt in addition to the urea initially present, and the total osmotic concentration in this region is higher than before (Figure 19.6c). Although there now exists a gradient for water to enter the interstitium from the ascending limb compartment, the walls of that compartment are relatively impermeable to water. But, because the collecting duct is water-permeable, water leaves the collecting duct until its total osmolality becomes the same as that of the adjacent interstitium. In the final equilibrium state (Figure 19.6d), the ascending limb compartment will have its original volume but will have lost salt. It has indeed become diluted. The collecting duct compartment has lost water in excess of solute and become concentrated. The water lost from the collecting duct is now in the interstitial compartment, whose volume is increased.

FIGURE 19.6 Generation of osmotic gradients from pure solutions by selective permeability. (a) Two compartments are shown corresponding to the thin ascending limb (TAL) and collecting duct (CD). The TAL is permeable only to salt, whereas the CD is permeable to urea and water. (b) These compartments are now shown separated by an interstitial region (I) and filled with solutions of equal osmolarity but different composition. (c) Sodium chloride can diffuse down its gradient from ascending limb to interstitium, but it cannot enter the collecting duct. (d) The interstitial region now has higher osmolarity because it contains both urea and salt. Because the ascending limb is nearly water-impermeable, water enters the interstitial region from the collecting duct, decreasing duct volume and increasing the concentration of urea remaining. [Modified from Stephenson, 1972]

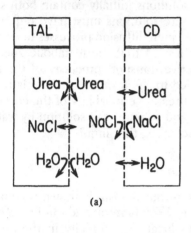

(a)   (b)

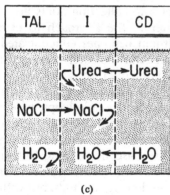

(c)   (d)

Thus, by using two pure solutions and allowing diffusion across specialized membranes, it is possible to create both a diluted ascending limb through salt loss and a concentrated collecting duct by water loss. Osmotic work has been performed, because new osmotic gradients have been created. The energy needed for this work is present in the initial pure salt solution, because this salt loses energy through dilution into the larger volume represented by the interstitium. To put it another way, it is clear that once the salt has diffused out of the ascending limb, it would take work to squeeze it back again. (In thermodynamic terms this is expressed as a change in the chemical free energy: $\Delta G = RT \ln C_1 - RT \ln C_2$.) Because physiological solutions initially contain both salt and urea, creation of the relatively pure solutions must have required metabolic energy. In the application of such diffusion processes to countercurrent multiplication, it is hypothesized that the metabolic energy input to the system occurs in the active transport processes of the ascending thick limb, and this is coupled to the formation of a relatively "pure" urea solution by virtue of the urea impermeability of the cortical and outer medullary collecting duct, and a "pure" salt solution by water removal from a salt-impermeable descending thin limb.

The new passive countercurrent model

One form of the passive countercurrent mechanism is shown schematically in Figure 19.7 (Kokko and Rector, 1972; Jamison and Maffly, 1976). Ascending limb salt transport creates local hypertonicity in the outer medulla and inner cortex. At this level, the collecting ducts are urea-impermeable, so that as they lose water in the approach to osmotic equilibrium, the urea solution within them becomes concentrated. At the border between outer and inner medulla the duct becomes urea-permeable. Urea diffuses out of the collecting duct, thus raising the urea content of the inner medullary interstitium. Water also enters the inner medulla by way of the thin descending limbs of Henle loops, which are relatively impermeable to urea and salt, but are permeable to water. Thus inner medullary urea (contributed by collecting ducts) creates an osmotic driving force favoring water removal. Because descending thin limbs contain relatively little urea compared with salt, water removal could create a relatively pure salt solution in the descending limb, but with the same osmotic concentration as the adjacent interstitium which contains *both* urea and salt. As the salt in the tubule must then be at higher concentration than in the interstitium, a driving force favoring salt removal exists, but the thin descending limb is impermeable to salt. After the fluid rounds the bend of the loop, however, the permeability of the ascending limb allows salt to diffuse out down its own concentration gradient. Some urea may diffuse into the ascending limb, but because this segment is much more permeable to salt than urea, and because flow along the segment prevents the establishment of equilibrium conditions, salt loss exceeds urea entry, and net dilution of the ascending limb fluid occurs. Such a process has been directly demonstrated in isolated perfused thin ascending limb segments (Imai and Kokko, 1976). Thus, the thin ascending limb may act as a passive salt source within the inner medulla. The small amount of urea entering the ascending limb

FIGURE 19.7 Schematic representation of the new passive countercurrent multiplier. Note that *two* functional loops are involved: the medullary loop of Henle and the cortical pathway comprised by the ascending limb, distal tubule, and cortical collecting duct. (1) Active salt reabsorption in the outer medulla dilutes ascending limb fluid and may cause outer medullary interstitial hyperosmolarity. (2) Water and salt removal in the cortical "loop" leaves a concentrated urea solution in the collecting duct. (3) Urea leaves the inner medullary duct passively, raising the urea content and osmolarity of the interstitium. (4) (Left side) Interstitial osmolarity (now urea plus salt) removes water from the (salt-impermeable) thin descending limb, thereby increasing the luminal salt concentration to a level higher than interstitial salt.'(5) As loop fluid ascends, the tubule becomes salt-permeable, allowing salt to diffuse passively down its concentration gradient. Because this segment is relatively impermeable to urea entry, such salt movement results in a net dilution of the ascending limb fluid, and creates a passive inner medullary salt source. [From Jamison and Maffly, 1976]

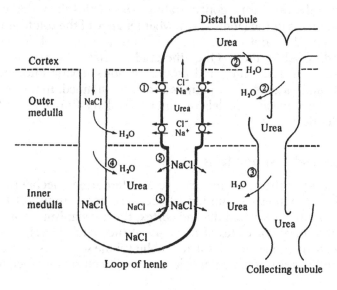

Distal tubule

Loop of henle

Collecting tubule

will be recycled via the distal convolution and reconcentrated in the collecting ducts of the outer medulla.

This new countercurrent hypothesis must satisfy mass balance considerations. During establishment of an osmotic gradient, net solute flow toward the papilla tip must exceed the flow away from the tip. To the extent that salt that comes down the descending thin limb diffuses laterally out of the ascending thin limb, this requirement is satisfied. Urea flow downward and out of the sides of the collecting ducts augments this loop solute source. As steady state is reached, these solutes, together with water from collecting ducts and descending thin limbs, must be continuously removed from the inner medulla by way of the vessels that pass through the region. As blood descends in these vasa recta, the plasma must approach osmotic equilibrium with the adjacent interstitium through a process in which solute entry is most important, consistent with a countercurrent isolation function. The water loss from vasa recta that does occur increases the plasma protein concentration and hence the colloid osmotic pressure in the capillary (Jamison et al., 1979). Hence classical Starling exchange mechanisms, involving the balance of hydrostatic and osmotic driving forces across the vasa recta walls, probably play an important part in the mechanism of inner medullary fluid removal. Direct measurement of pressure and protein concentration in vessels of the inner medulla has shown that intravascular protein provides an important driving force for water entry into the capillary, even if salts and urea have nearly come to equilibrium across the vessel wall (Sanjana et al., 1975). In steady state the net rate of water entry rate into the blood vessels must be equal to water removal from inner medullary collecting ducts during the final urine concentrating process plus the water leaving the thin segments of the Henle loops. Maintenance of the solute gradient in steady state does not require a continuous imbalance of descending and ascending solute flows. Once the gradient has been established, a strict equality of descending and ascending flows, taking into account both solute and water sources, permits it to be maintained. In this respect, both active and passive models of inner medullary function are functionally identical.

Advantages and new problems

The superiority of the two-solute passive countercurrent mechanism over the classical hypothesis is substantial. Because active transport is necessary only in the outer medulla and cortex, the mechanism is consistent with the structural correlates of transport at the cellular level, and with the permeability properties of isolated rabbit tubule segments. In the new model, urea has a fundamental role in the concentrating mechanism.

Finally, extensive computer analysis shows that such a mechanism can create an inner medullary solute gradient and generate concentrated urine. Accordingly, the new mechanism is rapidly becoming accepted as the best current model to explain generation of the inner medullary hypertonicity essential to the formation of concentrated urine.

It must be kept in mind that this mechanism is a working hypothesis and is not yet firmly established. Many important issues remain to be resolved. Among these are inconsistencies between in vitro observations on isolated rabbit tubules and micropuncture experiments performed in vivo; the nearly total lack of knowledge regarding the composition and flow rate of fluid in tubular segments entering and leaving the medulla; and the simplicity of physical structures in the hypothesis as contrasted with the actual structural complexity of animal kidneys. For example, although active sodium reabsorption has not been found in thin ascending limbs of isolated rabbit tubules studied in vitro, micropuncture studies in other species have been interpreted as indicating active transport from this segment (Marsh and Azen, 1975). The reason for the different findings is not clear.

The study of isolated tubules in vitro requires that investigators make assumptions regarding the composition of the fluid naturally surrounding the tubule. These assumptions may be partly wrong. It is also possible the data obtained may reflect damage to the tubule during the microdissection process. On the other hand, interpretation of papillary micropuncture experiments is usually based on the assumption that the concentration of solutes in the interstitium surrounding tubules is the same as that found in plasma samples from vasa recta. Even if this assumption is correct, it is not known that the vasa recta accessible to micropuncture on the surface of the papilla have the same composition as those deeper within the tissue (Andreoli et al., 1978). These questions are fundamental, because the classical hypothesis predicts lower sodium in the ascending limb than adjacent interstitium, as a result of active reabsorption, whereas the newer proposal predicts higher sodium in the tubule as the driving force for passive sodium reabsorption. It is clear that only direct measurement of concentrations in tubule lumens and interstitial compartments will allow driving forces and transport mechanisms to be established.

New techniques such as electron probe microanalysis of frozen papillary tissue sections now allow direct measurements in these previously inaccessible regions (Saubermann et al., 1981; Bulger et al., 1981). Similarly, the region deep within the kidney is entirely inaccessible to micropuncture and has rarely been studied by other techniques, so that the concentrations and flow rates in Henle loops and collecting ducts at the cortico-medullary border remain unknown. Computer models based upon "reasonable" assumptions about this generally succeed in predicting

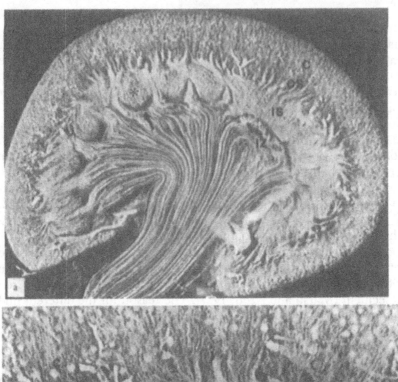

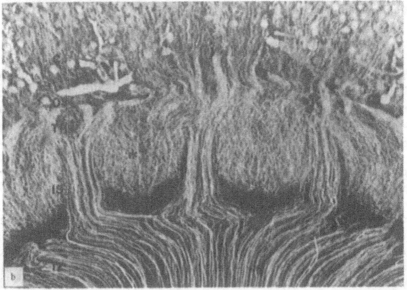

FIGURE 19.8 Photographs of blood vessels in *Psammomys* kidney made visible by a process of silicone injection followed by clearing. (a) The marked difference in vascular pattern in different regions is apparent. In the cortex (C) glomeruli are barely visible as white spots. The outer medulla contains two regions, the outer stripe (OS) and the inner stripe (IS). The inner medulla (IZ) extends to the papilla tip (out of picture

the formation of concentrated urine. However, depending on the assumptions made, different passive models of inner medullary events predict very different concentration profiles between the cortico-medullary border and the papilla tip. Because neither initial parameters nor the actual profiles of lumenal fluid compositions are known, the computer cannot provide a firm basis for selection among the different mechanisms proposed.

Finally, comparison of the physical arrangement of vascular and tubular compartments in hypothetical models with those of real kidneys shows that the hypotheses ignore most of the organizational detail. This may be entirely appropriate, for much of this detail may be entirely irrelevant to the basic mechanism. For example, although the correlation between urine-concentrating ability and papillary length has long been known, recent evidence indicates that an equally good correlation can be made with the numerical density of vasa recta in the outer medulla (Munkacsi and Palkovits, 1977). Grouped together in vascular bundles, these vessels undoubtedly serve as countercurrent isolators. In some species, however, these vascular bundles also contain thin descending limbs from nephrons having short loops, that is, loops turning within the outer medulla. Ascending limbs belonging to these loops are located in the interbundle region together with thin descending and thick ascending limbs of loops having turning points deep within the inner medulla (Kriz et al., 1972). In the desert rodent *Psammomys*, the outer medulla is quite remarkable because it includes two separate vascular compartments (Figure 19.8) (Bankir et al., 1979). One of these is composed of giant vascular bundles containing many more venous than arterial vasa recta, with all of the venous vessels originating in the inner medulla. These bundles also contain as many descending limbs of short loops as veins. The second vascular region, lying between the bundles, contains a dense capillary network that surrounds thick ascending limbs of both short and long Henle loops, together with thin descending limbs of long loops and collecting ducts (Kaissling et al., 1975). The apposition of thick ascending

FIGURE 19.8 (cont.)
to lower left). The arrow and asterisk indicate an outer medullary region shown at higher power below. (b) Arterial and capillary vessels and glomeruli at the corticomedullary border in *Psammomys*. Three bundles of relatively straight vasa recta separate the dense capillary networks of interbundle regions. Tubules are not visible in these preparations, but other studies show that the bundles contain many thin descending limbs, whereas thick ascending limbs and collecting ducts are located in the interbundle regions. The significance of this arrangement is not understood. [From Bankir et al., 1979]

limbs to collecting ducts might correspond to a proposal of the passive model that outer medullary salt transport could create local hypertonicity to enhance water removal from collecting ducts passing through this region. However, in contrast to the passive model, experiments in *Psammomys* indicate that the thin descending limbs of long loops appear to reach osmotic equilibrium with their surrounding interstitium largely through salt addition, rather than by water removal alone (Jamison et al., 1979). Because *Psammomys*, whose diet is very high in salt, may have evolved specialized mechanisms for salt excretion, these findings may not be relevant to the general problem of urine concentration. However, it seems clear that the trend in modeling is toward mechanisms that reflect actual details of outer medullary organization.

Summary

The basic mechanism for urine concentration is the flow of water down its osmotic gradient from the lumen of the collecting duct to the region of high solute concentration in the adjacent papillary interstitium. To this extent the process is simple. However, the process by which the papillary interstitium accumulates salt and urea is apparently quite complex. From mass balance considerations we know that there must be a solute source in the inner medulla. However, attempts to demonstrate active salt transport from the thin ascending limbs have been generally unsuccessful, and the ultrastructure appears inconsistent. Mathematical modeling techniques, combined with new data obtained from the study of isolated tubule segments in vitro, have led to the formulation of a new hypothesis: the passive countercurrent mechanism. In this model, active salt reabsorption from the outer medullary thick ascending limb is coupled, by structure and selective permeabilities, to the generation of a passive salt outflow from the thin ascending limb. This model is generally consistent with available data and accounts for the fundamental role of urea in the concentrating mechanisms. Additional findings, both structural and physiological, have clarified the role of the blood vessels in maintaining mass balance within the medullary region. Considered together, these observations and hypotheses form the foundation upon which future concepts of medullary function must be based. This chapter can be considered only as a brief introduction to the many exciting issues remaining to be resolved. For additional information the reader is referred to papers cited in the bibliography and to reviews that consider the major issues involved (Beeuwkes and Brenner, 1978; Beeuwkes, 1980; Jamison, 1981).

The author thanks J. Cant, S. Churchill, L. Kinter, and J. Shahood for their valuable criticism and suggestions. Original drawings for this chapter are by P. Battaglino. The manuscript was prepared by L. Batcheller and B. Creutz. The author's research was supported by USPHS Grants AM 18249, HL 19467, HL 02493, RCDA AM 00224 and by a grant from R. J. Reynolds Industries, Inc.

References

Andreoli, T. E., Berliner, R. W., Kokko, J. P., and Marsh, D. J. (1978) Questions and replies; renal mechanisms for urinary concentrating and diluting processes. *Am. J. Physiol.* 235:F1–F11.

Bankir, L., Kaissling, B., deRouffignac, C., and Kriz, W. (1979) The vascular organization of the kidney of *Psammomys obesus*. *Anat. Embryol.* 155:149–160.

Beeuwkes, R. (1980). The vascular organization of the kidney. *Annu. Rev. Physiol.* 42:531–542.

Beeuwkes, R., III and Bonventre, J. V. (1975) Tubular organization and vascular–tubular relations in the dog kidney. *Am. J. Physiol.* 229:695.

Beeuwkes, R., III and Brenner, B. M. (1978) Kidney. In *Peripheral circulation.* (P. C. Johnson, ed.) New York: Wiley.

Berliner, R. W., Levinsky, N. G., Davidson, D. G., and Eden, M. (1958) Dilution and concentration of the urine and the action of antidiuretic hormone. *Am. J. Med.* 24:730.

Bulger, R. E., Beeuwkes, R., and Saubermann, A. J. (1981) Application of scanning electron microscopy to x-ray analysis of frozen-hydrated sections. III. Elemental content of cells in the rat renal papillary tip. *J. Cell Biol.* 88:274–280.

Burg, M. B., Grantham, J. J., Abramow, M., and Orloff, J. (1966) Preparation and study of fragments of single rabbit nephrons. *Am. J. Physiol.* 210:1293.

Gamble, J. L., McKhann, C. F., Butler, A. M., and Tuthill, E. (1934) An economy of water in renal function referable to urea. *Am. J. Physiol.* 109:139.

Garner, J. B., Crump, K. S., and Stephenson, J. L. (1978) Transient behaviour of the single loop solute cycling model of the renal medulla. *Bull. Math. Biol.* 40:273–300.

Hargitay, B., and Kuhn, W. (1951) Das Multiplikationsprinzip als Grundlage der Harn Konzentrierung in der Niere. *Z. Elektrochem.* 55:539.

Imai, M., and Kokko, J. P. (1976) Mechanism of sodium and chloride transport in the thin ascending limb of Henle. *J. Clin. Invest.* 58:1054.

Jamison, R. L. (1981). Urine concentration and dilution: The roles of antidiuretic hormone and urea. In *The kidney*, Vol. 1 (B. Brenner and F. C. Rector, eds.), pp. 495–550. Philadelphia: Saunders.

Jamison, R. L., Bennett, C. M., and Berliner, R. W. (1967) Countercurrent multiplication by the thin loops of Henle. *Am. J. Physiol.* 212:357.

Jamison, R. L., and Maffly, R. H. (1976) The urinary concentrating mechanism. *N. Engl. J. Med.* 295:1059.

Jamison, R. L., Roinel, N., and deRouffignac, C. (1979) Urinary concentrating mechanism in the desert rodent *Psammomys obesus*. *Am. J. Physiol.* 236:F448–453.

Kaissling, B., deRouffignac, C., Barrett, J. M., and Kriz, W. (1975) The structural organization of the kidney of the desert rodent *Psammomys obesus*. *Anat. Embryol.* 148:121–143.

Kokko, J. P., and Rector, F. C., Jr. (1972) Countercurrent multiplication system without active transport in inner medulla. *Kidney Int.* 2:214.

Kriz, W., Schnermann, J., and Koepsell, H. (1972) The position of short and long loops of Henle in the rat kidney. *Z. Anat. Entwicklungsgesch.* 138:301–319.

Kuhn, W., and Ramel, A. (1959) Aktiver Salztransport als moglicher (und wahrscheinlicher) Einzeleffekt bei der Harnkonzentrierung in der Niere. *Helv. Chim. Acta* 42:628.

Kuhn, W., and Ryffel, K. (1942) Herstellung konzentrierter Losungen aus verdunnten durch blosse Membrane-wirkung: Ein Modellversuch zu Funktion der Niere. *Z. Physiol. Chem.* 276:145.

Levinsky, N. G., and Berliner, R. W. (1959) The role of urea in the urine concentrating mechanism. *J. Clin. Invest.* 38:741.

Marsh, D. J., and Azen, S. P. (1975) Mechanism of NaCl reabsorption by hamster thin ascending limbs of Henle's loops. *Am. J. Physiol.* 228:71.

Munkacsi, I., and Palkovits, M. (1977) Measurements on the kidneys and vasa recta of various mammals in relation to urine concentrating capacity. *Acta Anat.* 98:456–468.

Sanjana, V., Johnston, P. A., Troy, J. L., Deen, W. M., Robertson, C. R., Brenner, B. M., and Jamison, R. L. (1975) Hydraulic and oncotic pressure measurements in the inner medulla of the mammalian kidney. *Am. J. Physiol.* 228:1921.

Saubermann, A. J., Echlin, P., Peters, P. D., and Beeuwkes, R. (1981) Application of scanning electron microscopy to x-ray analysis of frozen-hydrated sections. 1. Specimen handling techniques. *J. Cell Biol.* 88:257–267.

Schmidt-Nielsen, K. (1979) *Animal physiology: adaptation and environment,* 2nd ed., p. 252 et seq. Cambridge University Press.

Stephenson, J. L. (1972) Central core model of the renal counterflow system. *Kidney Int.* 2:85.

Stephenson, J. L. (1978) Countercurrent transport in the kidney. *Annu. Rev. Biophys. Bioeng,* 7:315–339.

Ullrich, K. J., Kramer, K., and Boylan, J. W. (1961) Present knowledge of the counter-current system in the mammalian kidney. *Prog. Cardiovasc. Dis.* 3(5):395–431.

20

Insects: small size and osmoregulation

SIMON MADDRELL

Insects are remarkable animals in many of their physiological, behavioral, and ecological adaptations. One of the most striking things about them is that they particularly seem to thrive in osmotically stressful environments. Virtually none live in the sea where conditions are relatively stable, but they are common in fresh water, brackish water, and hypersaline lakes of osmotically and ionically varied composition. On land, arguably the most osmotically difficult environment because of the pronounced tendency for desiccation (Edney, 1977), insects really come into their own, claiming more than a million ecological niches. Whatever else may be true of insects, they certainly need to have a set of versatile and very effective osmoregulatory mechanisms. What adds to the impressiveness of their abilities in this direction is that insects are relatively small organisms whose size ought, because of the resulting increase in surface area to volume ratio, to add to their osmoregulatory stress. To emphasize this, let us compare the sizes of insects and vertebrates found in the terrestrial environment. The smallest terrestrial insects weigh only a few micrograms, whereas the smallest terrestrial vertebrates weigh more than 2 g, that is, nearly 10^6 times as much (Maddrell, 1981). Bearing in mind that the insect body form intrinsically is such as to expose a very large surface area (the very word *insect* derives from the Latin, meaning "cut into"), the figures imply an area/volume ratio that is close to 100 times larger than for terrestrial vertebrates. The difference is also reflected at other size ranges; the largest insects weigh no more than about 100 g, but terrestrial vertebrates can be 10^5 as large as this (an elephant can weigh more than 10^7 g). With so much apparently contributing to the difficulties that insects face in regulating their internal environment, how is it that they are able to survive in the harsh environments that they favor? The main object of this chapter will be to describe the effective osmoregulatory adaptations that research has so far shown insects to have. Where appropriate I shall try to point out the most important problems that remain for future research to resolve.

Insect cuticle in osmoregulation

The external surface

Insects could scarcely survive in the environments that they inhabit unless their external surfaces reduced to a minimum the movements of ions and water between the internal body fluids and the external world. The overriding importance of the wax-covered cuticle in providing just such a barrier has been recognized for a number of years (see Beament, 1961). The near impermeability of this barrier has been attributed to the presence at or near the surface of the cuticle of an oriented monolayer of lipid molecules. The existence of such a layer was deduced by Beament (1961) from his evidence for thermally induced phase changes in lipid material. Locke (1965) was later able to provide some evidence from electron micrographs of the outer layers of the cuticle that lipid material might occur in the predicted position (see Figure 20.1). Since these classic and important papers were published, more recent work has drawn attention to the possibility that the cuticle and its underlying epidermis might actually comprise at least two permeability barriers in series, an outer one being at the surface of the epicuticle and an inner one at the cell–cuticle interface. This would, of course, act so as to further decrease the rate of any ionic and water movements through the outer surface of the insect. This interesting idea was first advanced by Berridge (1970). In line with this idea is the proposal by Reynolds (1975) that the me-

FIGURE 20.1 Electron micrograph of a section through the surface layers of the cuticle of an insect (5th stage larva of *Calpodes ethlius*). [Micrograph courtesy of M Locke]

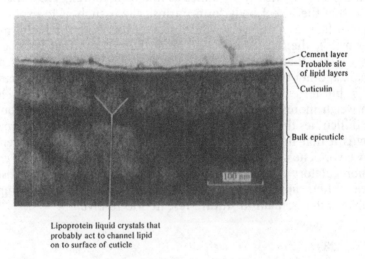

Cement layer
Probable site of lipid layers
Cuticulin

Bulk epicuticle

100 nm

Lipoprotein liquid crystals that
probably act to channel lipid
on to surface of cuticle

chanical properties of the abdominal cuticle in *Rhodnius* might depend on changes in intracuticular water activity dictated by control of pH within the cuticle.

Finally, it is worth mentioning that, in the cockroach, there is evidence that the permeability of the cuticle and epidermis to water may be under hormonal control (Treherne and Willmer, 1975). This may be important under the relatively rare conditions when the insect may need to excrete water. By increasing evaporation through the body wall, this can be achieved with little energy expenditure and without the inevitable small loss of ions that would occur if the same amount of water were eliminated through the excretory system.

It remains to be seen whether other insects can also control the water permeability of their integuments. In view of the possibility that overall permeability of the integument might depend on several barriers in series, it will be of interest to discover how its regulation is achieved.

Cuticle lining the hindgut

The importance of insect cuticle to osmoregulation does not end with its role as an almost impermeable coat. It has also a crucial role in the ability of insects to restrict water loss in excretion. The insect excretory system, as we shall see, has an unparalleled ability to recover virtually all of the water from excreted material. This rests very much on the highly developed water-absorbing activities of the cells of the insect hindgut. However, that these cells can carry this out so effectively almost certainly owes a great deal to the overlying cuticle which separates them from direct contact with the luminal contents. In the locust, the cuticle is only permeable to substances whose molecular diameters are less than about 0.7 nm (Phillips and Dockrill, 1968; Maddrell and Gardiner, 1980a). This allows the rectal cells to reabsorb water, and such small molecules as amino acids and monosaccharide sugars, but protects them from mounting concentrations of larger, potentially interfering substances in the lumen. In this respect it is significant that the excretion of metabolically produced acids is facilitated by their conversion to acylamides through condensation with amino acids (Smith, 1962). The resulting acylamides are actively taken up by Malpighian tubules and concentrated in the primary excretory fluid (Maddrell et al., 1974). When they reach the hindgut, their large size prevents them from penetrating the cuticle, and so they are more effectively eliminated than if they were able to escape passively at measurable rates from the hindgut. It seems quite possible that the possession of cuticle lining the part of the excretory system where water reabsorption occurs gives insects an important advantage over their competitors. Terrestrial vertebrates reabsorb water using epithelia that are in direct contact with the excretory material.

Cuticle lining the foregut

Insects have a layer of cuticle lining the foregut, and recent research has shown that in several insects this lining has a very limited permeability (Maddrell and Gardiner, 1980a). The importance this has for osmoregulation is that ingested material is effectively kept out of contact with the hemolymph, a great advantage if it is of very different osmotic concentration. The rate at which such potentially osmotically embarrassing fluid is presented to the midgut is controlled by regulating its passage from the foregut (Treherne, 1957). Overall, this adaptation allows insects rapidly to ingest considerable quantities of material of any osmotic concentration without suffering osmotic shock. One can also speculate that the possession of such an isolable region of the gut might allow digestive or fermentation processes to go on there that would be impossible and/ or damaging if they occurred in a part of the gut with permeable walls. It will be for future research to establish to what extent insects make use of this apparent possibility.

Respiration and osmoregulation

In vertebrates under desert conditions, the water content of the exhaled air is a major source of water loss (Schmidt-Nielsen, 1979). For example, a kangaroo rat fed dried barley and kept at 25 °C and 20% relative humidity loses water by evaporation, mostly from the respiratory tract at nearly three times the rate at which it loses water through the urine and feces (Schmidt-Nielsen, 1964). This is in spite of the fact that the kangaroo rat's nasal passages are so adapted as to cool the exhaled air, reducing its water content very significantly (Schmidt-Nielsen et al., 1970). The question naturally arises of how much water insects lose from their respiratory surfaces. Although some excellent early work has been done in this field, it has not been followed up, and it is probably fair to comment that our ignorance on this point is considerably greater than in any other area of insect osmoregulation.

The experiments that have been done (see Edney, 1977, for an excellent summary of this work) show that water loss from insects is greatly increased during activity, and that this phenomenon is almost certainly due to the greater length of time for which the spiracles are open. (The spiracles are the valves that control gas movements between the outside world and the lumina of the tracheal tubes.) Spiracular closure is also affected by the degree of hydration of the insect, the spiracles being kept closed for more of the time and opening less widely in dehydrated insects. Active ventilation of the tracheal system, which occurs in some insects, leads to much water loss. It is also known that, in active insects, respiratory water loss is considerably greater than the evaporative loss through the general body surface. Finally, insects with a low metabolic rate commonly display a peculiar phenomenon in which there is a periodic release of large amounts of carbon dioxide interspersed with longer periods when none is released. It seems that carbon dioxide is somehow stored in the insects' tissues as oxygen is absorbed, so that the internal pressure in the tracheal system is reduced. A small amount of air, equivalent in volume to the absorbed oxygen, can then be slowly drawn in through the partly open spiracles, and this inward movement may well slow diffusive loss of water through the spiracles. When the carbon dioxide content reaches a high level, the spiracles open wide and large amounts of CO_2 are released. This CO_2 mostly must come from the tissues, as the amounts released are much larger than the tracheal system could contain (Schmidt-Nielsen, 1979). This unusual cyclic method of gas exchange seems well adapted to curtail water loss by restricting it mostly to the periods of wide opening of the spiracle. Apart from this interesting method of controlling water loss, one can speculate that there are two general features of insects that may act to reduce water loss during respiration. The first is the obvious one that, because insects are ectotherms, their respiring tissues will be at or close to the temperature of the environment (active flight muscles being an exception to this). At low temperatures, then, not only will metabolism be slower so that the oxygen demand will be lower and the spiracles open for less time (or less wide), but the water vapor content of the tracheal air will be less so that the rate of diffusion out through the spiracles will be reduced.

The second feature is that because oxygen diffuses so very fast in air (300 000 times faster than in water) and because the tracheal system takes air very close to respiring tissues, there will be a steep concentration gradient acting to drive diffusion of oxygen through the relatively short fluid layer between the air and the tissue. It follows that the water-permeable area required for oxygen uptake need be relatively small, and that, as a consequence, evaporative water loss might be minimized.

It will be evident to the reader that much of what has been said in the above paragraphs suggests very strongly that insects need to be, and no doubt are, adapted to minimize the inevitable water loss arising from oxygen uptake. What is now urgently needed are some actual determinations of rate of respiratory water loss in insects at different temperatures. It is of major importance to see how the figures for terrestrial insects compare with those for vertebrates of broadly similar size, not only in respect of water loss for both groups of animals both at rest and in activity, but also in relation to the oxygen consumption. The interesting question is: How much water is lost by insects and by vertebrates when they take up similar amounts of oxygen? Given the strong selection pressure on animals living on land to conserve water, it will be of great interest to see if the insect tracheal system is more or less effective than the vertebrate system in minimizing the amount of water lost during respiration. At the risk of unnecessary repetition, it does seem that this is the major unexplored area in the field of insect osmoregulation.

The contribution of the excretory system to osmoregulation in insects

As discussed earlier in this chapter, insects expose to the environment a surface area that is much higher in relation to their volume than do vertebrates, for example. Particularly in the terrestrial environment, this relatively large surface area must increase their osmoregulatory problems. Although they are well protected by their wax-covered integument, terrestrial insects are so much smaller than terrestrial vertebrates that it is difficult to avoid the conclusion that insect tissues must have evolved a greater tolerance to varied conditions in the hemolymph.

Direct support for this view comes from experiments in which insects have been exposed to desiccating environments or have been fed different diets (see Edney, 1977, for references). In the stick insect, moderate dehydration can lead to a 20% increase in the osmotic concentration of the hemolymph. In *Tenebrio*, the hemolymph osmotic concentration was found to vary by more than 60% during exposure to moist or dry conditions. In *Schistocerca* fed hyperosmotic saline, the hemolymph osmotic concentration was higher by 29%, and the Na and K concentrations were greater by 46% and 73%, respectively. Similar large changes have been found in measurements of the osmotic concentration of the hemolymph in aquatic insect larvae living in osmotically varied environments. It is fair to add that, in most of these cases, the hemolymph volume changed very much more than its osmotic concentration, so that the concentration is at least partially regulated.

The relative ease with which in vitro preparations of many insect organs can be made may also be explained from the likelihood that insect tissues must have evolved a tolerance of changing conditions in their environment. Experiments with isolated Malpighian tubules from *Rhodnius* bear this out. They will continue to secrete fluid in salines that vary in osmotic concentration from 70 mOsm liter^{-1} to 700 mOsm liter^{-1} (Maddrell, 1969, 1980b), in pH from 6.0 to 8.8 (Maddrell and Gardiner, unpubl. results) and in an ammonium nitrate saline that contain no potassium, sodium, or chloride ions (Maddrell, 1969).

Some insect tissues owe their tolerance of unfavorable conditions to the properties of special protective and regulating epithelia that envelop them. The classic example is the perineurium that covers the central and much of the peripheral nervous system (Treherne, 1974). As witness to its properties, the axons of the isolated CNS of the cockroach will continue to conduct impulses for some hours in a Na-free solution, although conduction fails as soon as the perineurium is perforated. Appropriately enough, the perineurium and epithelia protecting other potentially sensitive tissues such as the eyes and testis are the only invertebrate epithelia known to have simple tight junctions (Lane and Skaer, 1980).

The insect tracheal system of respiration, in which the tissues have their oxygen supplies piped directly to them, may have allowed the evolution in insects of this tissue tolerance to varying extracellular conditions; the hemolymph need contain no respiratory pigment that could be adversely affected by changes in the hemolymph.

The evidence, then, is that insect hemolymph can vary in composition within wider limits than would be tolerable in many other animals. It follows that the insect excretory system need not operate as rapidly as it otherwise would have to, to control the composition of the hemolymph. Recent research has shown that the insect excretory system does indeed "filter" the hemolymph very much more slowly than might otherwise have been expected.

Reduced permeability of insect Malpighian tubules

Until recently, the accepted view of the insect excretory system was that the hemolymph is continously "filtered" by permeable Malpighian tubules. Useful substances are actively recovered from the filtrate both in the anterior hindgut and in the rectum. Any harmful substance appearing in the hemolymph passes into the primary excretory fluid in the lumina of the Malpighian tubules, and because it is not actively reabsorbed in the hindgut, it is eliminated. This constitutes a fail-safe mechanism, and its price is the continual committal of water, ions, and other useful substances to the flow of primary excretory fluid, from which they then have to be reabsorbed. All excretory systems, as far as is known, have to operate on this basis – of unselective "filtration" and selective reabsorption – in order to ensure the automatic excretion of novel toxins appearing in the extracellular fluid (Ramsay, 1958). The only variant is that in many animals, very large substances such as protein cannot passively enter the primary excretory fluid; their effective elimination is ensured by some further mechanism such as precipitation by antibodies. The discovery that has led to the question of whether the excretory system of insects operates in the same way as those of other animals is that the effective permeability of insect Malpighian tubules is very considerably less than that of the analogous filtration sites in the excretory systems of many other animals. In the "filtration" process of animals as diverse as rotifers, crayfish, millipedes, and vertebrates there is virtually free penetration of compounds up to the size of small proteins into the primary excretory fluid. In insects, the fluid leaving the Malpighian tubules contains even such small molecules as amino acids and sugars at concentrations that are only a fraction of those in the hemolymph (Ramsay, 1958; Maddrell and Gardiner, 1980b). In the rapidly secreting Malpighian tubules of *Rhodnius*, the fluid produced contains glycine and inulin at concentrations as low as 1% of those in the bathing fluid (Maddrell and Gardiner, 1974; Maddrell and Gardiner, 1980b). One might attempt to explain this difference as merely a consequence of the fact that the force driving hemolymph solutes into the lumina of the Malpighian tubules is only diffusion, not pressure as in most other animals.

Research on the Malpighian tubules of a millipede, *Glomeris marginata*, which is not an insect, shows that this explanation cannot be correct. As in insects, these Malpighian tubules secrete fluid into the lumen and do not filter it under pressure. Presumably most hemolymph solutes pass into the lumen by diffusion. In spite of this the fluid has virtually the same ionic composition as the bathing fluid even when it is varied over a large range. Solutes as large as inulin reach a concentration in the secreted fluid that is virtually the same as that in the medium, and even the passage into the lumen of dextrans of molecular weight up to 16 000 is slowed only a little (Farquharson, 1974).

Three questions now need answering: How can the apparent low permeability of insect Malpighian tubules be reconciled with the essential need for an excretory system to contain a permeable filtration site? How are the walls of insect Malpighian tubules organized to provide an effectively low permeability? Most important, what advantages is it supposed that insects might derive from an excretory system that allows hemolymph solutes only slow access into the primary excretory fluid?

Tolerance of slow "filtration" of the hemolymph

To answer the first question, it should first be pointed out that although such small solutes as amino acids cross the walls of insect Malpighian tubules slowly, the tubules still allow substances as large as inulin into the primary excretory fluid (Ramsay and Riegel, 1961; Maddrell and Gardiner, 1974). It follows that the apparent low permeability of the tubules is due to a limitation in the area of sites where solutes can cross into the lumen rather than to any restriction at these sites. This is crucial because it means that the Malpighian tubules can still fulfil the essential function of allowing most substances from the extracellular fluid passive and relatively unselective entry into the primary excretory fluid. The main difference from other animals is that such movements are only allowed to go on at a slow rate. As we saw earlier in this discussion, insect tissues are very tolerant of changes in the extracellular environment, so that it may not matter that novel potentially toxic or harmful substances are removed slowly as long as they are eventually eliminated. For toxic or useless substances that are often encountered, Malpighian tubules have an array of active transport mechanisms to hasten their elimination (Maddrell, 1977). Surprisingly, no such mechanism appears to have been evolved to cater to the excretion of nitrogenous wastes, the elimination of which one might think would be very important for an excretory system.

Insects excrete a range of nitrogenous compounds, but the most widely encountered, almost characteristic such substance is uric acid (Bursell, 1967). In spite of this, many insects seem to be unaffected not only by its slow elimination but even by its retention within the body. For example, the fat bodies of such insects as cockroaches (Tucker, 1977) and parasitic hymenopterous larvae (Corbet and Rotheram, 1965) contain considerable quantities of uric acid. In the cockroach, it is suggested that these deposits may act as nitrogen stores. In other insects uric acid is used as a white pigment in the body and wings (Berridge, 1965; Harmsen, 1966). Even in the blood-sucking insect *Rhodnius*, with its protein-rich diet, uric acid appears to cross the walls of the Malpighian tubules passively (Maddrell and Gardiner, 1974). Slow elimination of uric acid in insects is evidently tolerated rather well. This may have to do with the metabolic cost of the synthesis of this substance (Bursell, 1967).

Insect Malpighian tubules: structural considerations

The Malpighian tubules of insects behave as if their walls have permeable sites that are greatly limited in extent. In contrast, the very permeable Malpighian tubules of the millipede, *Glomeris*, allow much more rapid penetration of hemolymph solutes. Comparison of the structures of these two Malpighian tubules (Maddrell, 1980b) strongly suggests that it is the very different frontal areas of the intercellular clefts that explain their different permeabilities. In *Rhodnius*, for instance, it has been calculated that the area of the clefts occupies only 0.03% of the epithelial surface (Maddrell, 1980b). The similarly sized tubules of *Glomeris* look quite different (Farquharson, 1974); the cells interdigitate to a remarkable extent and, significantly, look very like the podocytes of the Bowman's capsule of the vertebrate kidney or of the antennal glands of crustaceans (Kümmel, 1973).

The relatively very large area occupied by the cells in the surface of insect Malpighian tubules raises the question of the function of these large flat cells. Or, to put it another way, could not a low rate of filtration of the hemolymph be simply achieved by tubules of smaller dimensions with more extensive intercellular clefts? The explanation is probably partly that insect Malpighian tubules may be important in various forms of intermediary and excretory metabolism and need to expose a large area to the hemolymph. In addition, it is arguably an advantage for insects to have the capability to excrete large quantities of fluid and/or to filter their hemolymph much more rapidly (Maddrell, 1981). This depends on a faster rate of fluid secretion, which in turn requires a more extensive cellular area (Maddrell, 1980b).

Advantages of slow "filtration" of the hemolymph

We have seen that because insects' tissues are tolerant of rather variable extracellular conditions, insects have been able to evolve with an excretory system that produces a primary excretory fluid containing substances from the hemolymph at rather low concentrations. It is instructive to reflect on the advantages that insects may thereby enjoy.

An obvious advantage is that because useful substances are only slowly removed from circulation, energy expenditure on reabsorbing them is minimized. Because production of the primary excretory fluid is also slow, there is a further evident saving of energy. A comparison of the rates of production of primary excretory fluid in insects and vertebrates shows that insects take 10 to 20 times longer than vertebrates to produce a volume of this fluid that is the same fraction of the total extracellular fluid volume (Maddrell, 1980c).

A direct indication of the different rates of operation of the excretory systems of insects and vertebrates comes from comparing the rates at which they both eliminate a test substance introduced into their extracellular fluids. Locusts eliminate injected inulin with a half-time of 8 days (Maddrell and Gardiner, 1974). The similar figure for man based on the known rate of glomerular filtration (Pitts, 1968) is 100 min.

A further major advantage to insects follows from the slow rate at which substances are passively removed from their hemolymph. It becomes practicable for them to maintain in the hemolymph high concentrations of useful substances even if these substances are of modest molecular weight. Flying insects often have more than 100 mmol of trehalose and/or diglyceride lipid in circulation as fuels for flight (Weis-Fogh, 1967). Many lepidopterans, hymenopterans, and coleopterans maintain 100 to 200 mmol of amino acids in the hemolymph (Maddrell, 1971). Finally, neurohormones in circulation will only slowly be removed and so can more easily function for extended periods of time. A potential advantage to those who would control insects with insecticidal compounds is that once such substances reach the hemolymph, they have longer in which to produce their damaging effects (Maddrell, 1980a).

Many insects have active transport systems in the Malpighian tubules that depend for their effectiveness on the relative impermeability of the walls. If the walls were much more permeable, the resulting gradients would rapidly be degraded by diffusion. Among such transport systems are active transport into the lumen of ions of magnesium, sulfate, and phosphate, as well as acylamides, sulfonates, alkaloids, and glycosides (Maddrell, 1977), and active reabsorption from the lumen of glucose (Knowles, 1975). In addition, fluid secretion by insect Malpighian tubules depends in nearly all cases on potassium transport. The resulting fluid is K-rich and allows herbivorous insects to eliminate excess K from the diet and form the primary excretory fluid with a single transport system. With more permeable tubules this would not be possible.

The hindgut in osmoregulation

That the hindgut, especially the rectum, has a key role in osmoregulation has long been known. During its passage through the hindgut from the Malpighian tubules and from the midgut, the primary excretory fluid and gut contents are subjected to extensive reabsorptive processes that very effectively recover useful substances (Wall and Oschman, 1975). In terrestrial insects, of course, it is the recovery of water that is crucial. All these reabsorptive processes are eased by the fact that the flow of fluid from the Malpighian tubules is a relatively slow process. This allows time for very effective absorption of all those substances that the insect needs to recover. We have already seen that the layer of cuticle that lines the hindgut is most important in protecting the reabsorptive cells from elevated concentrations of potentially disruptive substances in the lumen.

The action of the hindgut in these reabsorptive activities is widely known and well understood (see reviews by Wall and Oschman, 1975; Phillips, 1977) so that all that needs to be done here is to draw attention to some recent advances. The rectum is thought to play the key role in selective absorption in the hindgut, but the anterior hindgut is also involved. Two recent discoveries have thrown some light on the activities of the anterior hindgut. First, it has been found that recovery of potassium chloride during rapid diuresis in the cabbage white butterfly, *Pieris brassicae*, is localized in the anterior hindgut, the most direct demonstration so far of ionic regulation by this part of the gut (Nicolson, 1976). Second, Prusch (1972) has shown that the hindgut of larvae of *Sarcophaga*, which need to eliminate ammonia from their environment of carrion, can secrete NH_4Cl at a concentration of around 1 M.

Recent developments in research on the insect rectum have concentrated on the mechanisms whereby water is reabsorbed and how this is controlled, and have shown interesting differences in the abilities of rectal epithelia from terrestrial and aquatic insects.

In short, the rectum in terrestrial insects recovers water from the lumen by a process of solute recycling within the tissue. Ions are secreted into deep-set intercellular spaces, creating an osmotic gradient to draw water from the rectal lumen. Whether this occurs through the cell cytoplasm or via a paracellular route is an absorbing question for future research. Direct evidence for this mechanism comes from micropuncture samples from the rectal intercellular spaces of the cockroach (Wall et al., 1970) and from electron microprobe analysis of these spaces in deep-frozen sections of the rectal papillae of the blowfly (Gupta et al., 1980). In the locust, water can be absorbed from the luminal solution even if it contains only solutes such as xylose, which cannot themselves be absorbed. However, under these conditions uptake of water soon slows; continuous recovery of water requires the presence of absorbable ions in the lumen (Goh and Phillips, 1978).

The control of this process of water recovery probably rests on the action of two hormones. One, chloride transport stimulating hormone (CTSH), promotes active movements of chloride (and potassium as its counter ion) from the lumen (Spring and Phillips, 1980). The other hormone involved is an antidiuretic hormone (Mordue, 1969), which possibly acts by coupling water movements to ion movements. Further details of these processes and their hormone control seem likely to emerge from research that is now active in this field.

Secretion of hyperosmotic fluid in the rectum

Insects that live in hyperosmotic fluid environments are confronted by a similar osmoregulatory problem to that of terrestrial insects; they need to limit water loss and to recover water actively from the medium. However, saline water mosquito larvae have evolved a different solution to the problem (Bradley and Phillips, 1975). Instead of producing a concentrated urine by water reabsorption, they secrete a hyperosmotic fluid in an especially adapted posterior part of the rectum. The other element in their osmotic balance is that they ingest and absorb the saline water in which they live; it follows that the posterior rectum must secrete a fluid at least as concentrated as this.

Absorption of water vapor from the air

No account of osmoregulation in insects would be complete without mention of the remarkable ability of some terrestrial insects to absorb water vapor from the air. Selective pressure for effective osmoregulation, which is strong in all terrestrial animals, is certainly more severe in such small organisms as insects. We have seen that insects very effectively reabsorb water from fluid entering the hindgut. It seems that the process has been evolutionarily refined in several cases to such an extent as to allow the system to take up water vapor from moist air drawn into the hindgut. The insects in which this process has been studied in any detail are the mealworm, *Tenebrio molitor* (Ramsay 1971; Machin, 1978), and *Thermobia domestica* (Noble-Nesbitt, 1970, 1978). In *Tenebrio*, in which water can be recovered from air of 88% relative humidity, the mechanism depends on the ability of modified Malpighian tubules to secrete virtually saturated KCl solution into a chamber enclosing the rectum. More striking is the ability of *Thermobia* to absorb water vapor from air of only 45% relative humidity. Perhaps not surprisingly, how this remarkable performance is achieved is not yet known. The epithelial walls of the anal sacs where the uptake occurs have an ultrastructure dominated by a spectacular close-packed hexagonal array of mitochondria in cytoplasmic leaflets that protrude into the lumen (Noble-Nesbitt, 1973).

Nymphs and adult females of the desert cockroach, *Arenivaga investigator*, can take up water vapor from air at 83% relative humidity. They have developed a totally different site for water absorption, protrusible bladders on the hypopharynx on the underside of the head (O'Donnell, 1977). These bladders have a close pile of cuticular hairs, and water from the air condenses into a fluid layer maintained there. It seems likely that this condensation occurs because of a reduction in water activity at the fluid surface induced by physical means. Further work is needed to clarify this mechanism and to answer the important question of how the fluid is passed on to the mouth for ingestion.

Although there are still doubts as to the detailed mechanisms involved in these systems for uptake of water vapor, there can be no doubt that the uptake is eloquent proof of the extraordinarily well-developed adaptations of insects for life as small animals in the terrestrial environment.

Summarizing remarks

Insects are relatively small animals, yet they enjoy enormous success in a variety of osmotically stressful environments. It has been known for a long time that the insect integument is of central importance, and recent research has underlined this. The majority of research in the area in the last 20 years has been aimed at understanding the activities of the excretory system in osmoregulation. As we have seen, this has uncovered a great many elegant and effective adaptations. Relatively speaking, our knowledge of the effectiveness of the respiratory system of terrestrial insects in limiting water loss during oxygen uptake is still in a primitive state. We know from their prolonged survival under desiccating conditions that insects must have an effective system of control. What we need to know now is how, in detail, this control is exerted. It will also be of

302   SIMON MADDRELL

great comparative interest to discover how much extra water is lost during
the uptake of a given amount of oxygen and to compare this information
with similar figures for other terrestrial animals, particularly vertebrates.

References

Beament, J. W. L. (1961) The water relations of the insect cuticle. *Biol. Rev.
Cambridge Phil. Soc.* 36:281–320.
Berridge, M. J. (1965) The physiology of excretion in the cotton stainer, *Dys-
dercus fasciatus* Signoret. III. Nitrogen excretion and excretory metabo-
lism. *J. Exp. Biol.* 43:535–552.
Berridge, M. J. (1970) Osmoregulation in terrestrial arthropods. In *Chemical
zoology* (M. Florkin and B. T. Scheer, eds.), pp. 287–320. New York: Aca-
demic Press.
Bradley, T. J., and Phillips, J. E. (1975) The secretion of hyperosmotic fluid
by the rectum of a saline water mosquito larva, *Aedes taeniorhynchus*
(Wiedmann). *J. Exp. Biol.* 63:331–342.
Bursell, E. (1967) Excretion of nitrogen in insects. *Adv. Insect Physiol.*
4:33–67.
Corbet, S. A., and Rotheram, S. (1965) The life history of the ichneumonid
Nemeritis (*Devorgilla*) *canescens* (Gravenhorst) as a parasite of (*Anagasta*)
kuehniella Zeller, under laboratory conditions. *Proc. R. Entomol. Soc.
Lond.* A 40:67–72.
Edney, E. B. (1977) *Water balance in land arthropods.* Berlin: Springer-Verlag.
Farquharson, P. A. (1974) A study of the Malpighian tubules of the pill mil-
lipede, *Glomeris marginata* (Villers). III. The permeability characteristics
of the tubule. *J. Exp. Biol.* 60:41–51.
Goh, S., and Phillips, J. E. (1978) Dependence of prolonged water absorption
by in vitro locust rectum on ion transport. *J. Exp. Biol.* 72:25–41.
Gupta, B. L., Wall, B. J., Oschman, J. L., and Hall, T. A. (1980) Direct mi-
croprobe evidence of local concentration gradients and recycling of elec-
trolytes during fluid absorption in the rectal papillae of *Calliphora*. *J. Exp.
Biol.* 88:21–47.
Harmsen, R. (1966) The excretory role of pteridines in insects. *J. Exp. Biol.*
45:1–13.
Knowles, G. (1975) The reduced glucose permeability of the isolated Malpigh-
ian tubules of the blowfly *Calliphora vomitoria*. *J. Exp. Biol.* 62:327–340.
Kümmel, G. (1973) Filtration structures in excretory systems – a comparison.
In *Comparative physiology: Locomotion, respiration, transport and blood*
(L. Bolis, K. Schmidt-Nielsen, and S. H. P. Maddrell, eds.), pp. 221–240.
Amsterdam: North-Holland.
Lane, N. J., and Skaer, H. Le B. (1980) Intercellular junctions in insects tis-
sues. *Adv. Insect Physiol.* 15:35–213. New York: Academic Press.
Locke, M. (1965) Permeability of insect cuticle to water and lipids. *Science,
N.Y.* 147:295–298.

Machin, J. (1978) Water vapour uptake by *Tenebrio*: A new approach to studying the phenomenon. In *Comparative physiology: water, ions and fluid mechanics*. (K. Schmidt-Nielsen, L. Bolis, and S. H. P. Maddrell, eds.), pp. 67–77. Cambridge University Press.

Maddrell, S. H. P. (1969) Secretion by the Malpighian tubules of *Rhodnius*. The movements of ions and water. *J. Exp. Biol.* 51:71–97.

Maddrell, S. H. P. (1971) Inorganic ions and amino acids in haemolymph: Insects. In *Respiration and circulation* (P. L. Altman and D. S. Dittman, eds.), pp. 1917–1918. Bethesda, Md.: Federation of American Societies for Experimental Biology.

Maddrell, S. H. P. (1977) Insect Malpighian tubules. In *Transport of ions and water in animal tissues* (B. L. Gupta, R. B. Moreton, J. L. Oschman, and B. J. Wall, eds.), pp. 541–569. London: Academic Press.

Maddrell, S. H. P. (1980a) The insect neuroendocrine system as a target for insecticides. *Insect neurobiology and pesticide action*, pp. 329–334. London: Society for Chemical Industry.

Maddrell, S. H. P. (1980b) Characteristics of epithelial transport in insect Malpighian tubules. *Curr. Top. Membr. Transp.* 14:427–463.

Maddrell, S. H. P. (1981) The functional design of the insect excretory system. *J. Exp. Biol.* 90:1–15.

Maddrell, S. H. P., and Gardiner, B. O. C. (1974) The passive permeability of insect Malpighian tubules to organic solutes. *J. Exp. Biol.* 60:641–652.

Maddrell, S. H. P., and Gardiner, B. O. C. (1980a) The permeability of the cuticular lining of the insect alimentary canal. *J. Exp. Biol.* 85:227–237.

Maddrell, S. H. P., and Gardiner, B. O. C. (1980b) The retention of amino acids in the haemolymph during diuresis in *Rhodnius prolixus*. *J. Exp. Biol.* 87:315–329.

Maddrell, S. H. P., Gardiner, B. O. C., Pilcher, D. E. M., and Reynolds, S. E. (1974) Active transport by insect Malpighian tubules of acidic dyes and acylamides. *J. Exp. Biol.* 61:357–377.

Mordue, W. (1969) Hormonal control of Malpighian tubules and rectal function in the desert locust *Schistocerca gregaria*. *J. Insect Physiol.* 15:273–285.

Nicolson, S. W. (1976) Diuresis in the Cabbage white butterfly, *Pieris brassicae*: Water and ion regulation and the role of the hindgut. *J. Insect Physiol.* 22:1623–1630.

Noble-Nesbitt, J. (1970) Water balance in the Firebrat, *Thermobia domestica* (Packard). The site of uptake of water from the atmosphere. *J. Exp. Biol.* 52:193–200.

Noble-Nesbitt, J. (1973) Rectal uptake of water in insects. In *Comparative physiology* (L. Bolis, K. Schmidt-Nielsen, and S. H. P. Maddrell, eds.), pp. 333–351. Amsterdam: North-Holland.

Noble-Nesbitt, J. (1978) Absorption of water vapour by *Thermobia domestica* and other insects. In *Comparative physiology: Water, ions and fluid mechanics* (K. Schmidt-Nielsen, L. Bolis, and S. H. P. Maddrell, eds.), pp. 53–66. Amsterdam: North-Holland.

O'Donnell, M. J. (1977) Site of water absorption in the desert cockroach, *Arenivaga investigator*. *Proc. Natl. Acad. Sci. U.S.A. 74*:1757–1760.

Phillips, J. E. (1977) Excretion in insects: Function of gut and rectum in concentrating and diluting the urine. *Fed. Proc. 36*:2480–2486.

Phillips, J. E., and Dockrill, A. A. (1968) Molecular sieving of hydrophilic molecules by the rectal intima of the desert locust (*Schistocerca gregaria*). *J. Exp. Biol. 48*:521–532.

Pitts, R. F. (1968) *Physiology of the kidney and body fluids*. Chicago: Yearbook Med. Publ., 243 pp.

Prusch, R. D. (1972) Secretion of NH_4Cl by the hindgut of *Sarcophaga bullata* larva. *Comp. Biochem. Physiol. 41A*:215–223.

Ramsay, J. A. (1958) Excretion by the Malpighian tubules of the stick insect *Dixippus morosus* (Orthoptera, Phasmidae): Amino acids, sugars and urea. *J. Exp. Biol. 35*:871–891.

Ramsay, J. A. (1971) Insect rectum. *Phil. Trans. R. Soc. Lond. B 262*:251–260.

Ramsay, J. A., and Riegel, J. A. (1961) Excretion of inulin by Malpighian tubules. *Nature, Lond. 191*:1115.

Reynolds, S. E. (1975) The mechanism of plasticization of the abdominal cuticle in *Rhodnius. J. Exp. Biol. 62*:81–98.

Schmidt-Nielsen, K. (1964) *Desert animals*. Oxford: Clarendon Press.

Schmidt-Nielsen, K. (1979) *Animal physiology: Adaptation and environment*. 2nd ed. Cambridge University Press.

Schmidt-Nielsen, K., Hainsworth, F. R., and Murrish, D. E. (1970) Countercurrent heat exchange in the respiratory passages. Effect on water and heat balance. *Respir. Physiol. 9*:263–276.

Smith, J. N. (1962) Detoxication mechanisms. *Annu. Rev. Entomol. 7*:465–480.

Spring, J. H., and Phillips, J. E. (1980) Studies on Locust rectum II. Identification of specific ion transport processes regulated by corpora cardiaca and cyclic-AMP. *J. Exp. Biol. 86*:225–236.

Treherne, J. E. (1957) Glucose absorption in the cockroach. *J. Exp. Biol. 34*:478–485.

Treherne, J. E. (1974) The environment and function of insect nerve cells. In *Insect neurophysiology* (J. E. Treherne, ed.), pp. 187–244. Amsterdam: North-Holland.

Treherne, J. E., and Willmer, P. G. (1975) Hormonal control of integumentary water-loss: Evidence for a novel neuroendocrine system in an insect (*Periplaneta americana*). *J. Exp. Biol. 63*:143–159.

Tucker, L. E. (1977) The influence of diet, age and state of hydration on Na, K and urate balance in the fat body of the cockroach *Periplaneta americana. J. Exp. Biol. 71*:67–79.

Wall, B. J., and Oschman, J. L. (1975) Structure and functions of the rectum in insects. *Fortsch. Zool. 23*:193–222.

Wall, B. J., Oschman, J. L., and Schmidt-Nielsen, B. (1970) Fluid transport: concentration of the intercellular compartment. *Science, N.Y. 167*:1497–1498.

Weis-Fogh, T. (1967) Metabolism and weight economy in migrating animals, particularly birds and insects. In *Insects and physiology* (J. W. L. Beament and J. E. Treherne, eds.), pp. 143–159. Edinburgh and London: Oliver and Boyd.

PART FIVE
Movement and structure

Overview

Part Five considers the structural design of organisms.

In Chapter 21 McNeill Alexander discusses how the shapes of running and flying animals change with body size. He reviews various attempts to explain the changes in animal shape with size, and he concludes that the complexities involved in such scaling defy simple analysis. He suggests ways to proceed with future studies.

In Chapter 22 Stephen Wainwright discusses two major types of support systems utilized in the bodies of animals. He concludes that our knowledge of the morphology of animal bodies is keeping pace with our knowledge of their physiology, both in terms of remarkable discoveries and in numbers of important unanswered questions that arise.

In Chapter 23 Claude Lenfant and Shou-teh Chiang discuss the physiological consequences for humans and terrestrial animals of operating in the "weightless" environment of space. The effects of weightlessness, particularly on the musculoskeletal system, are severe and may limit the duration of exposure to a weightless environment. The data, however, are still sparse, and definite conclusions await future studies.

21

Size, shape, and structure for running and flight

R. McN. ALEXANDER

This chapter is about scale effects on the structure and functioning of limbs. It describes the relationships between properties of limbs (such as dimensions and frequency of movement) and body size. Two kinds of limbs, legs and wings, are discussed.

Although Galileo (1638) was probably the first scientist to discuss the scaling of limbs, Knut Schmidt-Nielsen is largely responsible for current interest in the subject. He published several important papers on scaling (including Schmidt-Nielsen, 1970, 1975) and convened the preliminary meeting that planned the Scaling Conference held in Cambridge in 1975 (Pedley, 1977).

The most useful tool in studies of scaling is the equation of simple allometry. Many physical laws have the form

$$y = ax^b \tag{1}$$

where x and y are variables, but the factor a and the exponent b are constants. Examples are Boyle's law,

$$p = kV^{-1}$$

the Stefan-Boltzmann law,

$$\Phi = \sigma T^4$$

and the equation that gives the area of a circle,

$$A = \pi r^2$$

(The meanings of the symbols in these equations are irrelevant to the present discussion.) The exponents in such laws are generally integers or the reciprocals of integers, but other fractional exponents occur in empirical equations of the same form, such as the equations that describe convective heat loss from solid bodies. Similarly, it is often convenient to use equation 1 to describe the relationship between two dimensions or properties, x and y, in animals of different size. So used, it is called the equation of simple allometry.

By taking logarithms, equation 1 is changed to

309

$$\log y = \log a + b \log x \tag{2}$$

so if x and y were precisely related by the equation, a graph of log y against log x (or a graph of y against x on logarithmic coordinates) would be a straight line of gradient b. Such precise relationships between dimensions cannot be expected in animals, and are not found. For instance, no straight line can pass through all the points for femur diameter shown in Figure 21.1. Nevertheless, the line drawn on the graph gives an excellent description of the general trend of the data. It was fitted by applying least-squares regression to the logarithms of the data (Bailey, 1959), and shows that bone diameter in millimeters is related to body mass in kilograms by

$$\text{diameter} = 5.2 \, (\text{mass})^{0.36} \tag{3}$$

This equation gives the best possible fit to the data shown but would need revision if more data were collected. Statistical analysis based on the assumption that the data used are a random sample of all possible data showed that the probabilities are 0.95 that the true factor (based on all possible data) lies between 5.0 and 5.4, and that the true exponent lies between 0.35 and 0.37. Exponents cannot generally be determined with such precision, unless data from animals of a very wide range of sizes are available. In this case the largest specimen (an elephant) had about one million times the mass of the smallest (a shrew).

Nearly all the data presented in this chapter apply to adult animals. The equations derived from them describe the trends observed when

FIGURE 21.1 A graph on logarithmic coordinates showing the lengths and diameters of the femurs of mammals plotted against body mass. Symbols: empty circles = insectivores; solid circles = primates; empty diamonds = rodents; solid diamonds = carnivores; empty squares = Bovidae; solid squares = other animals. [From Alexander et al., 1979a]

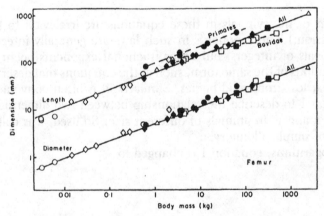

species of different sizes, within some higher taxon, are compared. In many cases quite different equations would be obtained if comparisons were made between different growth stages of the same species. For instance, the length of the hind foot (from heel to hoof) is proportional to (body mass)$^{0.21 \pm 0.05}$ in adult Bovidae, but in growth stages of one species (*Adenota kob*) it is proportional to (body mass)$^{0.09 \pm 0.03}$ (data of Sachs, 1967, and Ledger and Smith, 1964; the exponents given with 95% confidence limits). Antelope calves tend to have bigger feet than adult antelopes of the same mass.

Data

Tables 21.1 and 21.2 present allometric exponents which will be discussed in subsequent sections. All the data refer to adult animals except in the case of spiders (Table 21.1) where young animals were included. In Table 21.1 most of the data refer to legs. Wings and their muscles have been excluded from the data for birds in this table, except that wing bones are included with the rest of the skeleton in the data for skeleton mass.

The tables show that corresponding allometric exponents tend to be about equal for different groups of animals. However, the exponents for external lengths and bone lengths for Bovidae (antelopes, cattle, etc.) are lower than for other running animals. Also, the exponents for wing length and wing area for hummingbirds are higher than for other flying animals. The gradients of the lines in Figures 21.1 and 21.2 are the allometric exponents. Notice that the gradients of the lines for Bovidae and hummingbirds do not conform with the rest. The figures also show that primates have longer femurs than other mammals of the same mass, and butterflies have larger wings than other insects.

Similarity theories of scaling

Many attempts have been made to explain allometric data like those presented in Tables 21.1 and 21.2. This section reviews theories that postulate some rule of similarity.

Geometric similarity

Many discussions of the scaling of animals have started from the supposition that related animals of different sizes are geometrically similar and made of materials of the same density. This implies that all homologous linear dimensions should be proportional to (body mass)$^{1/3}$, all homologous areas to (body mass)$^{2/3}$, and the masses of homologous parts of the body to (body mass)^1. Table 21.1 shows that related running animals (other than Bovidae) tend to be quite near to geometric similarity.

TABLE 21.1 Exponents of allometric equations based on body mass, for mammals and some other running animals. Mean exponents are given when several equations are available; for instance, the exponents for bone lengths are means of exponents for several different bones. External length means quantities such as body length, leg length, or height at the shoulder. Circumferences of body segments are included with external diameters. Tail length, which is anomalous (exponent 0.04), has been excluded from the data for primates of 0.3–20 kg.

Group	Body mass (kg)	External lengths	Bone lengths	External diameters	Bone diameters	Skeleton masses	Muscle masses	Stride frequency
Insectivores[a]	0.003–0.6		0.38		0.39			
Primates[b]	0.3–20	0.27	0.31	0.36	0.38			
Primates[a]	0.6–60		0.34		0.39	1.05		
Rodents[a]	0.01–2		0.33		0.40			
Carnivores[c]	0.6–150	0.37						
Carnivores[a]	2–130		0.36		0.40			
Cats[d]						0.99	1.03	
Ungulates[c]	20–1000	0.29						
Ungulates[e]					0.34		0.98	−0.18
Bovidae[a,f]	4–500	0.26	0.26					
Cattle[g]		0.24		0.36				
All mammals[h]	0.02–7000					1.13		
All mammals[i]	0.006–7000					1.09		
All mammals[a]	0.003–2500		0.35		0.36			
All mammals[j]	0.02–700							−0.14
Running birds[k]	0.1–40		0.36[m]		0.40	1.07	1.01	
All birds[l,j]	0.003–80		0.36[m]			1.13		−0.18
Spiders[i]	3×10^{-5}–0.001							
Geometric similarity		0.33	0.33	0.33	0.33	1.00	1.00	
Elastic similarity		0.25	0.25	0.38	0.38	1.00	1.00	

[a] Alexander et al. (1979a).
[b] Stahl and Gummerson (1967).
[c] Radinsky (1978).
[d] Davis (1962).
[e] Alexander et al. (1977).
[f] Alexander (1977b).
[g] Brody (1945).
[h] Schmidt-Nielsen (1970).
[i] Anderson et al. (1979).
[j] Heglund et al. (1974).
[k] Maloiy et al. (1979).
[l] Prange et al. (1979).
[m] Leg bones only.

Most of the exponents for linear dimensions of bones are a little greater than 0.33, implying that larger animals have relatively larger bones. This is consistent with the tendency of skeleton mass to be proportional to (body mass)$^{1.1}$. The skeleton makes up about 4% of the mass of a shrew but 14% of the mass of an elephant. The exponents for muscle mass, however, are very close to 1.0. Muscle makes up such a large proportion of body mass in vertebrates that an exponent much different from 1 would imply an inordinately large proportion of muscle either in very large animals or in very small ones.

TABLE 21.2 Exponents of allometric equations, based on body mass, for flying animals. Birds 1, 2, and 3 are the groups defined by Greenewalt (1975), typified by the passeriforms, shorebirds, and ducks; they are birds that for their size have low, medium, and high (body weight)/(wing area), respectively. Normal hoverers are those hummingbirds and insects that are listed without asterisks in Weis-Fogh's (1973) Table 4.

Group	Body mass (g)	Wing length or span	Wing area	Aspect ratio	Wing muscle mass	Wing-beat frequency
Bats[a]	300–1400		0.66			
Megachiroptera[b]	20–1400	0.38	0.72	0.04		
Microchiroptera[b]	3–130	0.30	0.56	0.07		
Birds 1[a,c]	5–5,000	0.42	0.79	0.05	0.97	−0.36
Birds 2[a,c]	30–10,000	0.40	0.70	0.09	0.96	−0.19
Birds 3[a,c]	100–6,000	0.41	0.71	0.10	0.91	−0.24
Petrels[d]			0.59			
Hummingbirds[a]	2–20	0.57	0.96			
All birds[a]						−0.29
Diptera and Hymenoptera[a]	0.001–0.6		0.68			
Butterflies[a]	0.01–0.7		0.63			
Sphingid moths[a]	0.1–2		0.81			
Sphingid moths[e]	0.3–3	0.41	0.76	0.05	0.80[g]	−0.24
Saturniid moths[e]	0.2–2	0.39	0.70	0.06	0.83[g]	−0.12
Normal hoverers[f]	0.001–20	0.37	0.76[h]	−0.02[h]		−0.30
Geometric similarity		0.33	0.67	0	1.00	

[a] Greenewalt (1975).
[b] Norberg (1980).
[c] Rayner (1979).
[d] Warham (1977).
[e] Bartholomew and Casey (1978).

[f] Weis-Fogh (1973).
[g] Total mass of thorax. The exponent for Saturniids is not as printed in Bartholomew and Casey (1978), but has been recalculated from table of data.

[h] Calculated from wing length and chord.

In Table 21.2, all but one of the exponents for wing length are greater than 0.33, which implies that flying animals are not geometrically similar: The larger ones have relatively longer wings. Nevertheless, all the exponents for wing area (except for hummingbirds) are quite close to the value of 0.67 that they would have if the animals were geometrically similar. This is so because the wings of larger animals are relatively narrower as well as relatively longer, as the exponents for aspect ratio indicate.

Table 21.2 shows that wing muscle mass tends to be a smaller fraction of body mass in larger birds. This may perhaps reflect the tendency for wing mass to be a larger fraction of body mass in larger birds (Greenewalt, 1962, 1975). The exponents for thorax mass in moths, given in Table 21.2, are remarkably low, but refer only to moths of a narrow range of sizes.

The assumptions of geometric similarity and equal density do not by themselves lead to any predictions about the speeds and frequencies of movements. Hill (1950) and Wilkie (1977) made the additional assumption that homologous parts of animals have equal strengths: They are capable of withstanding (or, in the case of muscle, exerting) equal stresses. Their

FIGURE 21.2 A graph on logarithmic coordinates showing allometric equations relating total wing area to body mass for various groups of flying animals. The data are from the same sources as in Table 21.2. The point refers to the tiny wasp *Encarsia* (Weis-Fogh, 1973).

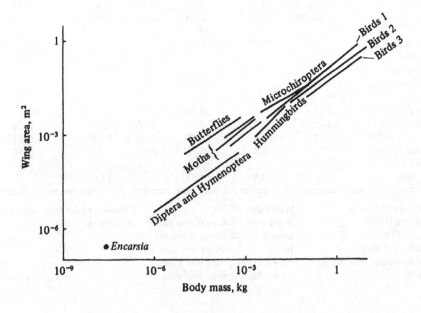

assumptions can be expressed as follows. Let a member of a group of geometrically similar animals have mass M. Let the mass of some part of its body be m, let some linear dimension in its body be l, and let the maximum stress developed in some part of its body be σ. Then for the whole group of animals:

$$m \propto M \qquad l \propto M^{1/3} \qquad \sigma \propto M^0 \qquad (4)$$

These assumptions imply that the maximum force that a muscle can exert is proportional to σl^2, and the distance it can shorten to l, so that the work that can be done in a single contraction is proportional to σl^3 and so to M. Hence the kinetic energy it can give to the body or to a limb is proportional to M, and the velocity to which it can accelerate the body or limb is proportional to M^0. Related animals of different sizes should be able to take off for jumps at the same velocity, and so be able to jump to the same height or distance. They should be able to accelerate their feet to the same speed, relative to the body, so that if this limits running speed, they should all be able to run at the same maximum speed. Hill (1950) presented evidence that mammals of a fairly wide range of sizes jump about equal heights and distances and run about equally fast. For instance, whippets, greyhounds, and racehorses all run at about 17 m s^{-1}. However, very large mammals (such as elephants) and very small ones (such as mice) neither run nor jump as well as medium-size ones (such as dogs, horses, and antelopes).

Proportionalities 4 were also proposed by Weis-Fogh (1977) as a model for the scaling of hovering animals. Most of the aerodynamic drag on the wings of all but the largest hovering animals is profile drag, which is about proportional to (wing area) $\times$ (speed)^2. If geometrically similar animals of different sizes moved corresponding parts of their wings at the same speed, the profile drag would be about proportional to wing area, and so to $M^{2/3}$. Because the cross-sectional areas of bones and muscles would also be proportional to $M^{2/3}$, the associated stresses would be independent of M as required by proportionalities 4. By the argument already applied to legs, the stresses associated with acceleration and deceleration of the wings at the ends of the wing beat would also be constant. If the wings beat through the same angle, equal speeds imply wing-beat frequencies inversely proportional to wing length. Geometrically similar hovering animals would have wing lengths proportional to $M^{1/3}$ and should beat their wings at frequencies proportional to $M^{-1/3}$. These requirements are not very different from the observed proportionalities to $M^{0.37}$ and $M^{-0.30}$ (Table 21.2).

This system of scaling would make the work done in each wing beat proportional to M. With wing-beat frequency proportional to $M^{-1/3}$, it would make the power output of the muscles proportional to $M^{2/3}$. This is unrealistic; measurements of the metabolic rates of hovering animals, from 1-mg mosquitoes to 5-g hummingbirds, show that power is approximately proportional to M (Weis-Fogh, 1973).

The approach to the scaling of hovering flight outlined in this section is oversimple, as Weis-Fogh (1977) realized. A more realistic analysis by Weis-Fogh is referred to in a later section (see under "Empirical similarities").

Elastic similarity

McMahon (1973, 1975a) proposed that terrestrial animals and trees should be scaled in such a way as to deform under gravity in geometrically similar fashion. The idea seems intuitively attractive for at least the horizontal branches of trees, which may be deformed more by their weight than by any other force. It seems less attractive for terrestrial mammals because the inertia forces involved in running and jumping are large compared to body weight, and are different multiples of body weight for animals of different sizes. The peak force that acts on a foot in running with duty factor β is about $\pi/2\beta$ times the weight the foot supports in standing (Alexander, 1977b). At their maximum speeds this is nine, six, and three times the standing force, in Thomson's gazelle, buffalo, and elephant, respectively (Alexander et al., 1977; Alexander et al., 1979b).

Though based on a doubtful premise, the theory leads to interesting and in some cases realistic conclusions. It postulates animals that have homologous parts made of materials of equal density and equal elastic moduli, and that deform in geometrically similar fashion under gravity. It does not postulate geometric similarity, but merely that all lengths (l) are proportional to each other, and all diameters (d) are proportional to each other, throughout the range of sizes of animals it describes. Horizontal beams with cross sections of the same shape have weights proportional to ld^2 and second moments of area of their cross sections proportional to d^4. Hence their deflections δ under their own weights are proportional to l^4/d^2 (Alexander, 1968, equation 15). Elastic similarity requires equal values of δ/l, and so of l^3/d^2. The same relationship is required between l and d for vertical pillars, if the criterion of elastic similarity is that their weights should be equal fractions of the loads that would make them fail by buckling. These considerations give

$$m \propto M \qquad l \propto M^{1/4} \qquad d \propto M^{3/8} \tag{5}$$

The data for Bovidae, in Table 21.1, agree well with proportionalities 5, but the data for lengths in other mammals do not. McMahon's (1975b) data on Bovidae also agree well with proportionalities 5. His data on Cervidae agree less well.

McMahon (1975a) assumed further that the stride frequencies of mammals of different sizes would tend to be proportional to the natural frequencies of vibration of the limbs. The mass and weight ot a beam ot length l, diameter d are proportional to ld^2. If its weight deflects the beam a distance proportional to l (as assumed in the theory of elastic similarity), the stiffness of the beam is proportional to $ld^2/l = d^2$. The natural frequency is proportional to $(\text{stiffness/mass})^{1/2}$ (see for instance Alexander, 1968), which is proportional to $l^{-1/2}$ or (using proportionalities 5) $M^{-1/8}$. Hence stride frequencies of elastically similar animals should be proportional to $M^{-1/8}$. Mammals use different frequencies at different speeds; so comparisons must be made at corresponding speeds, for instance, at the speed at which the animal changes from trotting to galloping. Heglund et al. (1974) measured the stride frequencies of mouse, rat, dog, and horse at this transition, and found them proportional to $M^{-0.14}$. This is very close to the theoretical prediction and might be regarded as strong confirmation of the theory of elastic similarity, but for three problems: First, mouse, rat, and dog belong to orders whose linear dimensions do not scale in accordance with elastic similarity (Table 21.1). Second, the kinetic energy associated with the forward swing of the legs, in mammals galloping fast, seems to be much less than the kinetic energy associated with the backward swing; so the swinging of the legs cannot be maintained elastically (Alexander et al., 1980). Third, a similar relationship between stride frequency and body mass is predicted by considerations of dynamic similarity, as will be shown in the next section of this chapter.

Dynamic similarity

Shapes are said to be geometrically similar when one can be made identical to another by a uniform change in the scale of length. Motions are said to be dynamically similar if they could be made identical by uniform changes of the scales of length and time. For instance, the motions of two pendulums of different length, swinging with the same angular amplitude, are dynamically similar. The motions of running (and of pendulums) involve interaction between gravity and inertia. Such motions can only be dynamically similar if they have equal Froude numbers u^2/gl, where u is a characteristic speed, g the acceleration of free fall, and l a characteristic length (Duncan, 1953). In making comparisons, u and l must be similarly defined for all the motions. In discussions of running it is convenient to make u the forward speed and l leg length (Alexander, 1976).

Though animals can run in dynamically similar fashion when they have the same Froude number, there is no necessity for them to do so; they might use different gaits, or different ratios of stride length to leg length. Nevertheless, observations of mammals of a wide range of sizes (from small rodents to horses) indicate that they tend to use similar gaits and equal values of (stride length)/(leg length), at any given Froude number (Alexander, 1977a).

Running in dynamically similar fashion requires stride lengths proportional to leg length l and speeds proportional to $l^{1/2}$. Stride frequency is (speed)/(stride length) and should therefore be proportional to $l^{-1/2}$. If l is proportional to $M^{0.25}$ (as for Bovidae, Table 21.1), stride frequency should be proportional to $M^{-0.13}$, as also required for elastic similarity. If l is proportional to $M^{0.35}$ (as for other mammals), stride frequency should be proportional to $M^{-0.18}$. The exponents for stride frequency given in Table 21.1 range from -0.14 to -0.18.

There is no inconsistency between elastic and dynamic similarity. Indeed, if the movements of animals depended entirely on elasticity (without inelastic extension and contraction of muscles), elastic similarity would be a necessary condition for dynamic similarity in running. In dynamically similar running motions, the forces on the feet of different animals are equal multiples of body weight. The elastic deformations due to these forces would be geometrically similar only if the animals were elastically similar.

The Froude number corresponding to maximum running speed tends to be smaller for large mammals (such as buffalo) than for small ones (such as gazelles; Alexander et al., 1977). Therefore, the motions of mammals of different sizes, at maximum speed, are not dynamically similar. This complication should be taken into account in any theory of the scaling of running.

The air movements involved in flight depend on the interaction of inertia and viscosity, so dynamic similarity would only be possible at equal Reynolds numbers ul/v (Duncan, 1953). Here u and l are a velocity and a length, as before, and v is the kinematic viscosity, which is a constant for air. To have equal Reynolds numbers, the wings of animals of different sizes would have to move at speeds u proportional to $1/l$. Also, dynamic similarity would require the wings of the animals to move at the same angle of attack and have the same lift coefficient, $2Mg/\rho Au^2$ (see, e.g., Alexander, 1968). Here ρ is air density and A wing area. For equal lift coefficients, u would have to be proportional to $(M/A)^{1/2}$ or, for geometrically similar animals, to $l^{1/2}$. Animals of grossly different size cannot be designed to fly with the same Reynolds number and the same lift coefficient. Flight is feasible over an enormous range of Reynolds numbers but only a narrow range of lift coefficients; stalling sets an upper limit to the range, and at low lift coefficients the ratio of drag to lift is high. Flying animals of different sizes operate at very different Reynolds numbers (Alexander, 1968; Weis-Fogh, 1973).

Empirical similarities

The complexities of scaling, both for running and for flight, defy simple analysis. The approach described in this section starts from empirical data rather than from general principles.

Alexander (1977b) assumed that the scaling of antelope legs depended mainly on the requirements for running at maximum speed. He obtained allometric equations describing their anatomy and movements and examined their implications for antelopes of different sizes running at maximum speed. He calculated that the maximum stresses in bones, muscles, and tendons were all more or less independent of body mass. The maximum stresses in bone and tendon were quite large fractions of the breaking stresses for these materials. The maximum stress in muscle was less than vertebrate striated muscle can exert in isometric contraction, but seemed quite a high value for a shortening muscle. Further, the mechanical work required of the muscles in each stride and the energy saved by elastic storage were both more or less proportional to body mass. Muscle mass was also about proportional to body mass. These calculations suggested that antelopes are scaled in such a way that at maximum running speed, bones, muscles, and tendons are all performing near the limits of their capabilities. If they were scaled otherwise, antelopes at at least one end of the size range could not run so fast.

This argument assumes that the performance of muscles is limited by the work they can do in a single contraction, rather than by their maximum power output. There seem to be physiological limits both to work and to power. Power output is (work per cycle) × (frequency); so power is likely to be limiting at high frequencies and work at low ones. Weis-Fogh and Alexander (1977) argued that work is limiting at frequencies below about 10 Hz (e.g., for running mammals) and that power is limiting above about 10 Hz (e.g., for hovering insects).

Weis-Fogh (1973) studied the dimensions and wing-beat frequencies of hovering animals. A wide range of these animals, from 1-mg mosquitoes to 20-g hummingbirds, use the technique he described as normal hovering. He calculated the average lift coefficient of their wings and obtained values in every case between 0.5 and 2.0. The highest values (for hummingbirds) are remarkably high, but the rest lie in the range known to be obtainable from simple aerofoils. He calculated the mechanical power used against aerodynamic drag and obtained values in every case between 13 and 47 W (kg body mass)$^{-1}$. In addition to doing work against drag, a hovering animal must do negative work, removing kinetic energy from its wings at the end of each stroke, and positive work, giving them kinetic energy for the next stroke. (This can be done either by muscles or, more economically, by an elastic mechanism.) The amounts of work involved in each stroke can be calculated from Weis-Fogh's data and multiplied by wing-beat frequency to obtain powers. These vary only between 24 and 57 W kg^{-1}, over the whole range of size from mosquitoes to hummingbirds. Hovering animals are scaled over this enormous range of sizes, so that they all work at fairly similar lift coefficients and have aerodynamic and inertial power requirements fairly nearly proportional to body mass. Measurements of metabolic rates confirm that power consumption is indeed about proportional to body mass.

This system of scaling, which keeps lift coefficients and power requirements per unit mass about constant, has interesting consequences for the elastic mechanisms of wings. The inertial power requirement is eliminated if the wings are mounted elastically and beat at their natural frequency. At this frequency, the kinetic energy of the wings in midstroke equals the elastic strain energy stored at the ends of the stroke. Geometrically similar animals made of similar materials would have natural frequencies proportional to $M^{-1/3}$. If they beat their wings through equal angles, the kinetic and elastic energies at the natural frequency would be proportional to M.

Hovering animals beat their wings at frequencies quite nearly proportional to $M^{-1/3}$, but they are not geometrically similar (Table 21.2). Because larger animals have disproportionately larger wings, the kinetic energy at midstroke is about proportional to $M^{1.3}$ (so that inertial power is about proportional to M). If the beating of the wings is to be maintained elastically, the elastic structures in the thorax must be capable of storing strain energy proportional to $M^{1.3}$.

There is probably a range of sizes of insects in which the flight muscles store the requisite quantity of elastic strain energy. In larger insects such as dragonflies and locusts, the flight muscles are supplemented as elastic energy stores by pieces of the elastic protein resilin. Some smaller insects, including Diptera, have a click mechanism that partly counteracts the elastic compliance of the muscles by storing elastic strain energy in midstroke (Alexander and Bennet-Clark, 1977).

Rayner (1979) discussed the scaling of forward flight in birds. He calculated that the minimum power required for flight is proportional to $M^{1.10}$ and that the flying speed for minimum power is proportional to $M^{0.16}$.

A constraint theory of scaling

A different approach to the scaling of hovering flight is most explicit in a paper by Lighthill (1977). He postulated geometrically similar animals beating their wings through the same angle, but required no other similarities. Instead he identified factors tending to limit the range of feasible wing-beat frequencies (n) for given linear dimensions (l).

Wings would not work well below some minimum Reynolds number, ul/v (probably about 100). Because the velocity u is proportional to nl, this implies

$$n \geq kl^{-2} \tag{6}$$

where k is some constant. Because the lift coefficient cannot exceed some maximum value, and equal lift coefficients imply $u \propto l^{1/2}$ for geometrically similar animals (see preceding section),

$$n \geq k_1 l^{-1/2} \tag{7}$$

The aerodynamic power requirement for hovering is the sum of two components, concerned with profile drag and induced drag. Neither of these may exceed the total power available from the muscles. Lighthill showed that this implied

$$n \leq k_2 l^{-2/3} \tag{8}$$

and

$$l \leq k_3 \tag{9}$$

FIGURE 21.3 A schematic graph of the logarithm of wing-beat frequency against the logarithm of a linear dimension, for hovering animals, showing the constraints identified by Lighthill (1977) and described in the text.

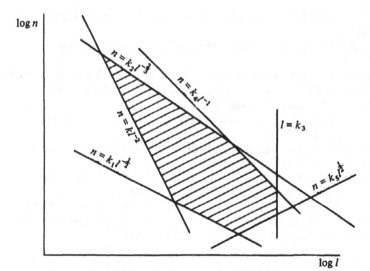

He identified two further constraints, due to the limited strength of the skeleton. All six constraints are shown in Figure 21.3, where they confine the set of possible pairs of (log l, log n) to the stippled hexagon. No quantitative predictions can be obtained from the diagram as it stands because the constants k, k_1, and so on, have not been evaluated, and the position and even the proportions of the hexagon are consequently unknown. Nevertheless, the diagram is consistent with the observations, first, that the wing-beat frequencies of hovering animals tend to be about inversely proportional to wing length (Table 21.2) and, second, that there seem to be fairly clear upper and lower limits to the range of sizes of hovering animals.

Optimality and allometry

Evolution by natural selection is a process of optimization. Its most fundamental tendency is to maximize the probability that the individual's genes will reappear in subsequent generations, but it is often possible to identify more tangible properties of the individual that have an important bearing on this probability. For instance, there has probably been strong selection in the evolution of flying animals tending to minimize the power required for flight. The concept of optimization has been mentioned in discussions of allometry (e.g., by Rosen, 1967, and Günther, 1975), but little use has been made of it.

References

Alexander, R. McN. (1968) *Animal mechanics*. London: Sidgwick and Jackson.

Alexander, R. McN. (1976) Estimates of speeds of dinosaurs. *Nature, Lond.* 261:129-30.

Alexander, R. McN. (1977a) Terrestrial locomotion. In *Mechanics and energetics of terrestrial locomotion* (R. McN. Alexander and G. Goldspink, eds.), pp. 168-203. London: Chapman and Hall.

Alexander, R. McN. (1977b) Allometry of the limbs of antelopes (Bovidae). *J. Zool., Lond.* 183:125-146.

Alexander, R. McN., and Bennet-Clark, H. C. (1977) Storage of elastic strain energy in muscle and other tissues. *Nature, Lond.* 265:114-117.

Alexander, R. McN., Jayes, A. S., and Ker, R. F. (1980) Estimates of energy cost for quadrupedal running gaits. *J. Zool., Lond.* 190:155-192.

Alexander, R. McN., Jayes, A. S., Maloiy, G. M. O., and Wathuta, E. M. (1979a) Allometry of the limb bones of mammals from shrews (*Sorex*) to elephant (*Loxodonta*). *J. Zool., Lond.* 189:305-314.

Alexander, R. McN., Langman, V. A., and Jayes, A. S. (1977) Fast locomotion of some African ungulates. *J. Zool., Lond.* 183:291-300.

Alexander, R. McN., Maloiy, G. M. O., Hunter, B., Jayes, A. S., and Nturibi, J. (1979b) Mechanical stresses in fast locomotion of buffalo (*Syncerus caffer*) and elephant (*Loxodonta africana*). *J. Zool, Lond.* 189:135–144.

Anderson, J. F., Rahn, H., and Prange, H. D. (1979) Scaling of supportive tissue mass. *Quart. Rev. Biol.* 54:139–148.

Bailey, N. J. T. (1959) *Statistical methods in biology.* London: English Universities Press.

Bartholomew, G. A., and Casey, T. M. (1978) Oxygen consumption of moths during rest, pre-flight warm-up, and flight in relation to body size and wing morphology. *J. Exp. Biol.* 76:11–25.

Brody, S. (1945) *Bioenergetics and growth.* New York: Reinhold.

Davis, D. D. (1962) Allometric relations in lions vs domestic cats. *Evolution* 16:505–514.

Duncan, W. J. (1953) *Physical similarity and dimensional analysis.* London: Arnold.

Galileo Galilei. (1638) *Discorsi e dimonstrazioni matematiche intorno a due nuove scienze.* Leiden: Elsevir.

Greenewalt, C. H. (1962) Dimensional relationships for flying animals. *Smithson. Misc. Collns.* 144(2):1–46.

Greenewalt, C. H. (1975) The flight of birds. *Trans. Am. Phil. Soc.* 65(4):1–67.

Günther, B. (1975) Dimensional analysis and theory of biological similarity. *Physiol. Rev.* 55:659–699.

Heglund, N., McMahon, T. A., and Taylor, C. R. (1974). Scaling stride frequency and gait to animal size: Mice to horses. *Science, N.Y.* 186:1112–1113.

Hill, A. V. (1950) The dimensions of animals and their muscular dynamics. *Sci. Prog.* 38:209–230.

Ledger, H. P., and Smith, N. S. (1964) The carcass and body composition of the Uganda kob. *J. Wildl. Mgmt.* 28:827–839.

Lighthill, J. (1977) Introduction to the scaling of aerial locomotion. In *Scale effects in animal locomotion* (T. J. Pedley, ed.), pp. 365–404. London: Academic Press.

McMahon, T. A. (1973) Size and shape in biology. *Science, N.Y.* 179:1201–1204.

McMahon, T. A. (1975a) Using body size to understand the structural design of animals: Quadrupedal locomotion. *J. Appl. Physiol.* 39:619–627.

McMahon, T. A. (1975b) Allometry and biomechanics: Limb bones of adult ungulates. *Am. Nat.* 109:547–563.

Maloiy, G. M. O., Alexander, R. McN., Njau, R., and Jayes, A. S. (1979) Allometry of the legs of running birds. *J. Zool., Lond.* 187:161–167.

Norberg, U. M. (1980) Allometry of bat wings. (Manuscript.)

Pedley, T. J. (ed.) (1977) *Scale effects in animal locomotion.* London: Academic Press.

Prange, H. D., Anderson, J. F., and Rahn, H. (1979) Scaling of skeletal mass to body mass in birds and mammals. *Am. Nat.* 113:103–122.

Radinsky, L. (1978) Evolution of brain size in carnivores and ungulates. *Am. Nat.* 112:815–831.

Rayner, J. M. V. (1979) A new approach to animal flight mechanics. *J. Exp. Biol.* 80:17–54.

Rosen, R. (1967) *Optimality principles in biology.* London: Butterworths.

Sachs, R. (1967) Liveweights and body measurements of Serengeti game animals. *E. Afr. Wildl. J.* 5:24–36.

Schmidt-Nielsen, K. (1970) Energy metabolism, body size, and problems of scaling. *Fed. Proc.* 29:1524–1532.

Schmidt-Nielsen, K. (1975) Scaling in biology: The consequences of size. *J. Exp. Zool.* 194:287–308.

Stahl, W. R., and Gummerson, J. Y. (1967) Systematic allometry in five species of adult primates. *Growth* 31:21–34.

Warham, J. (1977) Wing loadings, wing shapes and flight capabilities of Procellariiformes. *N. Z. J. Zool.* 4:73–83.

Weis-Fogh, T. (1973) Quick estimates of flight fitness in hovering animals, including novel mechanisms for lift production. *J. Exp. Biol.* 59:169–230.

Weis-Fogh, T. (1977) Dimensional analysis of hovering flight. In *Scale effects in animal locomotion* (T. J. Pedley, ed.), pp. 405–419. London: Academic Press.

Weis-Fogh, T., and Alexander, R. McN. (1977) The sustained power output from striated muscle. In *Scale effects in animal locomotion* (T. J. Pedley, ed.), pp. 511–525. London: Academic Press.

Wilkie, D. R. (1977) Metabolism and body size. In *Scale effects in animal locomotion* (T. J. Pedley, ed.), pp. 23–36. London: Academic Press.

22

Structural systems: hydrostats and frameworks

STEPHEN A. WAINWRIGHT

Thinking about "how animals work" over the years has been the self-appointed task of physiologists. Physiology rose out of and parallel to morphology, but, as is the nature of such evolutionary branchings, the branch tips continue to diverge, and specialization increases. Some Renaissance investigators still realize that no matter how interesting a function may be, it cannot exist or operate without a structural context. Knut Schmidt-Nielsen is such a person, and it is noteworthy that all of his major contributions in physiological research have depended on new understanding of the morphology surrounding them. With this in mind, a view of animal bodies as containers and carriers of all the physiologically interesting bits is offered here.

The present view is that a biological function can be understood only if the structure of the system at lower levels of integration is known. For example, how does the heart function? One cannot tell from looking only at the intact heart organ, but one must look at the tissues and see that the arrangement of tubes, valves, and muscle is responsible. Similarly, how does muscle tissue function? Only an analysis at microscopic and ultrastructural levels will give this answer. Therefore, to know how a body functions, we must define the structural system we call the body and then study its parts and how they are arranged to permit, control, and limit body function.

We recognize two major types of support systems in the bodies of animals: the jointed framework and the stretched membrane hydrostat. Most adult arthropods, echinoderms, and vertebrates have jointed frameworks, whereas hydrostatic support is found in polyps, worms, some molluscs, and larval arthropods and tunicates. Both framework and hydrostatic support systems occur commonly in bodies of echinoderms, molluscs, and vertebrates: A seastar walks on hydrostatic tube feet, but uses its framework to open the oyster; and we pump breath, blood, and breakfast through hydraulic systems in our bodies, whereas we play tennis with our frameworks.

Perhaps the most important difference between the two types of systems with regard to the mechanical function of animal bodies is that hydrostats are best powered by sheets of muscle that apply loads distributed over great areas of the body wall, whereas in jointed frameworks more nearly linear muscles often apply point loads to the rigid skeletal members. Point loads and long, rigid skeletal elements allow the use of leverage and mechanical advantage that in turn allow one muscle type to move the feet of differently shaped animals at different forces and speeds.

Equally important is the observation that the material formed by the animal in the hydrostat is the membrane (the body wall) whose passive function is to resist tensile forces, whereas the material formed by an animal with a framework includes both the tensile ligaments and the rigid bone or shell that resists compressive and bending forces. Hydrostatic animals have only to include some environmental water, or a suitably regulated derivative, in their bodies to create their compression-resisting component. This is consistent with the observation that animal life arose in the sea where incompressible liquid is cheap and abundant, and that only with the evolution of metabolically expensive, complex rigid materials did the arthropods and vertebrates come to dominate a wide variety of terrestrial habitats.

One feature shared by animals, plants, and man-made objects is that the physical integrity of their bodies is due to tension resistance of chemical bonds. Buckminster Fuller developed much of his theory for minimum weight structures from this notion of the tensile integrity of things. This helps us understand that bones alone do not support us: We must include the tensile ligaments as part of our integrated skeleton.

To mention only structural materials that resist tension and compression is to leave out the most unique of animal materials: the stretchable, flexible soft connective tissues that permit peristalsis in the gut and that allow skin to stretch over bent joints. As we shall see, the secret of these materials is that they are composite materials composed of strong, rigid fibers embedded in a gel-like matrix. The shearing properties of fibers and especially their interfibrillar matrices are vital to animals.

Thus we see that the materials of supportive systems are specialized to resist tension, to resist compression and bending, or to permit large, shearing deformations. Also, resistance to force is not an all-or-none property; structural materials within a species have a discrete range of such resistance that is tuned to the range of forces the system actually experiences. By and large, animal structures are stronger and stiffer than they need to be for daily operation, and the excess strength can be thought of as a safety factor outside whose limits natural selection acts (Wainwright, 1980).

This account will study the hydrostatic skeleton in some detail. Excellent accounts of frameworks by Alexander (1968), Liem (1977), and Gray (1968) are enthusiastically recommended.

Design of the hydrostatic skeleton

Generally speaking, animal bodies are cylindrical in shape. Somewhere in the body wall of every hydrostatic animal are multiple layers of fibrous connective tissue. In animals of most phyla, notably excluding the Arthropoda, the fibers are of collagen and are wrapped in helical array around the animal's body. All fibers in a layer are parallel to each other, and fibers in alternating layers describe right- and left-handed helices around the animal (Figure 22.2). Engineers routinely use strong, light, cheap fibers in crossed helical array to reinforce gas and water tanks, but animal bodies and plant cells have been doing it for millions of years.

In the pressurized, cylindrical body the presence of crossed helical fibers in the wall material has the dramatic function of controlling the changes in shape the body can make. One might well ask "Why be fiber-wound?" The answer is easy to see when you inflate a cylindrical balloon. The first puff of air just fills the balloon without stretching it much and allows it to stand supported by the air pressure inside (Figure 22.1). The next puff of air produces an aneurism – a bulge at one point along the cylinder. Such a bulge is a dangerous instability that arises from a principle of pressurized cylinders that depends only on the cylindrical shape itself. This principle is that in a pressurized cylinder the circumferential stress in the wall is twice the longitudinal stress: $\sigma_C = 2\sigma_L$. Another way to think of this is that the tendency for the cylinder to explode and split lengthwise is twice the tendency for the ends to blow off. So the answer to the question is that fibers can prevent disastrous anisotropic expansion of the body wall that might result from pressure increases due to muscle contraction.

FIGURE 22.1 A cylindrical rubber balloon (a) inflated just to support itself and (b) with an additional pulse of air that causes an aneurism.

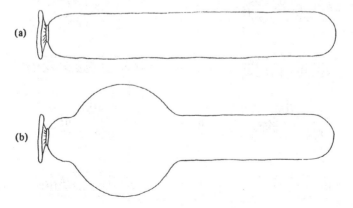

Now one might ask "Why put them in helical array?" There are two possible ways to arrange such reinforcement: as circumferential belts and longitudinal bindings or in crossed helical array. Figure 22.2 compares the ability of orthogonally and helically wound cylinders to stretch, bend, and twist. Helically wound bodies can change length and can bend a lot without kinking, and they resist torsion. Orthogonally wound bodies have fixed length, bend very little without kinking, and may twist easily. Each particular deformation is prevented only when it depends on stretching the relatively unstretchable fibers or, as in the kinking orthogonally wound cylinder, on compressing fibers axially; one simply cannot push on a string. Because virtually all the stretchy, contractile, squirming animals we know are helically wound, this is an enormously important principle in animal design. The answer to the question is that the crossed helical array allows bodies to stretch and shorten and to bend without kinking.

The crossed helical fiber array also limits changes in length and diameter of closed, pressurized cylinders and changes in volume in open, nonpressurized cylinders. A sea anemone can change its body volume a great deal by opening its mouth and contracting both longitudinal and circumferential muscles. With its mouth shut, it can, like most worms

FIGURE 22.2 Comparison of the properties of fiber-wound pressurized cylinders with fibers in rectilinear array on the left and in crossed helical array on the right. (Explained in the text.)

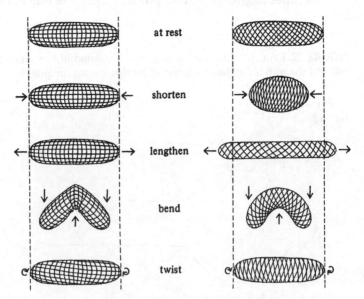

of many phyla, use longitudinal muscles to become short and fat (a common way to avoid predators) and the antagonistic circumferential muscles to become long and thin. These antagonistic muscle sets operate using the incompressible volume of the hydrostat as a fulcrum; when one set contracts, it causes the other to extend.

Consider an open-ended cylinder with a crossed helical fiber array. When it is maximally shortened, length and thus volume approach zero. When its diameter approaches zero at maximum elongation, the volume again approaches zero. In between these two extremes, the volume is maximum when the fiber angle is 54°44' (Figure 22.3) (Clark and Cowey, 1958). We presume that a sea anemone would follow the curve in Figure 22.3 as it changed volume.

Most animals are constant-volume, pressurized cylinders and therefore cannot follow the curve of volume change, but must change length and diameter on a single horizontal line in Figure 22.3. The particular horizontal line any worm species occupies is determined by the degree to which the worm's cross section becomes elliptical owing to low pressure between the extreme short and long shapes. Flatworms lie low on the graph (line AB), whereas highly pressurized nematode worms operate at a higher level (C and D). There are various other morphological constraints besides the helical reinforcement that prevent a particular worm

FIGURE 22.3 Graph showing the relation between the volume of an open-ended cylinder and the angle its reinforcing fibers make with the cylinder's axis. Ordinate on the right shows the ratio of major to minor axes of the elliptical cross section that closed cylinders would have at minimum pressure (animals at rest). Line AB: relationships measured in the nemertine Lineus longissima. C: values for the nematode Ascaris at rest. D: values for the squid Loliguncula at rest. Circles and ellipse indicate cross-sectional shapes and areas of the cylinders.

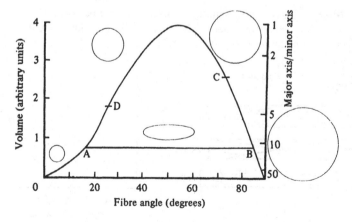

type from extending from one side of this curve to the other. The crossed helical array does put the ultimate limit on such changes in shape; no body shapes outside the curve in Figure 22.3 are known (Clark and Cowey, 1958).

Another principle of thin-walled, pressurized cylinders is that the stress in the wall equals the pressure times the radius of the cylinder divided by the thickness of the wall: $\sigma = pr/t$. Animals can control their internal pressure by muscular activity, and, because body size and wall thickness change very little, they can thereby change the skin stress. As the stress in an animal's body wall is changed indirectly by muscular action, so is its stiffness. (Remember that stiffness, $E = \sigma/\epsilon$ where σ is stress in pascals, and ϵ is strain in percent. Pressure-induced change in skin stress is accompanied by very little strain.)

For most familiar skeletal materials such as bone and shell, stiffness is a constant that depends on interatomic bonding. In fiber-reinforced materials and in soft connective tissues stiffness is strain-dependent. The reason for this is that the relatively stiff fibers are in an array that changes as the material is stretched so that the fibers become aligned progressively in the direction of the applied stress (Figure 22.4). The amount of stress in the material assumed by the fibers is a function of the fiber angle (the angle the fibers make with the stress direction), an angle that at any time depends on the original fiber angle and the percent change in length or strain.

So we see that animals with pressurized, cylindrical bodies have two ways to control the stiffness of their body walls: through the increased parallel alignment of the fibers with increased stretch or through the increase in internal hydrostatic pressure.

An example of such an animal is the large parasitic nematode, *Ascaris lumbricoides*, that thrashes about in the guts of pigs, horses, and man.

FIGURE 22.4 Stress–strain relationships in a piece of fiber-reinforced material such as skin. The slope of the curve is the strain-dependent stiffness of the material. (Explained in the text.)

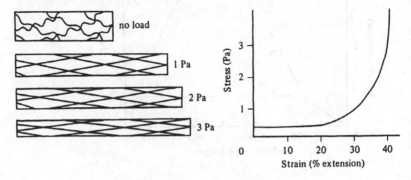

Ascaris's body is a very highly pressurized cylinder of circular section with tapered ends. A peculiar characteristic of all nematodes is that they have only longitudinal body muscles and completely lack an antagonistic set of circumferential muscles. The thick cuticle of *Ascaris* contains several layers of collagen fibers wrapped in crossed-helical array at a fiber angle of 75°. This high fiber angle makes the fibers nearly into belts that strongly oppose the tendency of the animal to increase its diameter while allowing a change in length when the longitudinal muscles contract (Harris and Crofton, 1958).

Collagen fibers are very stiff and linearly elastic, and they are relatively inextensible, that is, they break at 8 to 10% extension. As they are stretched forcefully over this short extension, they store the energy of the stretch as elastic energy. When the stretching force is removed, they shorten again and release the stored elastic strain energy. This means that when the longitudinal muscle on one side of *Ascaris's* body contracts, the animal bends and shortens and the internal pressure goes up, causing a slight circumferential stretching of the cuticle. When the muscle relaxes, the pressure drops and the elastic cuticle snaps back, returning the worm to its resting shape. Thus we see that the storage of elastic energy in a helically wound body wall serves to antagonize the action of the longitudinal body musculature, and we thereby understand how this worm and other nematodes "work" – in the Schmidt-Nielsen sense – without a set of antagonistic circumferential muscles. This principle is illustrated by the body wall structure in several phyla of worms, whereas the inverse situation occurs in the bodies of squids, where locomotion depends on forceful circumferential muscles, and there are no longitudinal muscles to return the body to resting shape. In this case the crossed fibrillar array starts at the low fiber angle of 28° (Figure 22.3, *D*), which allows the desirable great changes in diameter. Muscle contraction compresses the constant-volume mantle tissue, causing pressure in the tissue to rise and slightly stretching the collagen fibers that store strain energy. When muscle relaxes, pressure falls, and we believe that released strain energy returns the squid to resting shape, ready for the next power stroke of jet propulsion (Ward and Wainwright, 1972).

So we see that in polyps and wormlike, cylindrical animals the directional stiffness of fiber-reinforced body-wall material is controlled by the fiber angle. This arrangement can limit the extension in length and diameter of animal bodies, and it can replace entire muscle sets and the neural mechanisms necessary for their function. This is an interesting type of distribution of energy that merits further thought.

A way that changing wall stiffness helps animals is seen in the swimming shark. Sharks have hydrostatic pressure in the massive muscles under the skin, and the skin itself has many layers of collagen fibers in crossed-helical array. The internal pressure increases with swimming speed. With increased pressure, the skin stress increases and so does the skin stiffness. As the shark swims slowly, the connections of muscle to the pliant skin and to the backbone are enough to bend the body in swimming. As the shark swims faster, the increased stiffness of the skin allows it to serve as a whole-body tendon and thus to transmit the force and the displacement of anteriorly placed contracting muscle to the tail (Wainwright et al., 1978). Because of the precise fibrillar array in the skin, it can simultaneously transmit these large forces and allow the 15% extensions and contractions observed in fast swimming at constant stiffness. This is marvelous indeed – a tendon of controllable stiffness that acts in series with its muscle and changes length during its function!

Another feature of the shark skin tendon is that – again because of the helical fiber array – it makes use of the thickening or transverse work of the shortening muscle that is wasted in locomotion of running and flying creatures. The force of lateral expansion of the shortening muscle causes the collagen fibers in the skin to pull along their trajectories and thus to shorten the skin very forcefully in the same direction in which the muscle is pulling.

When a squid swims by jet propulsion, the force of a jet stroke is proportional to the volume of water displaced per unit time from its helically wound, cylindrical mantle cavity. From the designer's point of view, because the volume of a cylinder is $\pi r^2 L$, one can see that a unit change in radius or circumference will have a far greater effect on volume than will a unit change in length. It therefore is not surprising to find that squid jet strokes are powered by circumferential muscles (Ward, 1972). Whereas longitudinal muscles can decrease the volume of a cylinder, the use of circumferential muscles increases the amount of locomotor power the muscle system can generate per contraction.

In the squid, it is evident that any change in length of the mantle tissue during a jet stroke will decrease locomotor efficiency. The squid's mantle is wound by collagen fibers at 23° to the body axis. In such a fiber-wound cylinder, a 10% shortening by circumferential muscle will allow only a 2% increase in mantle length. In addition, because mantle tissue volume is constant (Ward, 1972), a 10% decrease in circumference with a 2% increase in length means there must be a very large increase in thickness. Increased thickness further decreases the diameter of the mantle cavity, displacing even more water and thus adding to the thrust of the jet stroke. Here again the fiber angle and cylindrical shape are morphological characters that control function.

Fibers in crossed helical array in the body wall thus convey emergent properties to the body wall itself and to the body. These properties are part of the basis of the posturing behavior and locomotor performance of cylindrical animals of many phyla. Differences in fiber angle alone will help to explain the diversity among many hydrostatic species that differ in such a major feature as the presence or absence of an entire set of antagonistic muscles in the body. And though hydrostatic animals are limited in their use of mechanical advantage to increase the force, displacement, or speed of a locomotor organ, they can undergo a much wider range of dimensional changes than can animals with framework skeletons.

Consider two examples. First, tentacles of the large hydra *Pelmatohydra oligactis* function from full extension at lengths of 20 mm to complete contraction to lengths of 0.5 mm. No one has yet shown how the epitheliomuscular cells or the extracellular collagenous connective tissue of these tentacles permit a 40-fold change in length, repeatedly, without rupture. Second, the caterpillarlike onychophoran *Peripatus* can perform whole-body peristalsis that allows it to walk its body through a hole one-ninth the diameter of its body at rest (Manton, 1958). Guts, heart, musculature, and exoskeleton must all suffer extreme deformations without rupture and without undue crowding of central nervous system, endocrine glands, gonads, and so on.

Analogy of hydrostat and backbone

Current dogma distinguishes sharply between hydrostatic and rigid-framework skeletons. By a philosophical sleight of hand – applying the concept of analogy – we shall see more clearly the general problem of axial support in all animals at the same time as we see that many framework and hydrostatic mechanisms are similar in both function and structure. Morphologically, legs and bodies of vertebrates and arthropods and arms of seastars and ophiuroids are beams made up of alternating rigid elements and flexible joints (Figure 22.5), the function of which organs is to bend without changing length. This makes engineering sense if it is pointed out that if the goal is to bend a body, any energy put into shortening it will be wasted. Although many lower invertebrates trade heavily on becoming shorter as means to withdraw and thus escape predators, this earthworm type of motion has been limited to creeping and burrowing animals. In any case, hydrostatic wormlike creatures gave rise to arthropods and vertebrates on different branches of the evolutionary tree.

All the living animals we think of as lower chordates have notochords. (A notochord is basically a rod in the middle of an animal body; the rod acts as a backbone by allowing bending without shortening.) The notochords of sturgeons, cyclostome fishes, amphioxus, and tunicate larvae are hydrostatic rods, wrapped in crossed-helical arrays of collagen fibers.

And the bony vertebral backbones of higher vertebrates are also wrapped in a fabric containing collagen in crossed helices. The crossed-helical array allows bending without kinking and helps prevent lateral displacement of vertebrae, whereas the rigid vertebrae and the soft intervertebral tissues resist shortening. Could the genetic instructions for forming the helical winding for a notochord be identical to the instructions for forming the helical winding in an external cuticle? Placement of the flexible but longitudinally rigid rod inside the body relieves the body wall of most of its support functions. For hydrostatic animals this is advantageous because puncture wounds in the body wall no longer interfere with posturing and locomotion. An axial skeleton in the early chordate was the

FIGURE 22.5 Three designs for the support of bodies: a, a few long rigid elements and flexure points in an elephant's hind leg; b, many short rigid elements and continuous bending in a backbone; c, walking leg of the horse-shoe crab, *Limulus*.

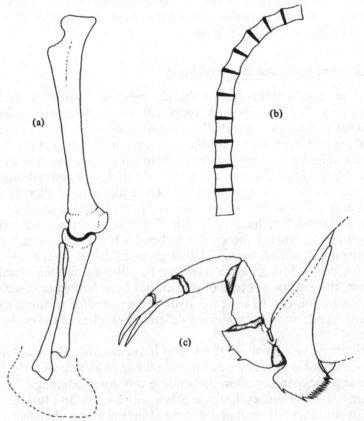

foundation to which mechanically advantageous lever-type appendages could later attach. The evolution of the backbone from a hydrostatic notochord in tiny protochordates such as larval ascidians (sea squirts) and amphioxus to the vertebral columns of the great whales, dinosaurs, and giraffes was permitted by the evolutionary invention of bone as a rigid material that replaced the water in the hydrostat with a more controllable, adaptable compression-resisting material.

Once long, rigid elements evolved, animals showed the mechanically advantageous features of lever systems. For example, through mechanical advantage muscle of a single type and range of shortening velocities and tensions can allow badgers to dig forcefully and deer to run swiftly (Smith and Savage, 1956). Recently another feature of long, thin appendages of framework animals that helps muscle to produce locomotion more efficiently has been discussed (Alexander and Bennet-Clark, 1977), namely, that collagenous Achilles tendons of men, dogs, and wallabies actually extend by 18, 25, and 11 mm, respectively, when the animals run or hop rapidly. This extension represents the storage of strain energy that drives the elastic recoil of the tendon as the foot lifts off the ground. This recoil energy helps power the step and thus is not lost to the locomotor efficiency of the animal. Because the ability of a tendon to store strain energy is proportional to its length, the longer the legs an animal has, the greater is its potential use of this structural feature.

A simple column is the easiest and cheapest support to build, and junctions between rigid and pliant materials are difficult to design and expensive to build and maintain against wear and fatigue; so how can we justify the vertebral columns of vertebrates and many echinoderms (Figure 22.5)? Once again we see design for flexibility without change in length. The ability to transmit force of muscle contraction across joints along a flexible structure without causing kinking or failure is important to us all. I know of no study aimed at elucidating this principle, but I shall call your attention to the most extreme example yet studied, namely, the seastar that opens oysters by brute force (Eylers, 1976).

Seastars have the following interesting combination of abilities: They can bend their arms into tight bends at any point along the arm, and they can stiffen the arm in any posture. Figure 22.6 shows how they are thought to accomplish this. The support system of each arm is made up of hundreds of tiny calcite ossicles that are interconnected by collagen. In parallel with the collagen ligaments between ossicles lie muscles. The stiffness and extensibility of collagen ligaments in echinoderms are controlled by the animal (e.g., Wilkie, 1978). Ordinary longitudinal muscle in the arm contracts locally, causing the arm to bend. The total deformation is taken up a tiny bit at each interossicular junction. This could be the mechanism of deformation.

At any bent shape, if the interossicular muscles are contracted and the ligaments made rigid, the ossicles meet over large, flat bearing surfaces that are now held together with such force that the entire arm is rigid. This is the suggested mechanism of stiffening.

FIGURE 22.6 A suggested mechanism by which the many tiny ossicles (less than 1 mm in length) might be articulated to allow the 10-cm-long arm to bend at any point and to stiffen in any position. (a) Sagittal section through one column showing ossicles and interconnecting muscle (thinner line) and catch ligament (thicker line). Single mechanism fully extended in two different positions (b and c). (d) See-through view of a single mechanism in bending, seen from above. (e) See-through view of a column in bending.

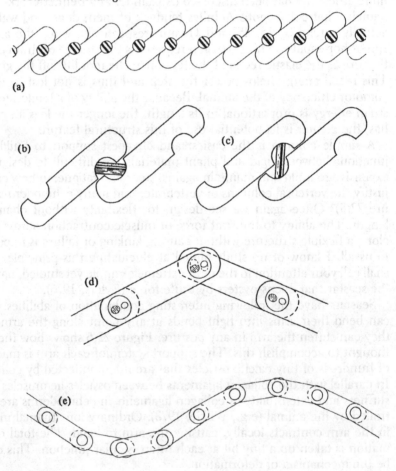

This design, using many small compression elements to support a long, flexible, forceful appendage, is unique among animal phyla and has not been much used by engineers. This mechanical system is fundamentally different from the long, hollow, cylindrical limb bones and arthropodan leg segments, and it merits much further attention. Of particular interest should be the mechanism by which the ligaments shorten. The force of shortening may be the interossicular muscles or may be small, intraligamental muscles such as those found in sea urchins (Wainwright et al., 1980). But how does the collagen behave? Do the molecules actually extend and contract? Or do they slide past one another? Whatever the mechanism is, it is sure to be something new in our experience of biostructural materials and systems.

The future

The recent realization that the skin of sharks is functionally involved in locomotion is causing us to completely reevaluate what we know about how sharks swim. Bony fishes are also fiber-wound. Does this have functional significance other than permitting bending without kinking? Whales, porpoises, and seals are fiber-wound, too. Whatever does this mean?

In the complex wall of the human aorta and large arteries, there is a layer containing a crossed-helical array of collagen fibers. Its fiber angle and its function are unknown. Cardiologists should care about what controls the elastic properties of these vessels.

Backbones of all kinds merit study of their taken-for-granted structure and function. At the extremes of bending, fish backbones are rigid and resilient. What is the material or systems basis for this interesting safety factor? Does the backbone of any animal store strain energy and use it in locomotion? The intervertebral disc appears to be a short segment of a fiber-wound cylinder still acting as a hydrostat in sharks and mammals. In sharks and some bony fishes it seems to be a very low-friction joint for the gentle bends involved in continuous aerobic swimming. Have engineers something to learn from this aspect of organism design?

In spite of the fact that the function of skeletons is not measured in terms of enzyme kinetics or neuronal electrodynamics, this function is vital to the organism and is a long way from being thoroughly understood. Our knowledge of the morphology of animal bodies is keeping pace with our knowledge of their physiology, both in the number of new discoveries of remarkable design features and in the number of unanswered but important questions that arise.

References

Alexander, R. M. (1968) *Animal mechanics*. London: Sidgwick and Jackson.

Alexander, R. M., and Bennet-Clark, H. C. (1977) Storage of elastic strain energy in muscle and other tissues. *Nature* 265:114–117.

Clark, R. B., and Cowey, J. B. (1958) Factors controlling the change of shape of certain nemertean and turbellarian worms. *J. Exp. Biol.* 35:731–748.

Eylers, J. P. (1976) Aspects of skeletal mechanics of the starfish *Asterias forbesii*. *J. Morphol.* 149:353–367.

Gray, J. (1968) *Animal locomotion*. London: Weidenfeld and Nicholson.

Harris, J. E., and Crofton, H. D. (1957) Structure and function in the nematodes: Internal pressure and cuticular structure in *Ascaris*. *J. Exp. Biol.* 34:116–130.

Liem, K. F. (1977) Musculoskeletal system. In *Chordate structure and function* (A. G. Kluge, ed.), 2nd ed., pp. 179–269. New York: Macmillan.

Manton, S. M. (1958) Habits of life and evolution of body design in arthropods. *J. Linn. Soc. (Zool.)* 44:58–72.

Smith, J. M., and Savage, R. J. G. (1956) Some locomotory adaptations in mammals. *J. Linn. Soc. (Zool.)* 42:603–622.

Wainwright, S. A. (1980) Adaptive materials: A view from the organism. *Symp. Soc. Exp. Biol.* 34:437–453.

Wainwright, S. A., Smith, D. S., and Baker, G. (1980) Echinoid catch apparatus: A contractile mechanism. *Am. Zool.* 20:845.

Wainwright, S. A., Vosburgh, F., and Hebrank, J. H. (1978) Shark skin: Function in locomotion. *Science* 202:747–749.

Ward, D. V. (1972) Locomotor function of the squid mantle. *J. Zool. (Lond.)* 167:487–499.

Ward, D. V., and Wainwright, S. A. (1972) Locomotor aspects of squid mantle structure. *J. Zool., Lond.* 167:437–449.

Wilkie, I. C. (1978) Nervously mediated change in the mechanical properties of a brittlestar ligament. *Marine Behav. Physiol.* 5:289–306.

23 Weightlessness: from Galileo to Apollo

CLAUDE LENFANT AND SHOU-TEH CHIANG

About 400 million years ago, some vertebrates first completed the transition from an aquatic to a terrestrial environment. That move required a great number of functional and structural modifications, many necessitated by the adaptation to gravity forces. An aquatic vertebrate lives in a "weightless" environment, supported by the surrounding water. A terrestrial vertebrate, however, must support itself against the forces of gravity, a situation that has a major effect on the design of its musculoskeletal system. Gravity also has major effects on the functioning of the cardiovascular system, and secondary effects on the excretory and respiratory systems. With the advent of space flight, we confront anew a "weightless" environment. In this chapter, we ask how well physiological systems built to operate under gravity forces can function when these forces are removed.

Musculoskeletal system

The study of gravity began with Galileo when he first introduced ballistics in 1634. Having recognized gravity as a physical entity, he went on to point out that large animals have relatively thicker weight-supporting bones than do smaller animals in order to compensate for the fact that the gravitational force exerted on the animal (i.e., its weight) increases as the cube of the body dimension, whereas the strength of the bone increases only as the square of the dimension (Galileo, 1634, in S. Drake's translation, 1974). A number of experimental approaches have been used to study what happens to the musculoskeletal system when gravity forces are altered or removed, and all yield similar results.

Effects of reduced gravity on locomotion

Gravity plays an important role in the mechanisms humans and other terrestrial vertebrates use to move along the ground. On the moon, gravity forces are one-sixth the value on earth, and in space flight they decrease to zero. How can these changes be compensated for, and what will be their long-term effects on the structure of the musculoskeletal system?

339

On earth, each stride by humans, other mammals, and birds involves an alternate transfer between gravitational potential energy in the vertical direction and kinetic energy in the forward direction (Cavagna et al., 1977). In this respect, walking is similar to a swinging pendulum or an egg rolling end over end. On the moon, decreased gravity forces greatly reduce the amount of potential energy stored in each step. In order to maintain the energy transfer, the kinetic energy must be correspondingly decreased, with a consequent reduction in forward speed. Indeed, the speed of walking on the moon does not seem to exceed 2 km h^{-1} (Margaria and Cavagna, 1964; Margaria, 1976).

During running, the feet push the body upward and forward simultaneously. Thus, no kinetic–gravitational exchange of energy can take place. Energy is conserved, however, by another mechanism – an elastic bounce of the body. On the moon, the vertical thrust greatly exceeds the forces of gravity, causing a shift from a "running" to a "jumping" mode of locomotion (Cavagna et al., 1972). When gravity forces are entirely absent, totally new modes of locomotion must be developed.

Effects of reduced gravity on the structure of bone and muscles

The stresses placed on bones and muscles are dramatically changed by a reduction of gravity forces. Experience with bedridden patients shows that the musculoskeletal system responds quickly to reductions in forces with bone demineralization, muscle atrophy, and deconditioning. In fact, rest and immobilization have been extensively used to simulate weightlessness, sometimes for periods exceeding a year. Because bed rest is not feasible for experiments with animals, immobilization with body casts has been used to simulate effects of weightlessness on the musculoskeletal system. The experiments, which sometimes have lasted for months, have produced results comparable to those obtained from human bed-rest studies.

Although no exposure to weightlessness in space has been of sufficient duration to assess the total extent of the effect of absence of gravity on the structure of the musculoskeletal system, enough observations have been made to recognize that this system is directly affected by weightlessness, even of relatively short duration. During space flight, just as during bed rest or immobilization, calcium increases in the urine almost immediately (Figure 23.1). It remains elevated throughout the entire flight and does not decline until the recovery phase begins. Fecal calcium also increases, a fact that further contributes to the negative calcium balance. These results are indicative of an alteration in bone structure, as confirmed by an increase in urine hydroxyproline, a degradation product of bone collagen (Whedon et al., 1976; Yagodovsky et al., 1976).

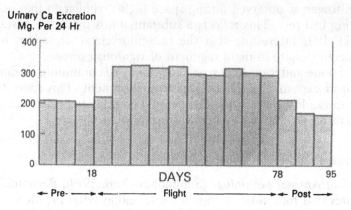

FIGURE 23.1 Effect of weightlessness on urinary Ca excretion. Data from one subject, during *Skylab 3* space flight. [From Whedon et al., 1976]

FIGURE 23.2 Effect of weightlessness on nitrogen balance. Intake is downward from zero baseline, excretion upward from intake lines; thus shaded areas above zero baseline indicate negative balance. Data from one subject, during *Skylab 3* space flight. [From Whedon et al., 1976]

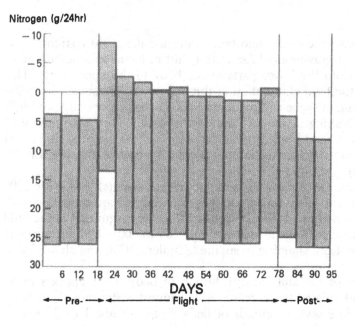

Muscle loss also occurs in response to reduction in forces. An increase in urinary nitrogen is observed during space flights, similar to that occurring during bed rest. This reflects a substantial loss in muscle tissue (Figure 23.2). It is interesting that the musculoskeletal alterations in astronauts occur despite in-flight regimens of vigorous exercise.

The loss of bone and muscle in space appears to be continuous during the duration of exposure to the weightless environment. This suggests that serious musculoskeletal impairment would be likely to occur during space trips lasting several years, such as trips to other planets.

Cardiovascular system

As discussed in *Animal physiology* (Schmidt-Nielsen, 1979), Bernoulli's theorem states that total fluid energy in a circulation system is the sum of the gravitational potential energy, the internal pressure potential energy, and the kinetic energy of the moving fluid. Reducing gravity, therefore, has dramatic effects on the pressures and the various forms of energy within the circulatory system. Circulatory changes similar to those that occur with reduced gravity forces occur during a normal day with changes in posture (e.g., when one changes from a standing to a reclining posture).

Volume changes

Undoubtedly, the most important cardiovascular event resulting from the removal of gravitational forces is a shift in blood volume and other body fluids from the lower parts of the body to the upper parts. This stems from the total elimination of the hydrostatic pressures. Since the very beginning of space exploration astronauts have reported "head fullness" and "flushing of the face and of the neck." They even described a movement of fluid from the legs to the thorax and the head that could be partly reversed by application of lower body negative pressure (LBNP) (Figure 23.3).

Quantitative measurements have shown that as much as 1500 to 2000 ml of blood and other fluids may be displaced upward during weightlessness (Kirsch et al., 1979). By comparison, the magnitude of the fluid shift into the intrathoracic space is only 400 to 500 ml when changing body position from standing to supine (Sandler, 1976), and about 700 to 800 ml during immersion.

The effect of the shift in distribution of body fluids appears to be somewhat attenuated by increased water output. Astronauts have been reported to lose several pounds of body weight in just 1 or 2 days. It should be noted that part of this weight decrease is due to a reduction in fluid intake and to negative nitrogen balance (resulting from muscle atrophy). However, most of it appears to stem from the activation of central (cardiac) mechano-receptors that in turn modulate hormone levels, especially causing suppression of the antidiuretic hormone (ADH). Although diuresis seems not to occur in space flights, it has been found to increase markedly in immersed subjects at the outset of the exposure. Further studies to understand the mechanism of water losses are needed, especially studies designed to assess water balance during the first hours of zero-G exposure.

Heart rate

Heart rate has been the most consistently and continuously measured parameter during space missions. The *Skylab* experiments (with the exception of *Skylab 2*) have demonstrated an increase in heart rate during weightlessness. There is, however, evidence of a downward trend in heart rate as the mission continues. The return to preflight values occurs in the first 2 to 3 days of the postflight recovery period.

Stroke volume and cardiac output

Measurements of stroke volume and cardiac output before and after space flight show a consistent and important decrease of both parameters

FIGURE 23.3 Effect of weightlessness and lower body negative pressure on resting volume of calf segment. This figure shows a conspicuous decrease of calf segment's volume during weightlessness, with partial compensation during LBNP. [From Johnson et al., 1976]

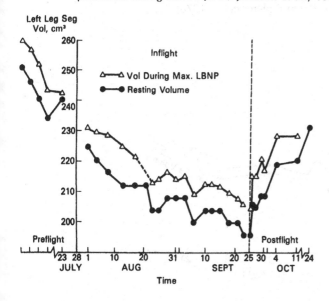

immediately after flight, with a return to normal preflight values within a few days. These data imply a fall of both the cardiac output and the stroke volume during weightlessness. The cardiac output, however, is thought to decrease much less than the stroke volume because of the compensatory increase in heart rate noted above (Buderer et al., 1976; Figure 23.4). In dogs instrumented for direct measurements, both cardiac output and stroke volume decreased during flight.

Systemic blood pressure and peripheral resistance

Mean systemic arterial blood pressure appears unaffected by weightlessness in space flight. The stability of blood pressure with a concomitant decrease in cardiac output implies an increase in the peripheral resistance (Figure 23.5), which seems to occur at rest as well as during exercise (Johnson et al., 1976).

Because blood pressure in the brain is determined by the opposite effects of heart-generated pressure and of gravity (i.e., vertical distance between heart and head), weightlessness results in an increase in cerebral blood pressure that in turn seems to cause (temporary) cerebral edema, at least in simulated conditions (Guell et al., 1979). Fortunately the adverse effect would be minimal in man where the heart–head distance is small (30–35 cm) (Figure 23.6). However, space travel would indeed be

FIGURE 23.4 Cardiac output and stroke volume before and after weightlessness exposure in subjects exercising at 50% of maximum oxygen consumption. Mean data from the three subjects who conducted the *Skylab 4* mission. [From Buderer et al., 1976]

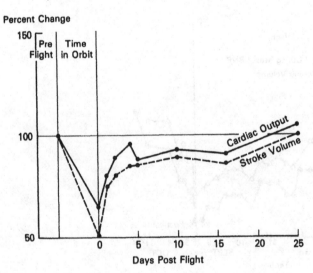

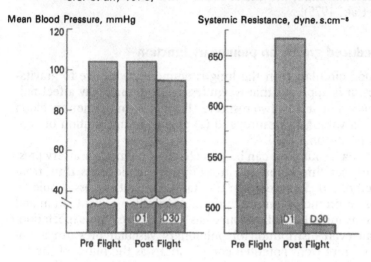

FIGURE 23.5 Mean arterial blood pressure and systemic resistance before and after weightlessness exposure in subjects exercising at 50% of maximum oxygen consumption. Mean data from the three subjects who conducted the *Skylab 4* mission (D = day). [Based on data from Buderer et al., 1976]

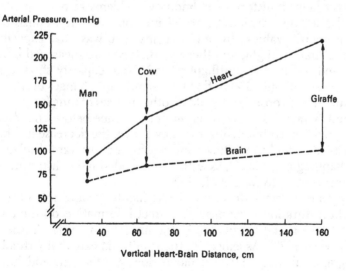

FIGURE 23.6 Mean arterial pressure at heart and at brain level in man, cow, and giraffe as a function of the vertical heart–brain distance. Note that the pressures at brain level are almost similar in all three species, whereas they differ greatly at heart level. [From Patterson et al., 1975]

a lethal event for the giraffe, whose heart–head distance is 175 to 200 cm. Should the corresponding hydrostatic pressure be removed for an extended period of time, the giraffe could suffer massive cerebral edema (Patterson et al., 1975)!

Effects of reduced gravity on pulmonary function

Because blood circulation in the lung is eminently sensitive to gravitational forces, it is apparent that weightlessness will markedly affect pulmonary function in at least two respects: (1) increase in pulmonary blood volume and in vascular pressure; and (2) change in distribution of ventilation and perfusion.

On theoretical grounds, it can be expected that pulmonary artery pressure becomes even throughout the lung during weightlessness; thus, from the viewpoint of blood oxygenation the state of weightlessness should be ideal because it promotes a more uniform distribution of ventilation and perfusion in the lung. Recent evidence has substantiated this expectation. However, an even distribution of pulmonary vascular pressure is not necessarily exempt from potential complications as the filling of the cephalad parts of the lung during weightlessness is analogous to the situation existing during high-altitude exposure or that present in patients with mitral stenosis. Thus it is possible that fluids may leak through the vascular wall, causing pulmonary edema to occur.

Relatively few measurements of pulmonary function have been made during the series of space flights thus far completed, probably because there has never been a history of pulmonary problems in participants. Data were collected only during the *Skylab* missions, and they consisted of simple spirometric values. In only one instance was vital capacity repeatedly measured in flight; all other parameters were measured only before flight and during the postflight period. Vital capacity showed a significant decrease (about 10%) in flight; no significant changes between pre- and postflight were observed for all the other indices of lung function, including chest X rays that were taken in one instance before and after flight. Although not corroborated by other evidence, the decrease in vital capacity observed in flight may have resulted from an upward displacement of the diaphragm at zero G, as well as from body-fluid redistribution into the thoracic cavity (Sawin et al., 1974).

Recent experiments made during parabolic flights causing zero-G trajectories of short duration have provided direct information on the distribution of perfusion and ventilation in the lung during weightlessness (Michels and West, 1978). As expected, the results indicate that virtually all the topographical inequalities of blood flow, ventilation, and lung volume that exist when gravitational forces are present are nearly abolished in a weightless environment. One would anticipate that uniformity of ventilation and perfusion throughout the lung, plus the fact that all pulmonary capillaries are open during weightlessness, would improve blood oxygenation. Unfortunately, as yet, no blood oxygenation data are available. However, evidence (in birds) showing that arterial P_{O_2} decreases with increasing G suggests that the opposite would be true when gravitational force is suppressed (Wolgenback et al., 1979).

Conclusions

The data on the physiological effects of weightlessness are still very sparse. The *Skylab*–space-shuttle missions will, we hope, fill many of the gaps that exist. However, the information in hand clearly substantiates Galileo's contention that gravity influences the design of terrestrial organisms.

References

Buderer, M. C., Rummel, J. A., Michel, E. L., Mauldin, D. C., and Sawin, C. F. (1976) Exercise output following skylab missions: The second manned skylab mission. *Aviat. Space Environ. Med.* 47:365–372.

Cavagna, G. A., Heglund, N. C., and Taylor, C. R. (1977) Mechanical work in terrestrial locomotion: Two basic mechanisms for minimizing energy expenditure. *Am. J. Physiol.* 233:R243–R261.

Cavagna, G. A., Zamboni, A., Faraggiana, T., and Margaria, R. (1972) Jumping on the moon: Power output at different gravity values. *Aerospace Med.* 43:408–414.

Galileo Galilei. (1634) *Two new sciences* (S. Drake, trans.). Madison: University of Wisconsin Press (1974).

Guell, A., Bes, A., Braak, L., and Barrere, L. (1979) Effect of a weightlessness simulation on the velocity curves measured by doppler sonography at the level of carotid system. *Physiologist* 22:S25–S26.

Johnson, R. L., Nicogossian, A. E., Bergman, S. A., and Hoffler, G. W. (1976) Lower body negative pressure: The second manned skylab mission. *Aviat. Space Environ. Med.* 47;347–353.

Kirsch, K., Röcker, L., and Wicke, H. J. (1979) Methodological aspects of future cardiovascular research in space. *Physiologist* 22:S11–S14.

Margaria, R. (1976) *Biomechanics and energetics of muscular exercise*. Oxford: Clarendon Press (Oxford University Press).

Margaria, R., and Cavagna, G. A. (1964) Human locomotion in subgravity. *Aerospace Med.* 35:1140–1146.

Michels, D. B., and West, J. B. (1978) Distribution of pulmonary ventilation and perfusion during short periods of weightlessness. *J. Appl. Physiol.: Respir. Environ. Exercise Physiol.* 45:987–998.

Patterson, J. L., Goetz, R. H., Doyle, J. T., Warren, J. V., Gauer, O. H., Detweiler, D. K., Said, S. I., Hoernicke, H., McGregor, M., Keen, E. N.,

Smith, M. H., Hardie, E. L., Reynolds, M., Flatt, W. P., and Waldo, D. R. (1975) Cardiorespiratory dynamics in the ox and giraffe, with comparative observations on man and other mammals, *Ann. N.Y. Acad. Sci.* 127:393–413.

Sandler, H. (1976) Cardiovascular effect of weightlessness. In *Progress in cardiology*, vol. 5 (P. N. Yu and J. F. Goodwin, eds.), pp. 227–270. Philadelphia: Lea and Febiger.

Sawin, C. F., Nicogossian, A. E., Schachter, A. P., Rummel, J. A., and Michel, E. L. (1974) Pulmonary function evaluation during and following Skylab's space flight. *Proc. Skylab Life Sci. Symp.*, Lyndon B. Johnson Space Center, Houston, 2:763–774.

Schmidt-Nielsen, K. (1979) *Animal Physiology: adaptation and environment*, 2nd ed. Cambridge University Press.

Whedon, G. D., Lutwak, L., Rambant, P. C., Whittle, M. W., Reid, J., Smith, M. C., Leach, C., Stadler, C. R., and Sanford, D. D. (1976) Mineral and nitrogen balance study observations: The second manned skylab mission. *Aviat. Space Environ. Med.* 47:391–396.

Wolgenback, S. C., Burger, R. E., and Smith, A. H. (1979) Effects of high-G on ventilation/perfusion in the domestic fowl. *Physiologist* 22:S59–S60.

Yagodovsky, V. S., Triftanidi, L. A., and Gorokhova, G. P. (1976) Space flight effects on skeletal bones of rats (light and electron microscopic examination). *Aviat. Space Environ. Med.* 47:734–738.

Index

absolute metabolic scope, 132
Acanthocybium solandri, 219
  see also wahoo
acetate, 145
acetyl CoA, 144
Achilles tendon
  dogs, 335
  men, 335
  wallabies, 335
acid–base balance, 207
acid–base control
  blood, 73ff
  within cells, 87
acid–base status, change mechanisms, 76
acidosis, metabolic, 77
Ackerman, R. A., 84
acylamides, 291, 299
Adaro, F., 60
Adenota kob, hind foot length, 311
Aepyornis
  body mass, 126
  egg weight, 126
aerobic metabolism
  cold-exposure measuring technique, 163
  exercise measuring technique, 163
  mammals, 31ff
Agalychnis, 242
aging, 109ff
  cause, 116
  commitment theory, 118
  definitions, 110
  fibroblast model, 119
  "pituitary clock" theory, 115
  rates, 116
air-breathing organs, fish, 93
air convection requirement, 79, 80, 82
air sacs, 8
alanine, 145
alanopine, 140
albacore, 216
  see also *Thunnus alalunga*
aldosterone, 249, 250
Alexander, R. McN., 168, 312, 316, 317,
    318, 319, 320, 326, 335
alkaloids, 299

alkalosis
  during exercise, 23
  metabolic, 77
Allen, J. A., 261
alligator, pH control, 78
allometric equations
  exponents: mammals, 312; running
    animals, 312; birds 313; moths, 313
allometry, 161
allosteric parameter, 228
Alopias vulpinus, 218
Alopias superciliosus, 218
Alopiidae, 218
alpha-imidazole, 75, 76
altitude–barometric pressure relationship,
    18
alveolar–arterial P_{O2}, 23, 24, 25
alveolar lungs, 10, 11
alveolar surface area, 42
Amadon, D., 126
Amblyrhinchus cristatus, blood pH, 84
amphibian
  cardiovascular system, 96
  cutaneous circulation, 98
  skin: diffusion limitation index, 62; as
    gas-exchange organ, 49, 54–9
Amphibolurus, blanching, 248
amphioxus, 333, 335
Amphipnous cuchia, 93
  circulation, 94
  heart, 95
  oxygen affinity, 94
anaerobic metabolism, 138ff
Anderson, J. F., 312
Anderson, J. R., 141
Anderson, M. E., 228
Andreoli, T. E., 283
Anguilla, 245
anoxia, metabolism, 141
Anser, 131
antelope, 311, 319
antidiuretic hormone, 246, 267, 269, 277,
    300, 343
apocrine glands, 249
Appleyard, R. F., 194

Kiceniuk, J. W., 66
kidney, 144, 146
  human, 269
  osmotic and solute gradients, 270
  Psammomys, 284, 285, 286
  rabbit, 276, 283
  rat, 269
Kilgore, D. L., 190, 191
killer whale, 68
Kimball, G. C., 114
kinetic energy, 340
King, J. R., 124, 134
Kinney, J. L., 78
Kirkland, J., 115
Kirsch, K., 342
Kirsch, R., 245
Kirschner, L. B., 256
Kishinouye, K., 221
Kleiber, M., 125, 126, 135, 162, 165, 166,
  173, 178
Kluger, M. J., 212
Knight, C. H., 259
Knowles, G., 299
Koefoed-Johnson and Ussing hypothesis,
  257
Kokko, J. P., 276, 280, 281
Kooyman, G. L., 66, 67, 68, 151, 152, 153,
  154, 155, 156, 157, 158
Krebs, H. A., 145, 216
Krebs cycle, 141
Kris, W., 285
Krogh, A., 36, 240, 245
Krogh's diffusion constant, 50, 61
Kruysse, A., 66
Kuhn, W., 273
Kümmel, G., 297
Kusaka, M., 146
Kutty, M. N., 144

lactate, 145, 165
lactate dehydrogenase, 143
lactic acid pulse, 155
Lahiri, S., 19
Lamna, rete, 222
Lamna ditropis, 218
Lamna nasus, 218
  see also porbeagle
Lamnidae, 218
Lampetra, 240
lamprey, 242
  see also *Lampetra*
Lampris regius, 218
Lane, N. J., 295
Lasiewski, R. C., 126, 181, 190, 205
Latimeria, 241
Laurent, P., 92
Laybourne, R. C., 17

Lea, M., 114
Lechner, A. J., 163, 165
Ledger, H. P., 311
Lee, A. K., 247
Lee, P., 242, 248
Leitch, I., 124, 125, 126
Lenfant, C., 68, 100, 152
Lepidocybium flavobrunneum, 218
Lepisosteus osseus
  acid–base control, 83, 84
  oxygen uptake, 83, 84
Leptonychotes weddelli, see Weddell seal
Levinsky, N. G., 276
Licht, P., 200, 247, 248
Liem, K. F., 326
Lighthill, J., 320, 321
Lillywhite, H. B., 199, 206, 247, 248
limbs, scaling in structure and function,
  309–22
Limulus, 334
Lineus longissimus, 329
Linthicum, D. S., 223
Linzell, J. L., 254, 256, 257, 259, 260, 261,
  262, 263
liquid–gas interface, exchange, 243
liquid–liquid interface, exchange, 241
litter weight
  as function of adult metabolic rate, 125,
    128
  as function of maternal weight, 125, 127
Littorina, 145
liver, 144, 146
lizard
  blue tongue, 208; see also *Tiliqua*
  heliothermic, blood pH, 85
  varanid, 84, 85; cardiovascular system,
    101; systemic and pulmonary blood
    pressures, 102, 103
Locke, M., 290
locomotion, effects of reduced gravity, 339
locust, 291, 300, 320
Loliguncula, 327
Lomholt, J. P., 94, 95
longissimus dorsi, 34, 35, 36
loop of Henle, 266ff
Loveridge, J. P., 247
Lucia, S. B., 114
lung
  function, 39
  human: diffusion limitation index, 62;
    gas exchange, 49, 60–1
  mammalian alveolar, 3, 10
  parabronchial avian, 3, 8
lungfish, 93, 244
  African, 96; see also *Protopterus*
  Australian, 96
  cardiovascular system, 95

Printed in the United States
By Bookmasters